Simulation Strategies to Reduce Recidivism

Faye S. Taxman • April Pattavina
Editors

Simulation Strategies to Reduce Recidivism

Risk Need Responsivity (RNR) Modeling for the Criminal Justice System

Springer

Editors
Faye S. Taxman
Department of Criminology,
    Law, and Society
George Mason University
Fairfax, VA, USA

April Pattavina
School of Criminology and Justice Studies
University of Massachusetts Lowell
Lowell, MA, USA

ISBN 978-1-4614-6187-6   ISBN 978-1-4614-6188-3 (eBook)
DOI 10.1007/978-1-4614-6188-3
Springer New York Heidelberg Dordrecht London

Library of Congress Control Number: 2013938600

Printed on acid-free paper

Springer is part of Springer Science+Business Media (www.springer.com)

Preface

Why Do We Need the RNR Simulation Tools?

Over the last decade there has been a growing consensus among academics and practitioners about the importance of using evidence-based practices in the field of corrections. The consensus extends to the need to better align existing practices with known offender and program attributes that will generate better outcomes. Embedded in evidence-based practices is that organizations should: (1) use a valid risk and need assessment tool to identify those factors in an individual that are amendable to change; (2) use cognitive-behavioral programs to address these risk-need profiles; (3) administer programs that are high quality; and (4) focus on recidivism patterns. These are essential elements of the Risk-Needs-Responsivity (RNR) model. RNR is well recognized as a contemporary framework for determining the role of programs in correctional environments (institutional and community). In many ways, this model has revived attention to rehabilitation and the importance of good quality programming. The empirical support for the model along with the lack of progress on recidivism reduction results from punitive or incapactative models of punishment has led to a broad base of support for the RNR framework.

The RNR framework is more readily accepted but there are three missing components. First, the literature is based on the RNR philosophy that calls for integrating treatment and rehabilitation into the correctional system. What is lacking is an empirical demonstration designed to explore the effects of adopting RNR model on a large scale. Most findings have referred to single study experiments or meta-analysis but have not illustrated the impact on a large scale. Such a demonstration would assess what would happen to recidivism if we took the RNR model to scale—that is, if we expanded the use of risk-needs tools, placed people in programs designed to address risk-need factors, and offered high quality programming—what would be the impact. This "what if" analysis is ripe for simulation models because they provide the mathematical and statistical approaches well suited to illustrate the impact of policy options. In other words, they provide additional empirical support to the notion that it is worth changing sentencing and correctional decisions to incorporate RNR principles. Discrete event simulation model makes the point even more clear since it demonstrates how changes over times and events (different decision points) can affect the outcomes from the correctional system. It provides the empirical evidence that it is beneficial to change systems and practices through thoughtful consideration of the options available.

Second, the original RNR framework is built on assumptions regarding the relationships among individual criminal justice risk factors, individual needs, and recidivism. The empirical evidence supporting these assumptions is ever-evolving as more research is produced. Of course, not all studies are the same in terms of methodological rigor, measurement of key variables, and generalizability of the findings. But the emerging literature suggests that the original version of the RNR model may need modification as more scientific information becomes available. For example, the first evidence-based practice principle is that high risk offenders should be placed in correctional programming. This follows from the finding that better results are possible for higher-risk offenders. However, this statement does not consider the degree to which individual needs may "trump" criminal justice risk factors, the relevance of non-criminogenic factors such as mental illness and housing stability that may affect success in the community, and key demographic key factors (such as age and gender) that affect offending patterns. The RNR model should recognize it could be higher number of needs and clinically relevant factors that increase the need for more structured programming.

Third, in the "real world" the RNR process is more complicated than it appears. The complexity is probably one of the biggest factors that affect the likelihood that the model can and will be used in practice. The complexity has to do with making decisions that integrate complicated information about the individual offender as well as understanding the pros and cons of each program or service. In other disciplines, years of training are provided to build these diagnostic skills—in the justice system, this type of skill development is not provided. Rather it is assumed that experiential learning ("on the job training") will suffice. However, the skills to assess complex human condition such as those factors that influence offending behavior, intergenerational substance abuse or criminal behavior, antisocial personality or values, and so on are not necessarily easy to isolate. Answering the question, "what should we do about this?" requires yet another set of skills. Understanding the impact of one decision on processing at other decision points requires more complex information. The complexity of the tasks supports the need for simulation models that support the decisions that individuals or jurisdictions need to make about "who should go into what program?"

This book evolved from many discussions among scholars over the years that are devoted to improving criminal justice policy and practice through integrating evidence and evidence-decision criteria to practice. Faye Taxman has been working on many of the concepts in this book over the course of her career. The notion of exploring how to improve the use of assessment tools for identifying the factors that should be addressed during the period of correctional control has been a theme in her work on seamless systems of probation, reengineering probation, and using evidence-based practices. This is supplemented by the attention to more intricate decisions such as how to respond to negative behaviors and how to increase compliance with the conditions of correctional control. The growing emphasis on implementation processes (see Taxman & Belenko, 2012 on implementation models) has led many in the corrections field to promote the use of assessment tools and models to drive decisions regarding placement in appropriate programs and services.

Energized by the evidence in support of the RNR model and interested in the possibilities for widespread adoption, the Bureau of Justice Assistance sought to learn more about the correctional impacts if the model were implemented on a larger scale. With their support, several related projects described in the chapters were undertaken with the purpose of determining the ways in which we could (1) build RNR-based tools to support RNR programming for agencies and (2) to understand how the application of these tools would collectively serve to reduce recidivism at the national level.

After the framework for the RNR program simulation tool was developed, the challenge was to determine how the use of this tool across agencies might contribute to significant reductions in recidivism on a national scale. April Pattavina's work in this area was informed by recent developments in computer simulation that would allow for investigation of this issue. Simulation models have recently been used for operations research in health care and criminal justice applications and were appropriate for our work. We seek to first give the reader an understanding of how these techniques may be applied in operational contexts and then use the techniques in our investigation into the recidivism reduction effects of the RNR model.

In conjunction with the evolution of more effective models and tools available for assessing offender risks and needs has been the advancement of information and computing technology that has created opportunities to realize the possibilities for these models beyond a single study or agency. Throughout much of this book, authors were able to take advantage of publically available datasets, such as the Survey of Inmates in State Correctional Facilities, the National Corrections Reporting Program, and the Bureau of Justice Statistics' classic recidivism study for our work in this book. Using the RNR model as a framework, it was possible to use the data sources to measure key RNR concepts, map the processes for treating offenders, and finally investigate the impact of a large-scale implementation of RNR programming on recidivism. We were also fortunate to have partnership with state and local correctional and substance abuse agencies to validate the tool and assumptions. Using the most current simulation techniques, it was possible to assess the impact over time and found that the RNR model holds considerable promise for reducing recidivism.

Simulation models are important tools that are underutilized in research and policy. The attraction of using different types of simulation models was that the issues are often too complicated to design in experiments. This book outlines how to put together a simulation tool, and then use the tool to assess various problems. The "how to" notion of this book is to help others consider the various steps to develop a simulation model. In the course of developing the simulation tools, we learned how to handle a number of challenging data, methods, and theory issues. These challenges are presented to foster a greater understanding of the mystic involved in creating a simulation model, and how to use data from one model to another. We hope this book helps foster a number of simulation models in the field of justice policies. A few years from now we will know whether we were successful in this goal.

This book was possible through the support from many people including the contributors. We are indebted to the authors of chapters for their hard work including James Byrne, Stephanie Ainsworth, Erin Crites, Michael Caudy, Joseph Durso, Avi Bhati, Andrew Greasely, Matthew Concannon, and David Hughes. Ed Banks from the Bureau of Justice Assistance offered endless assistance and his passion for helping improve the knowledge integration process inspired us. Thanks are also given to our colleagues that have moved this field forward including Edward Latessa, Todd Clear, and Redonna Chandler.

Fairfax, VA, USA   Faye S. Taxman
Lowell, MA, USA   April Pattavina

Reference

Taxman, F. S., & Belenko, S. (2012). Implementation of evidence-based practices in community corrections and addiction treatment. New York: Springer.

Contents

Part III   Simulation Applications

Part IV   Conclusion

Contributors

Stephanie A. Ainsworth Department of Criminology, Law and Society, George Mason University, Fairfax, VA, USA

Avinash Bhati Maxarth, LLC, Gaithersburg, MD, USA

James Byrne School of Criminology and Justice Studies, University of Massachusetts Lowell, Lowell, MA, USA

Michael S. Caudy Department of Criminology, Law and Society, George Mason University, Fairfax VA, USA

Erin L. Crites Department of Criminology, Law and Society, George Mason University, Fairfax, VA, USA

Joseph Durso Department of Criminology, Law and Society, George Mason University, Fairfax, VA, USA

Andrew Greasley Aston Business School, Birmingham, UK

David Hughes Human Services Research Institute, Cambridge, MA, USA

Jennifer Lerch Department of Criminology, Law and Society, George Mason University, Fairfax, VA, USA

April Pattavina School of Criminology and Justice Studies, University of Massachusetts Lowell, Lowell, MA, USA

Matthew L. Perdoni Department of Criminology, Law and Society, George Mason University, Fairfax, VA, USA

Liansheng Tang Department of Statistics, George Mason University, Fairfax, VA, USA

Faye S. Taxman Department of Criminology, Law, and Society, George Mason University, Fairfax, VA, USA

Part I
Advances in Simulation Modeling for Criminal Justice Planning and Management

Chapter 1
Planning for the Future of the US Correctional System

April Pattavina and Faye S. Taxman

Introduction

The United States correctional system is at a crossroads. Over the last 15 years, considerable attention has been drawn to the way that the US criminal justice system has been dealing with criminal offenders since the later part of the twentieth century. While disturbing, it is a well-known truth that the US incarceration rate is the highest among late modern democratic countries and the United States has 25 % of the world's prisoners (Lacey, 2010). What is especially noteworthy is the growth in scale of punishment since the late 1970s. From 1920 to the mid-1970s, the incarceration rate was stable at around 100 per 100,000 people, but from 1980 to 2008, the US incarceration rate increased from 221 to 726 per 100,000 people (Western and Petit, 2010). By the end of 2010, about 1 in every 104 adults was in the custody of state or federal prisons or local jails; 1 in 33 is under some type of correctional control (Glaze (2011)). Academics, policymakers, and practitioners have argued that this level of incarceration is unsupportable from both philosophical and humanitarian perspectives and is economically unsustainable. The pressing question is how do we plan to address the emphasis on mass incarceration? This is perhaps one of the greatest challenges confronting our contemporary society.

What is unique about today's correctional system is the massive size—over two million people incarcerated in prison and jail on any given day and another 5+ million on community supervision. And the two are not mutually exclusive. Failures on

A. Pattavina, Ph.D. (✉)
School of Criminology and Justice Studies, University of Massachusetts Lowell,
1 University Avenue, Lowell, MA 01854, USA
e-mail: april_pattavina@uml.edu

F.S. Taxman, Ph.D.
Department of Criminology, Law and Society, George Mason University,
10900 University Boulevard, Fairfax, VA 20110, USA
e-mail: ftaxman@gmu.edu

F.S. Taxman and A. Pattavina (eds.), *Simulation Strategies to Reduce Recidivism:*
Risk Need Responsivity (RNR) Modeling for the Criminal Justice System,
DOI 10.1007/978-1-4614-6188-3_1, © Springer Science+Business Media New York 2013

community supervision contribute to the size of the institutional population, and the size of the institutional population places demands on the need to expand community supervision. Yet, both institutional corrections and community corrections are stuck at the same place—the current array of institutional and community correctional programming is limited due to available resources, philosophies around the purpose of punishment, and historical attempts to remake and reshape the correctional landscape. That is, during the late 1980s and early 1990s when the war on drugs was waging strong and there was a surge in drug offenders with lengthier sentences, an attempt was made to remake the correctional landscape. The policy talk at that time (1990s) was focused on intermediate sanctions or the correctional interventions and programs that occurred between probation and prison. Morris and Tonry (1990), in their famous treatise *Between Prison and Probation: Intermediate Punishment in a Rational Sentencing System*, wrote:

> Our plea is for neither increased leniency nor increased severity; our program, if implemented, would tend toward increased reliance on punishments more severe than probation and less severe than protracted imprisonment. At present, too many criminals are in prison, and too few are the subjects of enforced controls in the community. We are both too lenient and too severe; too lenient with many on probation who should be subject to tighter controls in the community, and too severe with many in prison and jail who would present no serious threat to community safety if they were under control in the community. (p. 3)

Morris and Tonry envisioned a community punishment system that had programming which would occur between standard probation (face-to-face contacts) and prison (secured institutional setting). They discussed fines, community service orders, house arrest, three types of probation (intensive supervision, residential conditions, and treatment conditions), intermittent imprisonment, restitution and compensation, fees for service, electronic monitoring, and forfeiture. The integration of these correctional interventions within the existing sanction and treatment structures faced significant setbacks. First, some were tried and tested, and it became apparent that the "public community" (including legislators, stakeholders, citizens, correctional and probation agencies, and offenders) was not ready for this form of punishment. For example, the concept of day fines was tried with a number of implementation barriers that impeded progress toward institutionalizing them in the United States (see Hillsman, 1990). Second, individual evaluations and more contemporary meta-analyses and systematic reviews have found that some of these interventions do not reduce recidivism. Such is the case for control-oriented intensive supervision (MacKenzie, 2006). If an intervention does not improve recidivism rates, then it begs the question as to whether we should routinely employ this intervention. Third, with insufficient resources, some of these innovations are partially (or even barely) implemented which dilutes their potential effectiveness. This is the case for electronic monitoring, probation conditions, probation with treatment, day reporting programs, some treatments such as cognitive behavioral therapy or therapeutic communities, and other ideals. Collectively, the systematic and organizational resistance coupled with insufficient attention to program fidelity created hesitations to move forward to implement a continuum of punishments that expanded from probation from prison.

The work of Morris and Tonry laid a foundation that many jurisdictions struggle to realize. Today there are new innovations developed during the 1990–2000s that are gaining support in the field and worthy of including in this system of punishments. The first is the growing use of drug treatment and problem-solving courts in the United States. These courts integrate treatment with control conditions to create the type of community controls that Morris and Tonry envisioned. Second, there are a host of new technological innovations that are front and center in terms of the potential to exact controls on offender behaviors. These include drug testing, GPS, electronic monitoring, and now smart phone applications that allow for daily diaries, journaling, and location monitoring (Pattavina, 2009). Technological advances will continue to influence the development of new approaches to support and monitor offenders in the community.

Morris and Tonry struggled with a system for determining the appropriate sentencing or punishment level for an individual. They outlined the concept of interchangeability that was based on equity among certain punishments in terms of their level and type of controls but allowed the punishment to be tailored to the individual's situation. Hence, punishments could be "equivalent" in terms of severity, while substantively different. Similar to other sentencing schemes, the focus on assignment was based on severity of crime and criminal history, the two components of most sentencing guidelines. At the same time that Morris and Tonry were articulating this scheme, another set of scholars was advancing new concepts about offender management issues in corrections. Andrews and his colleagues offered a classification and programming scheme that focused more on the dynamic factors that affect offender outcomes. In their review of the literature, they proposed that correctional programming should be determined by the offender's risk (criminal history) and needs (factors that affect their continued involvement in the criminal behavior (see Andrews, Bonta, and Hoge 1990). This was further developed into a framework referred to as risk–needs–responsivity (RNR), which focused attention on placement decisions based on the factors designed to control the risk of the offender to the community and attending to the factors that are most likely to reduce the likelihood of further involvement in the justice system. Figure 1.1 below combines the two models—Intermediate Punishment and RNR—into a vision for a correctional landscape that would best serve to reduce recidivism at the individual level and to build a correctional system that is responsive to the various needs of offenders. The model is based on the premise that recidivism reduction requires tailoring programming and placements to minimize risk but also using the least restrictive environment to achieve this goal. This book describes the development of a simulation models that allow jurisdictions and individual actors to put into place an empirically driven framework for making correctional placement assignments.

The Predicament Arising from the Correctional Population Surge

The size and shape of the US correctional population has drawn serious attention since the recent economic recession. Political scientists and criminologists share the perspective that the dramatic growth of the prison population was largely achieved by policy changes that include the adoption of laws sending more drug and property offenders to

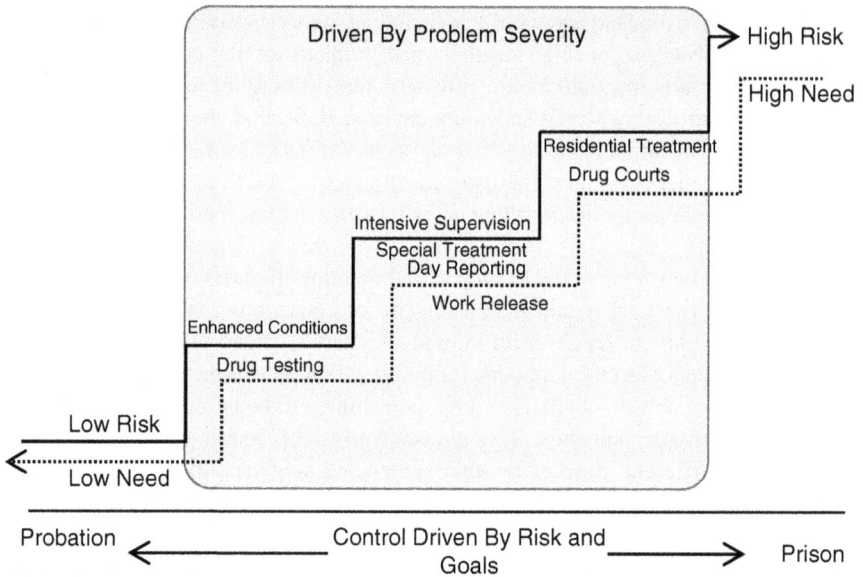

Fig. 1.1 Prototype risk–need model applied to various correctional settings

prison (rather than jail or probation), lengthening prison sentences for various crimes, and requiring prisoners to serve larger portions of their sentences before being released (i.e., mandatory minimums) (Simon, 2010). Taken together these policy changes reflect a punitive or punishment model that emphasizes retributive justice and incapacitation as a means to promote public safety (Auerhahn, 2003). These initiatives appear to come at a high cost with only a modest return on investment—that is, if we use a utilitarian calculus of the costs and benefits of this experiment, we must assess the outcomes from a different perspective. Weisburg and Petersilia (2010) report that the growth in state imprisonment rates since 1985 accounted for no more than 25 % of the decline in serious crime during the 1990s. Western (2008) is more skeptical and reports that prison populations accounted for 10 % of the drop in serious crime. The modest decline in serious crime from 1993 to 2001 was achieved by the $53 billion in additional correctional spending and added half a million new prisoners (Western, 2008).

The cost of correctional expansion warrants more than just a discussion of outcomes couched in cost–benefit terms. The policies supporting the growth in incarceration also led to a growing disproportionate number of minority people from distressed communities being sent to prison (Clear, Waring, & Scully, 2005; Lynch & Sabol, 2001). Research has consistently shown that the policies governing punishment have resulted in the incarceration of a disproportionate number of black males. Blacks are 7 times more likely to be incarcerated than whites, and large racial disparities can be seen for all age groups and at different levels of education. One in nine black men in their twenties is now in prison or jail (Western, 2008). The same appears to be occurring to Hispanic males, but not to the same degree. Finally, the concentration of people with lower socioeconomic status in the justice system, regardless of race or ethnicity, continues to be problematic.

Prison conditions have also worsened in the wake of the growing carceral population. Overcrowding has led to unsafe and unhealthy living environments for inmates. Research indicates that because of housing needs, space for recreation and work and rehabilitation programs are eliminated to allow all useable space to be converted to dormitories for additional prisoners. Correctional workers in crowded facilities experience more job-related stress and fear of inmates (Martin, Lichtenstein, Jenkot, & Forde, 2012). Crowding has become so severe that the Supreme Court declared California prison crowding unconstitutional and is forcing the state to radically change the way it houses criminal offenders. Judges have recently ordered the state to reduce its inmate population as a way to improve medical care (Biskupic, 2011).

What happens upon release from prison illuminates the cumulative social impact of incarceration. Each year, over 739,000 inmates return home from prison. Those coming home from prison face significant challenges. The problem of prisoner reentry has been well documented by leading scholars in the field (Petersilia, 2005; Travis, Solomon, & Waul, 2001; Travis & Visher, 2005). Most return to neighborhoods of concentrated disadvantage where support services are lacking. Men with prison records are often out of work (Visher & Kachnowski, 2007). The jobs they do find pay little and do not offer the benefits and earning potential necessary to support the socially valued roles of husband and provider (Uggen, Wakefield, & Western, 2005; Western, 2008). Petersilia (2005) provides a general profile of soon to be released prisoners based on inmate survey data. She found that 41 % reported that they did not have a high school diploma and 33 % were unemployed the month before arrest. Family disruption and substance abuse were also problems for many inmates. Approximately 27 % were divorced and 59 % reported using drugs in month before committing their crime. Nine percent had an overnight admission for a mental condition.

The prevalence of these life circumstances among inmates reflects the wide range of cognitive and behavioral deficits that will continue to challenge them when they are released back into the community. Yet, the correctional system seems unresponsive to these issues. Petersilia (2005) describes the current state of knowledge regarding offender needs and prison treatment programs. She laments that prison administrators are not able to inform researchers regarding the number of prisoners who need different types of programs or the extent to which offenders participate in programs. Even in cases where counts are available, details about the duration and intensity of programs are often lacking and programs likely to be evaluated attesting to greater concerns about the quality of the treatment programming (Gendreau, 1996; Lowenkamp & Latessa, 2005; Welsh & Zajac, 2004). Her analysis of survey data shows that less than half of those in need of drug and alcohol treatment had enrolled in a relevant treatment program. Moreover, she reports that those most in need of certain programs are not always the most likely to participate in them. Given that these programs operate on a volunteer basis, the participating inmates may reflect the pool of inmates that are lower in need but "savvy" in that they realize that program participation both consumes idle time in prison and appears to suggest the offender is preparing for release. These offenders consume valuable space in limited programs, leaving behind those that are of greater need. She found that participation in educational and vocational programs was similar across low, moderate, and high need levels. Phelps (2011) in a recent article reviewing the availability of programming in correctional institutions found that the type of programming has shifted from treatment to more "reentry" services that emphasize life skills. Typical rehabilitation programs are far and few between given the growth in more life skill building programs which the meta-analysis literature finds to be of little importance to the recidivism reduction efforts.

A survey of treatment programs in prisons by Taxman, Perdoni, and Harrison (2007) identified major gaps in the availability and delivery of treatment services to offenders. Moreover, the services that are available are generally of low to moderate quality (Friedmann, Taxman, & Henderson, 2007). They also found that many correctional facilities attempt to provide services but that the resources available limit the size of the programs. Thus, program capacity amounts to a small percentage of their daily population. The services available tend to be more oriented toward educational awareness and minimal counseling, as opposed to intensive clinical and treatment services, reinforcing the shift noted by Phelps (2011).

Faced with limited opportunities for a productive life after prison, many offenders are likely to eventually recidivate. The 1994 recidivism study by Langan and Levin (2002) found that within 3 years of release from prison, 68 % were rearrested, 47 % are subsequently convicted, and 25 % returned to prison for a new crime. These rates seem impenetrable given the rather consistent finding across studies (see Pew, 2011). High recidivism rates suggest that people released from prison appear unprepared for life on the outside and that they are being entrenched in the wheels of justice. Many return to prison numerous times in a process Lynch and Sabol (2001) refer to as churning where many offenders cycle in and out of prison serving short sentences, getting released, and returning a few months later on another charge or parole violation only to be released again in a few months. Some states are responding to this problem by eliminating harsh sentencing practices that lead to crowding (Mauer, 2011). Other states are reducing the use of incarceration for non-criminal technical violations occurring during the period of supervised release after a prison/jail term. Another set of states are exploring the state of correctional programming for offenders in efforts to promote offender change and rehabilitation.

The attention brought to the problem of reentry by leading scholars has helped to change the discourse on correctional programming by reasserting the importance of rehabilitation. The notion of rehabilitation is slightly different here in that the treatment programs are being discussed in the context of altering recidivism. The linkage between other philosophies of punishment—just deserts, incapacitation, retribution, and deterrence—and recidivism has been shown to be weak (see Cullen & Jonson, 2012) giving rise to a concern that in the utilitarian assessment of whether the surge in correctional populations has been fruitful for societal gains, the costs (fiscal, humanitarian, increased recidivism, etc.) outweigh the benefits.

Discussion among policymakers has begun to focus on what the corrections system is doing to help offenders prepare for life on the outside and what support was available to communities. The Council of State Governments created the Reentry Policy with a mission to develop a collaborative report recommending policies intended to improve outcomes for returning prisoners, their families, and communities (Travis & Visher, 2005). Then in 2008, President Bush signed the Second Chance Act, which funds literacy programs, drug treatment, and other services for prisoners and ex-prisoners. The Second Chance Act can be viewed as one achievement in the broader movement for improved prisoner reentry policy (Western, 2008) and lays the groundwork for a revitalization of the principled correctional and community correctional programming to address the unintended consequences of the surge in using the correctional system.

The Challenges Before Us in Creating a Continuum of Recidivism Reduction Programming

Despite the growing support for offender rehabilitation in the public discourse on crime and punishment, financial support for programs and services has been slow to materialize. The Department of Justice proposed a 100 million budget allocation for the Second Chance Act which amounts to barely .14 % of the 70 billion spent on corrections each year (Gottschalk, 2010). Moreover, even though Attorney General Holder recently stated in a speech that the administration would not focus on incarceration as the sole means to protect the public, the 2010 and 2011 budgets increased allocations for law enforcement and new construction. Clearly, there continues to be some political reluctance to fully support programs that promote offender rehabilitation. So the challenge remains for researchers and practitioners to find ways to improve our correctional system in ways that promote offender change without compromising public safety.

Perhaps the reluctance of policymakers to more fully support rehabilitative programming stems from the conflict among researchers within the corrections field regarding the impact of rehabilitation and treatment programs. It is well known among the corrections community that the value of a rehabilitative approach toward criminal offenders has been significantly challenged by research reviews claiming that evaluation studies of treatment programs largely failed to demonstrate successful offender outcomes (Martinson, 1974; more recently Farabee, 2005). By focusing on the issues of whether a program is "effective" (i.e., reduced recidivism, null results compared to the control condition), the discussion has been that rehabilitation programs do not "work." More recently some scholars contend that the effect size of the impact is small and may not be worth the investment (Clear & Austin, 2009). In response, advocates of rehabilitation such as Palmer (1992) suggest that we should interpret the work of Martinson and other detractors as a reminder that success is hard to come by and that correctional intervention has accomplished a great deal. The small effects can be increased by improving the quality of programming (Lowenkamp, Latessa, & Smith, 2005). The future of rehabilitation and correctional programming may have a more nuanced focus if we accept that it may not be possible to change *all* offenders, but we can devote our attention to addressing known criminogenic factors including substance dependence, social networks that include antisocial peers, and other targeted factors that affect offending activities.

To this end, the system should encompass the principle of ensuring that appropriate offenders are placed in appropriate programs instead of putting people in the "first available" program. As we plan for the future of correctional interventions, it might be wise to focus our efforts on who to target for deterrence, for incapacitation, and for retribution and who then we should target for rehabilitation-type programs. This will not be an easy undertaking because it requires addressing very complex questions that involve issues of cost, feasibility, justice, and public safety (Feeney introduction to Palmer, 1992) as well as more recent questions of responsivity, efficacy of interventions, and treatment matching.

Despite the criticisms launched against the potential legitimacy of correctional programming (due to program quality issues, see Cullen & Jonston, 2010; Lowenkamp et al., 2006), the search for the most effective ways to promote offender change has continued among a committed group of advocates, practitioners, and academics (Cullen, 2005). Researchers working in this area have generated an important body of literature devoted to advancing models of offender risk assessment and linking needs to appropriate offender treatment programs. This is an evolving area of work as more attention is paid to the question of "what works for whom." Even in the meta-analysis literature, a focus on moderator analyses to identify the patterns has emerged as scientists focus more attention on expanding our knowledge of maximizing our placement practices with evidence-based decision-making principles. While few studies have been able to isolate such patterns, researchers are committing to the use of moderators to better understand individual-level factors that account for positive (or negative) outcomes. In recent years, the RNR model (see below) has emerged as a dominant framework that emphasizes the importance of matching risk and need assessment with appropriate services that are consistent with the behaviors that drive their criminal activity (Ward & Maruna, 2007). It mirrors the movement in other fields—namely, substance abuse and mental health—where placement criteria have evolved to augment clinical decision-making processes to integrate evidence with clinical science to assist practitioners to improve individual-level outcomes.

According to Andrews and Bonta (2006, 2010), the RNR model integrates the psychology of criminal conduct into an understanding of how to reduce recidivism by targeting the unique individual factors that affect involvement in criminal behavior. The model proposes that correctional interventions should be structured according to three core rehabilitation principles: risk, need and responsivity. The risk principle specifies that offenders should be grouped by the criminal justice history that represents their "threat level" that a person may pose to society. Measures of static (historical) risk include age, criminal history, age at first arrest, number of prior probation violations, and other historical facts about individuals. The higher the level of risk, the greater the dosage or intensity of the treatment and controls should be. The need principle holds that the treatment of offenders should be targeted to specific dynamic risk factors (i.e., criminogenic needs) that are predictive of criminal behavior and that are amendable to change. Amenable to change infers the broad set of factors that contribute to offending behavior, but it does not include the demographic (i.e., age, gender, cultural) that may help explain recidivism rates but that there is little a person can do about these factors. These key dynamic risk factors are antisocial values and attitudes, antisocial peers and associates, criminal subculture, low self-control, substance use disorders, and dysfunctional family environment. The number and type of criminogenic need also drive the targeting decision in that offenders that exhibit more than one factor should benefit from more intensive services, treatments, and controls. The criminogenic need category focuses on presenting factors that can be addressed with proper attention. The third principle is responsivity, which involves the proper matching of correctional interventions in ways that consider contextual factors such as the ability and learning style of the offenders, the number of non-criminogenic destabilizing factors (i.e., mental health disorders, lower literacy rates, negative work history), and strengths such as stabilizing factors of a strong support network, good work and educational experiences, and positive social skills. Gender, age, and culture also affect responsivity since some programs, treatments, or approaches may be more beneficial to different demographic factors such as the gender-specific treatment, developmentally appropriate treatment, and culture competency in approaches.

When implemented correctly, the concept of service matching that is guided by principles of RNR is considered best practices for corrections (Taxman & Marlowe, 2006) and has been shown to significantly reduce recidivism in certain settings (Andrews, Zinger, Hoge, Bonta, Gendreau, and Cullen 1990). Research has also shown that nonadherence to RNR principles in service delivery, however, is not only ineffective but also detrimental to offender treatment outcomes (Lowenkamp & Latessa, 2005). Not treating offenders or placing offenders in inappropriate treatments can increase the risk of recidivism. Moreover, research suggests that program caliber is an important consideration when considering treatment delivery. Attending to implementation and quality is an important factor affecting the spread and utilization of treatment programming. There is a need to go beyond merely looking at the program components to assessing the quality of the delivery. Friedmann et al. (2007) reported that the attention to evidence-based practices is low in correctional programming including that treatment programs with the same name and identical treatment manuals vary in their overall program effectiveness from jurisdiction to jurisdiction (Latessa, Smith, Schweitzer, & Brusman Lovins, 2009). Taxman and Bouffard (2003) observed that there is great variation in substance abuse counseling regardless of the known program components.

A sizeable and growing body of literature devoted to each of the three RNR principles exists. For example, there is a growing body of literature on the development and application of relevant risk and need assessment instruments (see Pattavina & Taxman, 2007). A variety of risk assessment tools are available and many correctional agencies advancing in their use of such tools. While the risk tools may vary in content, they have the collective purpose of determining who is at higher risk of reoffending and identify the deficits and strengths of each inmate. This information is used to determine appropriate program planning. We are also learning more about which RNR-based programs are most successful at promoting offender change. Practitioners that have incorporated RNR elements into cognitive-based treatment plans and evidence-based reviews appear more satisfied that these programs significantly reduce recidivism. Evidence-based assessments of programs are necessary for determining which programs are most appropriate and which programs should be expanded or eliminated. Applying evidence-based research findings to the search for "best practices," evidence-based practices, or promising strategies benefits offenders, promotes public safety, and may be more cost-effective.

The commitment to evidence-based approaches serves to increase the demand for more rigorous research designs necessary to assess valid program evaluations and answer the question of what works for whom. The growing interest in determining "what works" has led to support for initiatives that promote evidence-based research. Examples include the Washington State Institute for Public Policy, sponsored by a state legislature to conduct program evaluation research; Crime Solutions, supported by the US Department of Justice (www.crimesolutions.org); and the Campbell Collaboration responsible for sponsoring a variety of evidence-based reviews on effective crime reduction strategies (Mauer, 2011).

It is not sufficient to rely only on the literature that addresses the respective elements of the RNR model separately as a way to determine the overall significance for correctional practice. What makes the RNR especially appealing is the focus on the interconnectedness among the three principles needed to achieve the most successful outcomes and the ability to provide more rationality to sentencing schemes and program placement criteria. While we continue to move forward with producing quality research studies with respect to program matching, quality, and effectiveness, it is equally important that we begin to examine the implementation implications of this model within an operational context (Ward & Maruna, 2007; Taxman and Belenko, 2012). A continuing need exists to expand the research base to assess how the connections among RNR dimensions operate in real-world correctional settings. Research shows that the current distribution of treatment services to offenders in prison, jail, and community corrections is inconsistent with the needs of the offender populations, as discussed in Chaps. 2 and 6 in this book. Significant improvements cannot be made unless this gap is closed. There is thus a pressing need to help jurisdictions develop guidelines as to how to allocate offenders into appropriate services. The list continues to grow of the facets of how to fine-tune the correctional system to integrate the rudiments of RNR-based evidence-based practices.

About This Book

We wrote this book to articulate an approach to implementing RNR into practice in justice, correctional, and health organizations that serve people involved in the criminal justice system. In the chapters that follow, the authors present research from projects designed to collectively inform the comprehensive development of a model that is grounded in principles of the RNR model and attempts to make connections among the principles in a way that maximizes matching that will produce reductions in recidivism. Ultimately the model can also include the cost-effective possibilities. The work we present will establish validated estimates of key parameters that describe the national corrections population and appropriate treatment services. These parameters will be used to build simulation models designed to examine how varying levels of RNR implementation affect offender recidivism. At the national level, simulation results can be used to inform the debate on the integration of RNR as an emergent framework into practice. At the local level, simulation inputs are translated into an expert system designed to assist in the day-to-day decisions correctional staff make about the best program options for offenders available in specific jurisdictional settings. Much of this book is about building the RNR model to incorporate the major research findings and then demonstrating how the model can work as a static model and a discrete-event model. The components of the RNR Simulation Tool expert system are described in this book.

Taxman and colleagues will present a case study for the issues related to treatment gaps in Chap. 2. The purpose of this chapter is to establish clearly the issues that confront the RNR model—that is, a methodology for examining how to assess current level of programming, a range of programming available, and the gap between need and programming. It also shows the assumptions that are plausible and needed in an RNR model. The case study focuses on substance abuse treatment to establish some of the key components of models. But it makes the case that provides the approach for a broader range of criminogenic needs.

How to build a useful simulation model is addressed in Chap. 3. Greasley introduces the reader to the stages of simulation model development and makes recommendations on how to build and validate models. Simulation models are often built in stages and rely heavily on model conceptualization and process mapping. He outlines the technical features associated with constructing a working model and discuss methods used to test alternatives.

In Chap. 4, Taxman, Caudy, Pattavina, Byrne, and Durso present the empirical basis for the RNR model that will be used as the basis for simulation models presented in the subsequent book chapters. They will identify important assumptions relevant to the RNR model in measuring risk and needs, as well as the issues related to responsivity. Their interpretation of RNR model will allow us to transcend the "what works" mantra to the more focused question of what works for what kind of offender and under what circumstances (Brennan, 2012). The assessment and treatment needs that derive from their summary of the RNR framework will be used as a basis for measurement of RNR concepts in the chapters that follow.

The RNR framework presented in Chap. 4 serves as the guide for the design of contemporary data-driven techniques that will be used in subsequent chapters to create and validate measures of offender risk and criminogenic needs, build the link between risk and needs and appropriate treatment groups, and identify the programs that work for each treatment group. The offender risk/need profile distributions and matched treatment options will then be used as inputs for a nationally based simulation model that examines recidivism outcomes for offenders through RNR adoption in a prison setting. The distributions will also be adjusted to reflect locally based offender inputs for use in an expert system to guide local jurisdictions in implementing RNR-based program model. A specific simulation model designed to estimate the cost-effectiveness and public safety outcomes from programs that divert special populations from jail to community alternatives is also presented. The offender risk and need profiles developed and validated in Chap. 5 will be used by the authors in Chap. 8, to inform the creation of a synthetic data set designed to reflect the profile distributions and associated recidivism estimates of an inmate population.

Creating and validating offender risk and need measures is an essential first step in mapping the RNR process. An important component of the proper use of risk assessment instruments is the practice of validating the instrument for the particular sample on which it is to be applied. Offender populations vary across jurisdictions according to age, gender, race, ethnicity, and type of crimes. It is therefore a best practice to measure an instrument's validity for a particular sample before using it in that setting to make treatment placement decisions. In Chap. 5, Ainsworth and colleagues present the construction of a nationally representative database that merges publically available data on offender risk, needs, and recidivism. This is followed by a discussion of static risk and criminogenic need factors and the various procedures that were used to create and validate risk and need scales created using synthetic data. The resulting risk and need profiles and distributions created from these data will be used as the standard inputs in subsequent chapters that examine program matching and RNR outcomes for offenders at the national level.

In Chap. 6, Crites and colleagues present a method used to incorporate responsivity concepts into treatment planning. She uses the risk and need profile parameter estimates developed in Chap. 3 to identify the appropriate program content and dosage that meet offender risk levels and needs. Six program levels are described representing a continuum of care using increasing intensities of programming targeting different levels of needs. For individuals, key contributors to program-level assignment are risk level based on criminal history, dependence on hard drugs, multiple criminogenic needs, and presence of multiple destabilizing factors (e.g., unstable housing, dysfunctional family, low education).

In addition to identifying the appropriate programming level for individuals, Crites chapter also describes a program classification tool that is designed to identify which program level a specific program or intervention fits into based on characteristics such as target, dosage (clinical hours), content, and staff credentials. Once individuals and programs have been classified, individuals can be match to programs within the appropriate program level to meet their needs. In response to the need for program fidelity, a special program assessment tool is developed that will allow jurisdictions to evaluate available programs along four dimensions including setting, duration, content, and caliber. This chapter concludes with a discussion of pilot tests of this model using data from state and local criminal justice agencies.

Connecting individual risk and needs (described in Chap. 5) with appropriate program levels (described in Chap. 6) is the first stage of the model. Next, the simulation model must determine which specific correctional interventions are most effective at reducing recidivism in each level. In Chap. 7, Caudy and colleagues discuss how evidence-based reviews and meta-analytic findings from the field of criminology can be used to inform simulation model that estimates the impact of adherence to the principles of the RNR model on recidivism. Because meta-analytic research syntheses provide summary effect sizes which reflect the numerous primary studies that have been done on a given topic, they are particularly well suited for informing policy and practice. This chapter illustrates the utility of evidence-based reviews and meta-analyses for identifying the most effective correctional treatment programs. Successful programs can be added to program-level inventories and used as resources for practitioners when selecting appropriate programs for their jurisdictions as well as assisting in the treatment matching process. Caudy and colleagues subsequently illustrate different approaches to measuring system outcomes and simulate the impact on recidivism. Ultimately the model should model should help us identify what impacts might we expect on recidivism if we are able to effectively transfer the RNR principles of effective treatment into actual correctional settings? This is an important concern, and structuring correctional treatment protocols to be consistent with RNR principles in real-world settings may be easier said than done. Research in this area is limited, but a recent study in a local prison setting conducted by Bourgon and Armstrong (2005) found that when properly implemented within an RNR framework, treatment significantly reduces recidivism. Caudy et al found that treating 4 offenders with RNR programs will prevent one recidivism event. This is in contrast to punishing 33 people in order to prevent one recidivism event.

Given that the offender risk and need profiles constructed in Chap. 5 are created from nationally based data sources, they are most useful for supporting RNR implementation on a national level. The profile distributions will require adjustments to reflect locally based populations to support state and local jurisdictions wishing to adhere to an RNR-based offender treatment protocol. In Chap. 8, Bhati and Taxman describe the design and use of a synthetic database for this purpose as the foundation for a simulation model. The model borrows the parameters discussed in Chaps. 5, 6, and 7. Synthetic databases have, at their core, theoretically possible attribute profiles. The profiles are weighted (or re-weighted) to reflect different aggregate properties. The properties may reflect such features as means, rates, variances, covariances, and correlations of various attributes. In effect, once constructed, the synthetic database can be analyzed in much the same way as a real sample from the population of interest. In other words, the synthetic databases can be customized to reflect the characteristics of a local jurisdiction, thereby making it more relevant for localized policy simulations. This chapter describes the methodology used in constructing and re-weighting synthetic databases and demonstrates the procedure with real data from several jurisdictions. This chapter provides an overview of the RNR Simulation Tool expert system and discusses its potential applications to the field. The tool is comprised of three portals that operate collectively to guide the application of the RNR principles in a variety of correctional settings. This innovative web-based simulation tool provides decision support for agencies at the individual, program, and jurisdictional level.

The preceding chapters have used empirical evidence supporting RNR to create links among the principles that model a delivery framework that can be used to guide implementation in correctional settings. This framework presents an opportunity to use simulation techniques to investigate the potential impacts of implementing RNR practice without requiring changes to existing system. Simulation generally refers to a computerized version of a model, which is run over time to study the outcome implications of defined interactions. For our purposes, simulation can be used to show the effects of RNR implementation in a virtual setting.

Simulation techniques have been used to model criminal justice system operations dating as far back as the 1970s (Nagel, 1977). Despite a long history in the criminal justice field, simulation modeling was not widely used due to large resources that were necessary to build and maintain complex integrated models, along with the lack of available data to validate model outcomes. Advances in computer technology and simulation software have made access to simulation model development easier, and the availability of archived criminal justice data sets has provided important resources that can be used to build and validate model inputs. Simulation models have become useful tools for investigating the impacts of various sentencing strategies on the corrections system (Auerhahn, 2003) forecasting prison populations (Austin, 1990) and more recently have been introduced to examining the long-term effects of drug addiction (Zarkin, Dunlap, Hicks, & Mamo, 2005). The application of simulation techniques to assess the RNR model is appropriate given that we are interested in understanding the impact of adopting RNR as a model of correctional treatment. Simulation allows us to create an operational computerized version of the RNR model and then explore various "what if" scenarios regarding RNR implementation and compare the outcomes without disrupting the existing system Chap. 10 describes the building and application of a discrete event simulation model to examine the impact of several treatment scenarios on recidivism at the national level. The results show that RNR programing substanially reduces the number of returns to prison.

Special populations present unique challenges for correctional treatment delivery. This may be particularly true for patients with serious mental illness. Some of these offenders would be better served by being placed in specifically designed treatment programs in the community rather than in jail where serves are lacking. Simulation modeling can be useful for investigating the impacts of jail diversion on these populations and on system outcomes such as cost and public safety. Chapter 9 provides details on a simulation model for projecting the costs and benefits of comprehensive and evidence-based services for mentally ill offenders. The development of the model had two main objectives: (1) to develop the model using operations research methods to simulate the impacts of jail diversion programs and (2) to test that model to obtain projections of the fiscal and client outcome implications of implementing a jail diversion program for the criminal justice system, the mental health system, and the total system expenditures in a community. The model results quickly allowed a comparison of diverted and not diverted groups on several key variables, including costs to mental health and substance abuse systems, costs to the criminal justice agencies, jail days, and individual outcomes (i.e., functional level improvement).

The Mental Health/Jail Diversion Simulation Model provides a tool for communities to use in the process of planning a jail diversion program with a fiscal impact assessment. The model addresses an important public policy consideration: specifically, whether and to what extent jail diversion achieves current and future cost savings. The model confirms the pattern of cost shifting from the criminal justice to the mental health system observed in prior studies. Moreover, the results of the simulations provide stakeholders responsible for designing jail diversion programs with insight into how eligibility criteria affect the pool of individuals who can be intercepted, as well as the overall fiscal impact of the interception itself.

In the last chapter of this book, Taxman, Caudy and Pattavina discuss the future of RNR modeling and simulation for the US correctional system. Whether the goal is to better understand how the criminal justice system works or to examine the possible outcomes of anticipated or planned changes in criminal policies or practices related to correctional treatment, the authors will draw upon the work in this book that demonstrates simulation models can be useful tools for building knowledge about the operation and improvement of the criminal justice system.

The particular focus of this book has been on correctional treatment and planning using an RNR treatment approach. The application and use of simulation tools hold much promise for the future of corrections because policymakers and practitioners are looking for improved means to manage correctional populations in ways that can help offenders lead productive lives. Moreover, academics have provided important treatment frameworks and evidence-based studies to inform the search for improved treatment options. Simulation can be used to test the effects of changes in treatment delivery on a national scale and to serve as a basis for an expert system designed to aid local practitioners. Despite the promise that simulation holds for advancing correctional goals, challenges remain. This chapter concludes with a discussion of the challenges and suggest opportunities for advancing simulation work in criminal justice.

References

Andrews, D. A., Bonta, J., & Hoge, R. D. (1990). Classification for effective rehabilitation: Rediscovering psychology. *Criminal Justice and Behavior, 17*(1), 19–52.

Andrews, D. A., Bonta, J., & Wormith, J. S. (2006). The recent past and near future of risk and/or need assessment. *Crime & Delinquency, 52*(1), 7–27.

Andrews, D. A., & Bonta, J. (2010). The psychology of criminal conduct. Elsevier Science Ltd.

Andrews, D. A., & Bonta, J. (2010). Rehabilitating criminal justice policy and practice. *Psychology, Public Policy, and Law, 16*(1), 39–55. doi:10.1037/a0018362.

Andrews, D. A., Zinger, I., Hoge, R. D., Bonta, J., Gendreau, P., & Cullen, F. T. (1990). Does correctional treatment work? A clinically-relevant and psychologically informed meta-analysis. *Criminology, 28*(3), 369–404.

Auerhahn, K. (2003). *Selective incapacitation and public policy: Evaluating California's imprisonment crisis.* Albany, NY: State University of New York Press.

Austin, J. (2009). Reducing America's correctional populations: a strategic plan. Washington, D.C.: National Institute of Corrections.

Austin, J. (1990). America's growing correctional-industrial complex. National Council on Crime and Delinquency. San Francisco, CA.

Biskupic, J. (2011, May 24). Supreme Court stands firm on prison crowding. *USA Today.*

Bourgon, G., & Armstrong, B. (2005). Transferring the principles of effective treatment into a "real world" prison setting. *Criminal Justice and Behavior, 32*(1), 3–25.

Brennan, T. (2012). Signaling and meta-analytic evaluations the presence of latent offender groups: The importance of coherent target group selection. *Criminology & Public Policy, 11*(1), 61–73.

Caudy, M. (2013). Reducing recidivism through correctional programming: using meta-analyses to inform the RNR simulation tool. In simulation strategies to reduce recidivism: risk need responsivity (RNR) modeling in the criminal justice system. Springer.

Clear, T., & Austin, J. (2009). Reducing Mass Incarceration: Implications of the Iron Law of Prison Populations. *Harvard Law & Policy Review. 3*, 307–324.

Clear, T., Waring, E., & Scully, K. (2005). Communities and reentry: Concentrated reentry cycling. In J. Travis & C. Visher (Eds.), *Prisoner reentry and crime in America* (pp. 179–208). New York, NY: Cambridge University Press.

Cullen, F. T. (2005). The twelve people who saved rehabilitation: How the science of criminology made a difference. *Criminology, 43*(1), 1–42.

Cullen, F. T., & Jonson, C. L. (2011). *Correctional theory: Context and consequences.* SAGE Publications, Inc.

Farabee, D. (2005). *Rethinking rehabilitation: Why can't we reform our criminals?* Washington DC: American Enterprise Institute Press.

Friedmann, P. D., Taxman, F. S., & Henderson, C. E. (2007). Evidence-based treatment practices for drug-involved adults in the criminal justice system. *Journal of Substance Abuse Treatment, 32*(3), 267–277. doi:10.1016/j.jsat.2006.12.020.

Gendreau, P. (1996). The principles of effective intervention with offenders. In A. T. Harland (Ed.), *Choosing correctional options that work: Defining the demand and evaluating the supply* (pp. 117–130). Thousand Oaks, CA: Sage Publications.

Glaze, L. (2011). *Correctional population in the United States, 2010.* Washington, DC: US Department of Justice.

Gottschalk, M. (2010). Cell blocks & red ink: Mass incarceration, the great recession & penal reform. *Daedalus-Journal of the American Academy of Arts and Sciences, 139*(3), 62–73.

Hillsman, S. T. (1990). Fines and day fines. *Crime and Justice, 12*, 49–98.

Lacey, N. (2010). American imprisonment in comparative perspective. *Daedalus-Journal of American Academy of Arts and Sciences, 139*(3), 102–114.

Langan, P. A. & Levin, D. J. (2002). *Recidivism of prisoners released in 1994.* Bureau of Justice Statistics, Special report. Washington, DC: US Department of Justice.

Latessa, E. J., Smith, P., Schweitzer, M., & Brusman Lovins, L. (2009). *Evaluation of selected institutional offender treatment programs for the Pennsylvania Department of Corrections.* Prepared for the Pennsylvania Department of Corrections in Harrisburg, PA.

Lowenkamp, C. T. & Latessa, E. J. (2005). Developing successful reentry programs: Lessons learned from the "what works" research. *Corrections Today, 67*(2), 72–74; 76–77.

Lowenkamp, C. T., Latessa, E. J., & Smith, P. (2006). Does correctional program quality really matter? The impact of adhering to the principles of effective intervention. *Criminology & Public Policy, 5*(3), 201–220. doi:10.1111/j.1745-9133.2006.00388.

Lynch, J., & Sabol, W. (2001). *Prisoner reentry in perspective* (Crime policy report, Vol. 3). Washington, DC: Urban Institute.

MacKenzie, D. L. (2006). What works in corrections: reducing the criminal activities of offenders and delinquents. New York, NY: Cambridge University Press.

Martin, J. L., Lichtenstein, B., Jenkot, R. B., & Forde, D. (2012). They can take us over any time they want: Correctional officers' responses to prison crowding. *The Prison Journal, 92*(1), 88–105.

Martinson, R. (1974). What works? Questions and answers about prison reform. *The Public Interest, 35*, 22–54.

Mauer, M. (2011). Sentencing reform: Amid mass incarcerations-guarded optimism. *Criminal Justice, 26*(1), 27–36.

Morris, N., & Tonry, M. (1990). *Between prison and probation: Intermediate punishments in a rational sentencing system.* New York, NY: Oxford University Press.

Nagel, S. S. (Ed.). (1977). *Modeling the criminal justice system.* Beverly Hills, CA: Sage.

Palmer, T. (1992). *The re-emergence of correctional intervention.* Newbury Park, CA: Sage.

Pattavina, A. (2009). Use of electronic monitoring as persuasive technology: Reconsidering the empirical evidence on the effectiveness of electronic monitoring. *Victims & Offenders, 4*(4), 385–390.

Pattavina, A., & Taxman, F. (2007). Community corrections and soft technology. In J. Byrne & D. Rebovich (Eds.), *The new technology of crime, law and social control*. Monsey, NY: Criminal Justice Press.

Petersilia, J. (2005). From cell to society. In J. Travis & C. Visher (Eds.), *Prisoner reentry and crime in America* (pp. 15–49). New York, NY: Cambridge University Press.

Pew Charitable Trusts (2011). State of recidivism: The revolving door of America's Prisons. Washington, DC.

Phelps, M. S. (2011). Rehabilitation in the punitive era: The gap between rhetoric and reality in U.S. prison programs. *Law & Society Review, 45*(1), 33–68.

Simon, J. (2010). Clearing the "troubled assets" of America's punishment bubble. *Daedalus-Journal American Academy of Arts and Sciences, 139*(3), 91–101.

Taxman, F. S., & Bouffard, J. A. (2003). Substance abuse counselors' treatment philosophy and the content of treatment services provided to offenders in drug court programs. *Journal of Substance Abuse Treatment, 32*(3), 267–277.

Taxman, F., & Marlowe, D. (2006). Risk, needs, responsivity: In action or inaction? *Crime & Delinquency, 25(2), 75–84*.

Taxman, F., Perdoni, M., & Harrison, L. D. (2007). Drug treatment services for adult offenders: The state of the state. *Journal of Substance Abuse Treatment, 32*(3), 239–254.

Taxman, F. S., & Belenko, S. (2012). Implementing evidence-based practices in community corrections and addiction treatment (2012th ed.). Springer.

Travis, J., Solomon, A., & Waul, M. (2001). *From prison to home: The dimensions and consequences of prisoner reentry*. Washington, DC: Urban Institute.

Travis, J., & Visher, C. (2005). Introduction: Viewing public safety through the reentry lens. In J. Travis & C. Visher (Eds.), *Prisoner reentry and crime in America* (pp. 1–14). New York, NY: Cambridge University Press.

Uggen, C., Wakefield, S., & Western, B. (2005). Work and family perspectives on reentry. In J. Travis & C. Visher (Eds.), *Prisoner reentry and crime in America* (pp. 209–243). New York, NY: Cambridge University Press.

Visher, C., & Kachnowski, V. (2007). Finding work on the outside: Results from the "returning home" project in Chicago. In S. Bushway, M. Stoll, & D. F. Weiman (Eds.), *Barriers to reentry? The labor market for released prisoners in post-industrial America* (pp. 80–144). New York, NY: Russell Sage.

Ward, T., & Maruna, S. (2007). *Rehabilitation*. London: Routledge.

Weisburg, R. & Petersilia, J. (2010). The dangers of pyrrhic victories against mass incarceration. *Daedalus-Journal American Academy of Arts and Sciences, 139*(3), 124–133; 146–147.

Welsh, W. N., & Zajac, G. (2004). A census of prison-based drug treatment programs: Implications for programming, policy, and evaluation. *Crime & Delinquency, 50*(1), 108–133.

Western, B. (2008). Reentry: Reversing mass imprisonment. *Boston Review*. Retrieved July 12, 2012 from http://www.bostonreview.net/BR33.4/western.php.

Western, B., & Pettit, B. (2010). Incarceration & social inequality. *Daedalus, 139*(3), 8–19. doi:10.1162/DAED_a_00019.

Zarkin, G., Dunlap, L., Hicks, K., & Mamo, D. (2005). Benefits and costs of methadone treatments: Results from a lifetime simulation model. *Health Economics, 14*(11), 1133–1150.

Chapter 2
A Case Study in Gaps in Services for Drug-Involved Offenders

Faye S. Taxman and Matthew L. Perdoni

Introduction

The US correctional system is a de facto health service provider because at any given time, nearly eight million offenders are under its control (seven million adults and nearly 650,000 youth), and many of these offenders are in need of physical, mental health, and substance abuse services (Binswanger, Krueger, & Steiner, 2009; Glaze & Bonczar, 2006, 2008; Taxman, Young, Wiersema, Rhodes, & Mitchell, 2007). The seven million adults represent nearly 5 % of the adult population in the US (ages 18–65), and the 650,000 youth represents about 4 % of youth in the 13–18 age range. The prevalence of substance use disorders in this population is reported to be nearly 70 % (Glaze & Bonczar, 2006; Karberg & James, 2005; Mumola & Bonczar, 1998), and substance abuse disorders are 4 times more likely among offenders than the general population (SAMHSA, 2006a). Despite strong evidence that substance abuse treatment is an effective strategy to reduce drug use and increase public safety (Chandler, Fletcher, & Volkow, 2009), significant gaps exist in the service delivery system both within the justice system and the substance abuse treatment system at large. This is the purported problem, but there has been little documentation of what the service gap is or how this service gap affects potential outcomes.

The 2005 National Criminal Justice Treatment Practices (NCJTP) survey illustrates that a wide array of services are provided across the spectrum of correctional settings (Taxman, Perdoni, & Harrison, 2007),[1] but the capacity for serving the

---

[1] For a discussion on juvenile justice settings, see Young, D. W., Dembo, R. Henderson, C. E. (2007). A national survey of substance abuse treatment for juvenile offenders. *Journal of Substance Abuse Treatment*, 32:255–266.

F.S. Taxman (✉) • M.L. Perdoni
Department of Criminology, Law and Society, George Mason University,
10900 University Boulevard, Fairfax, VA 20110, USA
e-mail: ftaxman@gmu.edu

F.S. Taxman and A. Pattavina (eds.), *Simulation Strategies to Reduce Recidivism:*   21
Risk Need Responsivity (RNR) Modeling for the Criminal Justice System,
DOI 10.1007/978-1-4614-6188-3_2, © Springer Science+Business Media New York 2013

offender population is very low with an even smaller proportion of offenders having access to appropriate services. As noted by Tucker and Roth (2006), a population impact is a function of three factors: size of the target population, intervention utilization, and effect size. That is, outcomes will be enhanced by exposing the largest percentage of the population to the intervention. For example, a vaccine is not going to eradicate a disease if the vaccine is not delivered to a large enough proportion of the potential target population. Better, and more desirable, findings are known to occur when larger proportions of the population are treated. Few examples exist that assess either the degree to which the intended pool is receiving the appropriate type of services or the impact on desired outcomes like reduced substance abuse and reduced recidivism.

In this chapter, we use two types of simulation models to illustrate the impact of treatment matching on the ability of the justice system to improve recidivism outcomes. We begin with an illustration of the concept of treatment matching in substance abuse treatment where there is more agreement as to the types of substance use disorders that should be placed in different types of treatment programming. In the following case study, we illustrate the components of the matching process in a discipline where there is more clarity in terms of standards regarding placing substance abusers into appropriate levels of care. After we illustrate some of the criteria, we then focus on the gap analyses.

This chapter illustrates the concepts which provide the theoretical framework underlying the simulation model. It provides a model for responsivity for offending behavior that focuses on criminal justice risk, offending behavioral health, and placement criteria. The proposed system incorporates the constructs of the risk, need, and responsivity (RNR) model (Andrews & Bonta, 2003; Taxman & Marlowe, 2006) to maximize the reduction in recidivism by focusing on appropriate placement of offenders in treatment services.

A Case Study of Substance Abusers in the Justice System

The current criminal justice and health systems provide drug treatment services to approximately 10 % of the offender population in need of care (Taxman, Perdoni, & Harrison, 2007). Furthermore, individual level data shows that less than 20 % of offenders have been in treatment services during their period of incarceration (Beck, 2000); this percentage is lower still for offenders under probation or parole supervision (Mumola & Bonczar, 1998). A population impact could be achieved by increasing the service delivery to 30 or 40 % of the population, but only if offenders are targeted for the appropriate levels of care. Reducing recidivism is only possible by providing a larger percentage of the offender population with appropriate treatment services.

One challenge in modeling the intended benefits is that the existing data are often lacking. Existing studies examine offender participation in treatment on a daily basis—often referred to as "stock." However, "stock" measure provides only a limited perspective, because access is affected by the "flow" through treatment programs and services each year, not just the daily populations. This chapter focuses on the annual flow of offenders through the correctional system and their exposure to treatment services while under supervision. The annual flow is relevant because many offenders are involved in the justice system for less than 18 months (see Appendix A). A focus on the annual flow provides a more accurate depiction of the potential for treating offenders while under supervision and allows for an assessment of offenders' impact on and utilization of community health systems.

The Issues of Substance Abuse and Health Problems Among the Adult Offender Population

Research shows that the majority of adult offenders has diagnosable substance abuse disorders and at rates much higher than that of the general adult population. Although less than 10 % of the general population of adults are characterized as dependent or abusers (SAMHSA, 2006b), Mumola and Karberg (2006) report that 53 % of prison inmates meet such criteria. Over 80 % report prior drug use, and almost 60 % use in the month leading up to arrest (Mumola & Karberg, 2006). Nearly 70 % of jail inmates can be classified as either drug dependent or abusers, and over half (55 %) use in the month prior to arrest (Karberg & James, 2005). Half of the community-based offender population uses drugs regularly, and over 30 % use in the month leading up to their immediate offense (Mumola & Bonczar, 1998).

Substance abuse not only plays a role in the day-to-day lives of offenders but is often a factor in offending behavior. One-third of prison inmates were under the influence when committing their instant offense (Mumola & Karberg, 2006), as were approximately 30 % of jailees (Karberg & James, 2005) and 14 % of probationers (Mumola & Bonczar, 1998). Data from the Arrestee Drug Abuse Monitoring program (ADAM) shows that approximately 70 % of both males and females tested positive for one or more illicit substances upon arrest and that this rate remained relatively consistent over the study's 2 decades of data collection (ADAM, 2000). Taylor and colleagues (Taylor, Fitzgerald, Hunt, Reardon, & Brownstein, 2001) found that nearly one-quarter of male (23 %) and female (24 %) arrestees tested positive for two or more drugs. Roughly 20 % of prison and jail inmates report that they committed their immediate crime in order to purchase drugs (Karberg & James, 2005; Mumola & Karberg, 2006). James reports that one-quarter of persons in jail are confined for a drug offense (James, 2004), as are 20 % of prisoners (Sabol, Couture, & Harrison, 2007) and nearly 30 % of probationers (Glaze & Bonczar, 2007).

The offender population also displays higher rates of health disorders, which may complicate health issues. These problems are further complicated, and sometimes driven, by their substance use. Due in part to the exposure to substance abuse and violence in the communities to which they return, offenders released after periods of incarceration have higher than average fatality rates (Binswanger et al., 2007). These environments have also been found to couple with physiological factors to trigger drug cravings (Chandler et al., 2009; Grimm, Hope, Wise, & Shaham, 2001; Volkow et al., 2006), and the communities themselves are negatively impacted by offenders' health problems (Freudenberg, 2001). The adult offender population is far more likely than the general public to be infected with HIV/AIDS (Maruschak, 2004; Weinbaum, Sabin, & Santibanez, 2005) and other health problems like diabetes and high blood pressure (Hammett, 2001). With the exception of diabetes, angina or myocardial infarction, and obesity, offenders tend to have higher odds of other medical disorders than the general population with some variation based on the correctional setting (Binswanger et al., 2009). Over 30 % of the confined offender population is infected with Hepatitis C (Beck & Maruschak, 2004), as compared to less than 2 % of the general public (CDC, 2008). Approximately 60 % of prison and jail inmates suffer from mental health problems (James & Glaze, 2006), and it has been estimated that roughly 20 % of prisoners, jail inmates, and offenders under community supervision could classify as mentally ill (Ditton, 1999). Often, these mental health problems are accompanied by substance abuse disorders (Abram, Teplin, McClelland, & Dulcan, 2003). The offender population exhibits behaviorally and medically complex disorders that are difficult to treat.

Treatment Services in the Criminal Justice System

The nexus between substance abuse and criminal involvement provides evidence of the importance of addressing offender's risk behaviors while they are involved with the justice system. Glaze and Bonczar (2007) estimate that out of every ten parolees exiting supervision, four have been unsuccessfully terminated from supervision. Of every five exiting probationers, one is terminated from supervision. Estimates have also shown that roughly 70 % of prisoners recidivate within 3 years after release (Langan & Levin, 2002). Common to all of these figures is that reincarceration or violation of supervisory terms is often the result of positive drug tests or failure to comply with treatment plans.

Despite the overwhelming need for services (Binswanger et al., 2007; Chandler et al., 2009; Glaser & Greifinger, 1993; Hammett, Gaiter, & Crawford, 1998; Hammett, Harmon, & Rhodes, 2002), the criminal justice system does not recognize its role as a service provider (Taxman, Henderson, & Belenko, 2009). Studies have shown that the availability of and access to treatment while involved in the justice system is minimal. A recent study provides nationally representative estimates on the availability of treatment services across all correctional settings (Taxman, Perdoni, & Harrison, 2007). These findings show that substance abuse treatment services are sparse, and when they are provided, they tend to be inadequate for dealing with the severity of the problems presented. More importantly, the survey findings show that intensive treatment services (defined as more than a single counseling session a week) are rarely offered, and they are provided to few offenders in the system. Drug and alcohol education and outpatient counseling are most prevalent throughout the system. Three-quarters of prisons offer drug and alcohol education, as do 53 % of community agencies and 61 % of jails. Just over half (55 %) of prisons provide under 4 hours of group counseling per week, as do 47 % of community agencies and 60 % of jails. This distribution of services is inconsistent with the severity of the substance abuse disorders reported by offenders (Belenko & Peugh, 2005).

Intensive treatment services—intensive outpatient (5 or more hours a week), therapeutic community, and drug treatment courts—would be more appropriate for the justice-involved-dependent population given the severity of substance abuse. However, less than half of prisons offer 5 hours or more of group counseling per week, compared with 30 % of community agencies, and roughly one-quarter of jails (Taxman, Perdoni, & Harrison, 2007). Twenty percent of prisons provide therapeutic communities, as do 26 % of jails and 3 % of community correctional agencies. Furthermore, the capacity to provide services is relatively low, and only a small number of offenders can participate in them (Chandler et al., 2009).

The prominence of substance problems among the offender population and astonishing rates of recidivism support the notion that these issues are interrelated. Seventy-five percent of the overall offender population is supervised in community settings. As illustrated above, this segment of the offender population possesses the same risk and need characteristics as prison and jail inmates. Thus, by failing to address the mental and physical health and addiction problems displayed by the offender population, the criminal justice system exposes community health to unnecessary risks.

Methods of the NCJTP Survey Used in Modeling Treatment Placement

The NCJTP survey consists of a representative sample of prisons, jails, and community correctional agencies (Taxman, Young et al., 2007). We began with facilities listed in the Bureau of Justice Statistics' (BJS) 2000 census of prisons to generate the prison portion of the sampling frame (Stephan & Karberg, 2003). Federal prisons, prisons devoted to medical or mental health treatment, and prisons categorized as community corrections facilities were not included in the study. From the remaining facilities, prisons specializing in drug and alcohol treatment were sampled with certainty ($n=58$). The remainder of the sample was generated by selecting prisons from this frame using the methodology employed by BJS for their national surveys of prisons. In the first step of this method, the county was broken into regions representing the South, West, Midwest, and Northeast, and the four states with the largest correctional populations were classified separately, resulting in eight regions. Within these eight strata, facilities were chosen randomly with the probability of selection proportionate to the size of the facility's daily population. Ninety-two additional prisons were selected using this technique.

There is not a complete listing of community corrections agencies, since many are operated by state or local agencies. Therefore, we generated a sampling frame using a two-stage stratified cluster technique. The first stage was the selection of counties from within the 3,141 US counties or county equivalents. We used the same technique that was used for the selection of the prison sample by stratifying the eight national regions by the size of the county's population. County population included three categories: "small" (less than 250,000), "medium" (250,000–750,000), and "large" (more than 750,000). From the resulting 24 strata, we selected counties with populations over three million with certainty and again utilized the probability proportional to size technique to generate a sample of 72 counties. The second stage was the selection of correctional programs and services within the 72 counties, resulting in a sample of 644 facilities.

Survey instruments were mailed to wardens, chief probation and parole officers, and other facility administrators. The survey contained questions on daily facility operations, gathered demographic information on the administrator and the facility, and collected data on funding, treatment practices, attitudes and philosophies on treatment and service delivery, screening and assessment practices, integration with outside agencies, management techniques, and other questions on facility and offender management. A response rate of 70 % was attained for the prison sample and a response rate of 71 % attained for the community sample.

Sampling weights were also developed for the data. For the prison sample, we assumed that nonrespondents were missing at random. We developed weights based on the probability of selection, adjusted the sampling weights for nonresponses (Elliot, 1991), and trimmed excess values (Potter, 1988, 1990). A similar process was followed for the community sample, though values were not adjusted for nonresponse (Taxman, Young et al., 2007).

Modeling the Findings

Average daily populations: The NCJTP survey polled administrators on the types of treatment services available in their facilities, as well as the number of offenders in these various programs on any given day. We applied the sampling weights to the survey data to estimate the average daily population (ADP) of offenders in the various treatment services in all US correctional agencies. These estimates are represented in the column titled "Average Daily Population in Services" in Tables 2.1, 2.2, and 2.3.[2]

---

[2]For more detailed information on the daily populations and access rates reported in Tables 2.1, 2.2, and 2.3, see Taxman, F. S., Perdoni, M. L., & Harrison, L. D. (2007). Drug treatment services for adult offenders: The state of the state. *Journal of Substance Abuse Treatment*, 32:239–254. It should be noted that one adjustment occurred in these tables to account for an adjustment of the classification of one unit from a jail facility into a community corrections facility.

Table 2.1 Estimate of offender treatment needs and annual flow through treatment services in prisons

Average daily population of adults in prisons: 1,233,867

Service	Average daily population in services	Daily population in need of treatment	Est. population receiving services Conservative Model (1) (% population receiving appropriate treatment)	Est. population receiving services Liberal Model (2) (% population receiving appropriate treatment)	Est. % population flowing through services annually (range)
Alcohol and drug education	75,543	N/A	83,683 (N/A)	100,591 (N/A)	6.8–8.2
Low intensity	34,618	238,963	58,245 (24.4)	70,377 (29.5)	4.7–5.7
Medium intensity	64,475	228,574	47,919 (21.0)	56,102 (24.5)	3.9–4.5
High intensity	45,487	406,633	57,833 (14.2)	69,953 (17.2)	4.7–5.7
Total in clinical services*	144,580	874,170	163,997 (18.8)	196,431 (22.5)	13.3–15.9
Total adjusted for phased treatment structures			150,948 (17.3)	180,826 (20.7)	12.2–14.7

*Excludes drug and alcohol education

Estimating need for different levels of treatment services: We reviewed the literature on severity of substance use disorders among the correctional population to generate estimates on the number of offenders with substance abuse disorders in need of clinical treatment services. Belenko and Peugh estimate that 31.5 % of male prisoners are substance dependent, 18.7 % have a serious abuse disorder, 20.2 % are abusers, and 29.6 % have no substance abuse problem (Belenko & Peugh, 2005). Female prisoners have a higher rate of dependence, with 52.3 % falling within this classification, while 16.2 % have a serious abuse disorder, 8.3 % are classified as abusers, and 23.2 % have no substance abuse problem requiring treatment. These figures mirror those reported in other studies for jail inmates and community-based offender populations (BJS, 2004; Taylor et al., 2001). We used these estimates to determine levels of service need.

Given that there are gender differences in the severity of the problem, the next step was to account for gender breakdowns across all settings. According to Harrison and Beck (2006), 93 % of prisoners are male and 7 % are female, and 87 % of jail inmates are male and 13 % female. Glaze and Palla (2005) report that 77 % of probationers are male and 23 % female and that 88 % of the parolees are male and 12 % are female.

We then generated estimates of the number of offenders with some type of substance problem in the criminal justice system on any given day. First, sampling

Table 2.2 Estimate of offender treatment needs and annual flow through treatment services in jails

Average daily population of adults in jails: 745,765

Service	Average daily population in services	Daily population in need of treatment	Est. population receiving services Conservative Model (1) (% population receiving appropriate treatment)	Est. population receiving services Liberal Model (2) (% population receiving appropriate treatment)	Est. % population flowing through services annually (range)
Drug and alcohol education	46,071	N/A	130,217 (N/A)	157,874 (N/A)	17.5–21.2
Low intensity	43,334	139,374	102,241 (73.4)	125,484 (90.0)	13.7–16.8
Medium intensity	16,674	137,090	14,891 (10.9)	18,178 (13.3)	2.0–2.4
High intensity	11,578	254,616	14,350 (5.6)	16,833 (6.6)	1.9–2.3
Total in clinical services*	71,586	531,080	131,482 (24.8)	160,495 (30.2)	17.6–21.5
Total adjusted for phased treatment structures			73,670 (13.9)	90,046 (17.0)	9.9–12.1

*Excludes drug and alcohol education

weights were applied to the data, and estimates of the ADP in each setting were generated. These estimates were then split by the gender breakdowns listed above. Finally, the prevalence of substance abuse disorders reported by Belenko and Peugh was applied. This figure is represented in the column titled "Daily Population in Need of Treatment" in Tables 2.1–2.3.

Matching offender need to treatment services: The next step was to categorize treatment services by their levels of intensity. The risk, need, and responsivity (RNR) model suggests that offenders should be assigned to services based on the seriousness of their risk of recidivism and the severity of their problem behavior (such as substance abuse, mental health disorders, sexual deviance, histories of violence) (Taxman & Marlowe, 2006). This model resembles the Patient Placement Criteria (PPC) recommended by the American Society of Addiction Medicine (ASAM) for substance abuse (Graham, Schultz, Mayo-Smith, & Ries, 2003). The RNR and PPC models are built on the premise that the severity of the problem disorder should control the duration, design, content, and type of service delivered. Together, these models suggest the following service categorization. Individuals with dependent disorders should participate in more *intensive services* than those with threshold disorders. For those with dependent disorders, intensive services involve more frequent interaction with counselors and a therapeutic community setting. Intensive

Table 2.3 Estimate of offender treatment needs and annual flow through treatment services in community correctional facilities

Average daily population of adults in community corrections: 5,864,152

Service	Average daily population in services	Daily population in need of treatment	Est. population receiving services Conservative Model (1) (% population receiving appropriate treatment)	Est. population receiving services Liberal Model (2) (% population receiving appropriate treatment)	Est. % population flowing through services annually (range)
Drug and alcohol education	192,072	N/A	310,277 (N/A)	373,974 (N/A)	5.3–6.4
Low intensity	145,070	1,035,572	237,949 (23.0)	288,351 (27.8)	4.1–4.9
Medium intensity	40,520	1,065,295	38,189 (3.6)	44,431 (4.2)	0.7–0.8
High intensity	27,987	2,107,622	25,286 (1.2)	28,966 (1.3)	0.4–0.5
Total in clinical services*	213,577	4,208,489	301,425 (7.2)	361,748 (8.6)	5.1–6.2
Total adjusted for phased treatment structures			281,693 (6.7)	338,834 (8.1)	4.8–5.8

*Excludes drug and alcohol education

outpatient counseling services (offered for 5 or more hours per week), considered *medium intensity*, are more appropriate for individuals who do not use substances daily and whose use does not interfere with daily functioning. *Low-intensity* outpatient counseling services, including those providing infrequent counseling and some type of pharmacological medications (like methadone maintenance), are suited for individuals with low-threshold disorders. This categorization of service intensity is reflected in Tables 2.1–2.3.

Flow of offenders through treatment services: In this study, we develop two models for measuring the annual flow of offenders through treatment service. These models are based on the number of times per year a facility offers a particular treatment program and the retention rates in these programs. The number of times a program can be offered in a year is determined using the duration of the reported program.[3] As shown in Appendix A, Model 1, a more conservative model assumes that treatment programs are offered less frequently (fewer times per year), while Model 2, the more liberal model, assumes that programs are offered more frequently (more

[3] When respondents indicated multiple durations for a single program, we used the response indicating the shorter duration.

times per year than assumed in Model 1). We do not assume that these programs are offered in "closed group" formats, where all offenders enter and leave on specific and common days. Using estimates from prior studies, we assume that retention rates in residential programs are approximately 65 % (Joe, Simpson, & Broome, 1999; Martin, Butzin, Saum, & Inciardi, 1999) and retention rates in other services are approximately 55 % (Joe et al., 1999). The retention rate figures are used to reduce the estimates of offenders receiving care each year.

The models also adjust estimates for the number of offenders participating in various services as a *part* of a total treatment program. In these "phased treatment structures," offenders participate in more than one service at a time. Drug and alcohol education is often the first phase of a layered treatment program. Thus, if the agency administrator reported that the number of offenders in drug and alcohol education is equal to the number of participants in other services, then we assume it is a phased treatment structure. Furthermore, if the facility offered three or more services with the same enrollment, we assume each is part of a phased structure. We adjusted the annual population estimates by counting the enrollment in the individual services making up the phased treatment structure only once.[4] Overall, 13 % of facilities offer services through a phased structure.

Estimates of the annual flow through treatment services were generated by multiplying the number of times per year that the service can be offered by the number of offenders in the program and then reducing this estimate by the retention rate assumptions. The calculation was completed for each group of services (high-, medium-, and low-intensity classifications) using the criteria described above. The sampling weights were applied to the data to generate national estimates of the flow of offenders through the services. The model also generates estimates adjusted for the population participating in phased treatment structures.

Results from the Model

The following tables report the capacity of the correctional system to provide services through their facilities or in conjunction with outside agencies.

Prisons: As shown in Table 2.1, 874,170 of the 1.2 million offenders in prisons likely need some form of clinical substance abuse treatment, but the actual number receiving appropriate care on the average day is under 145,000. Between 163,997 (Conservative Model (1)) and 196,431 (Liberal Model (2)), prisoners complete treatment programs annually. After adjusting for phased treatment structures, the flow estimate ranges between 150,948 and 180,826 offenders, respectively.

---

[4]The count of services in each case was also factored into this process. If a facility reported that they provide four or more services, the threshold value was set at 3. However, when the facility reported three services, the criteria for determining phased programming was set at a minimum of two identical values, and when the facility reported two services, the criteria was at least one identical value.

Overall, most prisoners have access to less intensive services, geared for those with low-threshold disorders. Although a higher proportion of prisoners complete low- (under 5 hours of group counseling and methadone) and high-intensity (therapeutic communities) programs each year than in other settings, these estimates still represent only a small percentage of the population in need.

Jails: Table 2.2 provides flow estimates for jails. Over 531,000 of the 745,765 inmates in US jails are in need of some level of treatment services on any given day. However, the daily capacity for providing services is low, as only 71,586 (14 %) of jail inmates have access to treatment daily. The annual participation rate ranges from 131,482 (Conservative Model (1)) to 160,496 (Liberal Model (2)) jail inmates.

The annual flow estimates for jails drop considerably after the models are adjusted for phased programming. It is estimated that over 40 % of the population of jail inmates completing particular services each year participate as a part of a phased treatment structure. Furthermore, the post-adjustment estimate of annual completions in the Conservative Model (1) is roughly equal to the estimate of jailed people in clinical treatment programs on any given day.

Community corrections: Table 2.3 shows the estimated treatment need and annual completions in community correctional settings. Seventy percent of the offenders under community supervision have some type of substance use disorder, meaning that about 4.2 million need clinical treatment services. On any given day, only 213,577 receive such care (5 %). The annual flow through community-based programs ranges from 301,425 to 361,748 offenders. After adjusting for phased treatment structures, the estimates drop to 281,693 under the Conservative Model (1) and 338,834 under the Liberal Model (2). Like prisons, about 6 % of annual completions in community corrections participate in phased treatment structures.

Annual completion estimates are lowest for high-intensity services. Under the Conservative Model (1), an estimated 25,286 offenders complete high-intensity services annually, as opposed to 237,949 for low-intensity treatment services, and 310,277 for drug and alcohol education. Under the Liberal Model (2), an estimated 361,748 offenders complete clinical services each year, of which 288,000 complete low-intensity programming. Thus, high-intensity service completions account for roughly 8 % of the total annual completions in community-based settings, while low-intensity services account for nearly 80 %.

Correctional programming: Correctional programs, such as intensive supervision, work release, and day reporting, are designed to be graduated sanctions that intensify the supervision of offenders in the community. Often these programs are designed to address offenders that have a high-risk profile. Treatment services are a frequent component of these programs, although most of the provided treatment services fall within the range of educational and/or infrequent counseling type of services.

As shown in Table 2.4, half of the agencies providing intensive supervision (53 %) and transitional housing (50 %) programs incorporate treatment services as a part

Table 2.4 Prevalence of treatment services within correctional programs

Program	% Agencies	% With programs that include SA treat-ment services	% With services that include drug and alcohol education	% With services that include counseling services	% With services that include medium-intensity services	% With services that include high-intensity services
Day reporting	10.8	19.2	93.2	54.4	56.4	54.1
Intensive supervision program	41.1	52.6	92.3	81.0	55.2	11.8
Work release	22.2	39.0	90.1	82.1	21.2	62.9
Transitional housing	15.0	49.5	92.5	75.8	65.7	23.6
Vocational training	28.4	29.7	97.4	70.9	86.4	17.3
Education	48.5	17.0	94.7	51.9	41.7	32.0

of their program structure.[5] Much less common is the inclusion of services in other correctional programs, such as education (17 %), day reporting (19 %), and vocational training (30 %). Almost all facilities including treatment within correctional program-ming provide drug and alcohol education (over 90 % across all such programs), but rates drop drastically when the focus shifts to more intensive treatment services. Typically, as the intensity level increases, the less likely it is a service modality is incorporated into a program—much like the trends observed for treatment services in general discussed in the previous sections. While 81 % of the intensive supervision programs including treatment provide low-intensity services, 55 % incorporate medium-intensity services, and only 12 % include high-intensity services. Although 76 % of the transitional housing programs including treatment provide low-intensity services, 66 % include medium-intensity services, and 24 % include high-intensity services. Overall, the availability of high-intensity services within correctional pro-grams is low, ranging from 12 % (intensive supervision) to 63 % (work release).

The Importance of Modeling RNR and Bringing the System Closer to Best Practices

Effective public health and public safety strategies emphasize the importance of risk reduction as a primary goal. For public health, the desired reductions are focused on physical and psychological health, including reduced substance abuse. Public

---

[5]For information on the availability of treatment services within correctional programs by criminal justice setting, see Taxman, F. S., Perdoni, M. L., & Harrison, L. D. (2007). Drug treatment services for adult offenders: The state of the state. *Journal of Substance Abuse Treatment*, 32:239–254.

safety is concerned about reducing the odds of being involved in criminal behavior, particularly personal and violent crimes. The value of providing quality drug treatment services to offenders cannot be understated since all evidence points to ineffective programs having little to no impact on offender outcomes (Cullen, Myer, & Latessa, 2009).

Although it is well acknowledged that offenders in the correctional system need substance abuse treatment services, this study puts a face on the concerns about the dearth of services available. The study measures the current capacity of the correctional and public health systems to provide substance abuse treatment services, both in terms of overall capacity but more importantly the capacity to provide the appropriate level of care. First, using a clinically based definition for substance use disorders (defining the problem severity), this case study illustrates that half of the offenders do not have a substance use disorder that requires any intervention or services. While some might have low-threshold substance use behaviors, these behaviors do not warrant the need for treatment services. For the other half of offenders that are in need of substance abuse treatment services, the existing system is woefully inadequate. For these individuals with severe substance abuse disorders, the typical treatment programs available are geared towards those with low-threshold disorders. Second, the study finds that for the total offender population (i.e., regardless of setting) anywhere between 7.6 and 9.2 % of offenders, on an average day, can participate in some type of programming and/or treatment program. This low capacity illustrates one of the compelling problems for the correctional system: providing few services or having limited access to services illustrates that service delivery is not a priority, as determined either by correctional administrators or by funding agencies. Low capacity for providing access to services affects both the ability to deter or rehabilitate criminal behavior, and even more importantly it negatively impacts the ability of the correctional culture to embrace programs and services as part of the core operations. Low capacity means that services operate at the margin of the mission of correctional agencies perpetuating the competing correctional values of punishment vs. effective treatment services.

The analytical framework for this case study provides a roadmap for addressing service gaps in this difficult policy arena of providing treatment services for drug-involved offenders. The need to expand the array of services in the community has been a consistent theme for nearly 2 decades. Potter (1990) recommended a set of intermediate sanctions (e.g., day reporting centers, work release, intensive supervision) to provide better oversight and management of the offender population in the community. This occurred concurrently with the development and implementation of drug treatment courts which served to provide a novel approach for increasing access to treatment services for offenders, followed by the design and implementation of Residential Substance Abuse Treatment services in prisons (and jails) with continued care in the community.

In 2001, the Serious and Violent Offender Initiative (SVORI) was built on the premise that more services and programs were needed to reduce the risk of recidivism, yet there is little to suggest that providing such services had an impact (Lattimore et al., 2010). As shown in this case study, and companion studies,

treatment programs for the most part continue to exist at the margins, and even correctional programs such as intensive supervision and work release are providing
to a relatively small percentage of the population that could benefit from these services (see Taxman, Perdoni, & Harrison, 2007 for a discussion). The challenge
to the system is not only to provide substance abuse treatment services but as Potter
(1990) identified earlier to expand the variety of community programming.

The analytical framework used in this study should apply to other critical areas
where there is a need to better understand how the available programs match the
criminal risk of offenders or other psychosocial needs (i.e., mental health, sexual
deviancy, education deficits). As far as we know, there has been no systemic
analysis of criminal risk levels (propensity to commit criminal behavior) and the
type of appropriate programming that will serve to reduce the risk of recidivism.
Yet, meta-analyses exist that demonstrate the efficacy of drug treatment courts
(Mitchell, Wilson, Eggers, & MacKenzie, 2012), in-prison treatment and aftercare
(Wilson, Mitchell, & MacKenzie, 2006), work release and education programs
(Wilson, Gallagher, & MacKenzie, 2000), and intensive supervision (MacKenzie,
2006) that demonstrate the value of different programs, and there are a host of other
correctional programs that are available that theoretically reduce the likelihood
of recidivism such as electronic monitoring, home arrest, and transitional housing.
Unlike the ASAM criteria that deal with patient placement for substance abuse
disorders, no such industry standards exist regarding the appropriate level of supervision or programming (care) for offenders presenting different risk portfolios.

For programs that are designed to safely manage the offender in the community,
more research is needed to define the types of programs that are needed for different
risk portfolios of offenders. And, standards are then needed for the programs. It is
recognized that a need exists to implement a cadre of programming that serve
to address the risk factors along with the other areas that serve to propel people
to continue their criminal behavior. The evidence-based practices literature consists
of this theme that diagnostics should drive programming (see Andrews & Bonta,
2003; Taxman & Marlowe, 2006) and that by failure to do so illustrates that "quackery" guides action instead of professionalism (Cullen et al., 2009; Latessa, Cullen,
& Gendreau, 2002).

The roadmap that derives from a risk, need, and responsivity (RNR) framework
for service delivery is that the appropriate level of care will be a major consideration
given towards efforts to reduce the risk for recidivism and other important behaviors
such as drug use and risky behaviors. A focus on this model holds the promise
of advancing the adoption of quality services that are suitable to reduce the risk
of recidivism. In the criminal justice lexicon, risk is synonymous with punishment
severity; in the treatment arena risk should indicate the need for more intensive,
structured services that address multiple dimensions. For example, consider the following illustration of an RNR approach that links criminal justice risk to substance
abuse severity (Fig. 2.1 below). Here, the level of programming would be associated
with risk factors, and the available services would be used to address the severity
of the problem disorder (substance abuse) given the criminal justice risk factors.
Low-risk offenders would be offered less services unless they have a dependent

Substance Abuse/ Dependency	High Risk	Medium Risk	Low Risk
Dependent	Residential TX	Intensive outpatient	Intensive Outpatient Employment
Abuser	Criminal Values Therapy as part of Intensive Outpatient	Outpatient with Employment	Outpatient with Employment
None	Criminal thinking/values Employment	Criminal thinking/values Employment	Employment

Fig. 2.1 Illustration of RNR approach with substance use

disorder, and then the focus would be on the disorder. But low-risk offenders with no disorder would not be required to participate in treatment; instead the emphasis will be on other punishments (such as fines) and prosocial behaviors such as employment. This model serves to comingle the provision of adequate care into the equation of reducing the risk of recidivism.

Gladwell's (2000) concept of a "tipping point" suggests that change efforts take hold in an environment only after moving beyond a critical point. Findings from this study show that few offenders can access appropriate services, and with so few services available in the system, the system remains focused on security, enforcement, and punishment. The integration of treatment, including a therapeutic environment that promotes offender change, is barely present. Existing research shows the problems associated with implementing treatment in correctional settings (Taxman & Belenko, 2012; Farabee et al., 1999, Taxman & Bouffard, 2000), and much of these issues are due to the correctional culture that does not recognize its role as a service provider. Treatment programs are considered secondary to the primary mission of the agency, and as a result implementation problems arise from the misalignment of correctional goals.

Revamping the correctional mission is a critical issue given the spiraling incarceration populations and costs and the need to develop a community corrections system that prevents incarceration. Little attention over the last 2 decades has been devoted to the community correctional system. Federal funds have been used for Residential Substance Abuse Treatment Programs (in-prison treatment) and drug treatment courts that serve around 49,000 offenders every year—both make a small dent in the nearly eight million offenders under correctional control. Neither of these are part of an overall strategy to develop a community correctional system that is responsive to the various risk and criminogenic needs of offenders. Most of the solutions on the table—use of evidence-based treatment programs and supervision—do not address the need to expand the full range of programming and services geared to the needs of the offender profiles.

If there is a desire to move towards a more evidence-based approach, then there will be a need to adapt the correctional environment. The implementation of evidence-based practices and the use of appropriate assessment and placement protocol are keys to the successful expansion of correctional alternatives. Probation and parole supervision exposes the offender to an array of service outlets in the community. However, these agencies are not structured to effectively manage their clientele, as they implement only 4.6 (of 11) evidence-based practices (Friedmann, Taxman, & Henderson, 2007). Drug courts, a widely heralded alternative to traditional criminal justice involvement, implement an average of 5.9 EBPs and only 6.1 of the "10 key components" (Taxman & Perdoni, 2009). And, the use of risk and screening tools across all correctional settings leaves something to be desired (Taxman, Perdoni, & Harrison, 2007). With broader utilization of EBPs, universal adoption of standardized screening and assessment tools, and the implementation of an approach like the RNR model, offenders suited for lower levels of control and who pose the least risk to public safety can be assigned to more efficient and effective types of supervision.

Conclusion

Acting on the ideas presented here can be done fluidly and efficiently, because a common principle underlies everyone: do what works. With the majority of the criminal justice population suffering from various substance abuse, health, and mental health problems, it is clear that the system must assume responsibility for implementing the changes necessary to effectively address offender needs. However, findings from this case study show that less than 10 % of offenders successfully complete treatment programs sufficient to address their needs. Even a modest improvement on current practices would not yield drastic changes given the size of the correctional population and the unmet needs. A strategic approach will better enable correctional agencies to focus on implementation issues associated with better practices, and this includes using the RNR model to guide the types of services available to the offender population. This is what we will demonstrate in other chapters of this book.

Acknowledgements This study was funded under a cooperative agreement from the US Department of Health and Human Services, Public Health Service, National Institutes of Health, and National Institute on Drug Abuse (NIH/NIDA) to George Mason University (Grant U01 DA016213-01, Action Research to Advance Drug Treatment in the CJS). The funding for this cooperative agreement was supplemented by the Center for Substance Abuse Treatment, Bureau of Justice Assistance, Centers for Disease Control and Prevention, and National Institute on Alcohol Abuse and Alcoholism. The authors acknowledge the collaborative contributions by federal staff from NIDA and members of the coordinating center (George Mason University) and the nine research center grantees of the NIH/NIDA CJ-DATS 1 Cooperative. The contents are solely the responsibility of the authors and do not necessarily represent the official views of NIH/NIDA or other federal agencies or any other participants in CJ-DATS.

Appendix A: assumptions underlying models 1 and 2

	Model 1	Model 2
Duration assumption (number of times per year service is offered)		
Under 30 days	10	12
31–90 days	5	6.2
91–120 days	3	3.5
121–180 days	2	2.4
181–365 days	1.2	1.3
Over 365 days	0.8	1
Retention rate assumption		
Low-/medium-intensity programming	55 %	
High-intensity programming	65 %	

References

Abram, K. M., Teplin, L. A., McClelland, G. M., & Dulcan, M. K. (2003). Comorbid psychiatric disorders in youth in juvenile detention. *Archives of General Psychiatry, 60*(11), 1097–1108.

Andrews, D. A., & Bonta, J. (2003). *The psychology of criminal conduct.* Cincinnati, OH: Anderson.

Arrestee Drug Abuse Monitoring (ADAM). (2000). *Arrestee drug abuse monitoring annual report, 1999.* Washington, DC: Office of Justice Programs, National Institute of Justice.

Beck, A. J. (2000). *State and federal prison population tops one million.* Washington, DC: Office of Justice Programs, Bureau of Justice Statistics.

Beck, A. J. & Maruschak, L. M. (2004). *Hepatitis testing and treatment in state prisons.* (NCJ-199173C). Washington, DC: Office of Justice Programs, Bureau of Justice Statistics.

Belenko, S., & Peugh, J. (2005). Estimating drug treatment needs among state prison inmates. *Drug and Alcohol Dependence, 77*(3), 269–281.

Belenko, S., Taxman, F. S., & Wexler, H. K. (2008). *Technology transfer in correctional agencies for substance abuse treatment programs.* Washington, DC: National Institute on Corrections.

Bhati, A. S., Roman, J. K., & Chalfin, A. (2008). *To treat or not to treat: Evidence on the prospects of expanding treatment to drug-involved offenders.* Washington, DC: The Urban Institute, Justice Policy Center.

Binswanger, I. A., Krueger, P. M., & Steiner, J. F. (2009). Prevalence of chronic medical conditions among jail and prison inmates in the United States compared with the general population. *Journal of Epidemiology and Community Health, 63*(11), 912–919.

Binswanger, I. A., Stern, M. F., Deyo, R. A., Heagerty, P. J., Cheadle, A., Elmore, J. A., et al. (2007). Release from prison: A high risk of death for former inmates. *The New England Journal of Medicine, 356,* 157–165.

BJS. (2004). Bureau of Justice Statistics Correctional Surveys, as presented in *Correctional populations in the United States, annual, prisoners in 2004,* and *probation and parole in the United States, 2004.* Washington, DC: Office of Justice Programs, Bureau of Justice Statistics.

Centers for Disease Control and Prevention (CDC). (2008). *Viral hepatitis statistics and surveillance.* Retrieved February 23, 2009 from http://www.cdc.gov/hepatitis/Statistics.htm.

Chandler, R. K., Fletcher, B. W., & Volkow, N. D. (2009). Treating drug abuse and addiction in the criminal justice system: Improving public health and safety. *Journal of the American Medical Association, 301*(2), 183–190.

Cullen, F. T., Myer, A. J., & Latessa, E. J. (2009). Eight lessons from *Moneyball*: the high cost of ignoring evidence-based corrections. *Victims and Offenders, 4*(2), 197–213.

Ditton, P. M. (1999). *Mental health and treatment of inmates and probationers* (NCJ-174463). Washington, DC: Office of Justice Programs, Bureau of Justice Statistics.

Elliot, D. (1991). *Weighting for non-response: A survey researcher's guide.* London, England: Social Survey Division, Office of Population Census and Surveys.

Farabee, D., Prendergast, M. L., Cartier, J., Wexler, H., Knight, K., & Anglin, M. D. (1999). Barriers to implementing effective correctional treatment programs. *Prison Journal, 79*(2), 150–162.

Freudenberg, N. (2001). Jails, prisons, and the health of urban populations: A review of the impact of the correctional system on community health. *Journal of Urban Health, 78*(2), 214–235.

Friedmann, P. D., Taxman, F. S., & Henderson, C. E. (2007). Evidence-based treatment practices for drug-involved adults in the criminal justice system. *Journal of Substance Abuse Treatment, 32*(3), 267–277.

Gladwell, M. (2000). *The tipping point: How little things can make a big difference.* Boston, MA: Little Brown.

Glaser, J. B., & Greifinger, R. B. (1993). Correctional health care: A public health opportunity. *Annals of Internal Medicine, 118*(2), 139–145.

Glaze, L. E. & Bonczar, T. P. (2006). *Probation and parole in the United States, 2005* (NCJ-215091). Washington, DC: Office of Justice Programs, Bureau of Justice Statistics.

Glaze, L. E. & Bonczar, T. P. (2007). *Probation and parole in the United States, 2006* (NCJ-220218). Washington, DC: Office of Justice Programs, Bureau of Justice Statistics.

Glaze, L. E. & Bonczar, T. P. (2008). *Probation and parole in the United States, 2007* (NCJ-224707). Washington, DC: Office of Justice Programs, Bureau of Justice Statistics.

Glaze, L. E. & Palla, S. (2005). *Probation and parole in the United States, 2004* (NCJ-215091). Washington, DC: Office of Justice Programs, Bureau of Justice Statistics.

Graham, A. W., Schultz, T. K., Mayo-Smith, M. F., & Ries, R. K. (2003). *Principles of addiction medicine* (3rd ed.). Washington, DC: American Society of Addiction Medicine.

Grimm, J. W., Hope, B. T., Wise, R. A., & Shaham, Y. (2001). Neuroadaptation: Incubation of cocaine craving after withdrawal. *Nature, 412*(6843), 141–142.

Hammett, T. M. (2001). Making the case for health interventions in correctional facilities. *Journal of Urban Health, 78*(2), 236–240.

Hammett, T. M., Gaiter, J. L., & Crawford, C. (1998). Researching seriously at-risk populations: Health interventions in criminal justice settings. *Health Education & Behavior, 25*(1), 99–120.

Hammett, T. M., Harmon, M. P., & Rhodes, W. (2002). The burden of infectious disease among inmates of and releases from US correctional facilities, 1997. *American Journal of Public Health, 92*(11), 1789–1794.

Harrison, P. M. & Beck, A. J. (2006). *Prison and jail inmates at midyear, 2005* (NCJ-213133). Washington, DC: Office of Justice Programs, Bureau of Justice Statistics.

James, D. J. (2004). *Profile of jail inmates, 2002* (NCJ-201932). Washington, DC: Office of Justice Programs, Bureau of Justice Statistics.

James, D. J. & Glaze, L. E. (2006). *Mental health problems of prison and jail inmates* (NCJ-213600). Washington, DC: Office of Justice Programs, Bureau of Justice Statistics.

Joe, G. W., Simpson, D. D., & Broome, K. M. (1999). Retention and patient engagement models for different treatment modalities in DATOS. *Drug and Alcohol Dependence, 57*(2), 113–125.

Karberg, J. C. & James, D. J. (2005). *Substance dependence, abuse, and treatment of jail inmates, 2002* (NCJ-209588). Washington, DC: Office of Justice Programs, Bureau of Justice Statistics.

Langan, P. A. & Levin, D. J. (2002). *Recidivism of prisoners released in 1994* (NCJ-193427). Washington, DC: Office of Justice Programs, Bureau of Justice Statistics.

Latessa, E. J., Cullen, F., & Gendreau, P. (2002). Beyond correctional quackery— Professionalism and the possibility of effective treatment. *Federal Probation, 66*(2), 43–49.

Lattimore, P. K., Steffey, D. M., & Visher, C. A. (2010). Prisoner reentry in the first decade of the twenty-first century. *Victims and Offenders, 5*(3), 253–267.

MacKenzie, D. L. (2006). Aftercare following a correctional bootcamp may reduce recidivism. *Criminology & Public Policy, 5*(2), 359–362.

Martin, S. S., Butzin, C. A., Saum, C. A., & Inciardi, J. A. (1999). Three year outcomes of therapeutic community treatment for drug-involved offenders in Delaware: From prison to work release to aftercare. *Prison Journal, 79*(3), 294–320.

Maruschak, L. M. (2004). *HIV in prisons and jails*. Washington, DC: Office of Justice Programs, Bureau of Justice Statistics.

Mitchell, O. J., Wilson, D. B., Eggers, A., & MacKenzie, D. L. (2012). Assessing the effectiveness of drug courts on recidivism: A meta-analytic review of traditional and non-traditional drug courts. *Journal of Criminal Justice, 40*(1), 60–71.

Mitchell, O. J., Wilson, D. B., & MacKenzie, D. L. (2007). Does incarceration-based drug treatment reduce recidivism? A meta-analytic synthesis of the research. *Journal of Experimental Criminology, 3*(4), 353–375.

Mumola, C. J., & Bonczar, T. P. (1998). *Substance abuse and treatment of adults on probation, 1995*. Washington, DC: Office of Justice Programs, Bureau of Justice Statistics.

Mumola, C. J. & Karberg, J. C. (2006). *Drug use and dependence, state and federal prisoners, 2004* (NCJ-213530). Washington, DC: Office of Justice Programs, Bureau of Justice Statistics.

Potter, F. J. (1988). *Survey of procedures to control extreme sampling weights*. Alexandria, VA: Proceedings of the Section on Survey Research Methods.

Potter, F. J. (1990). *A study of procedures to identify and trim extreme sampling weights*. Alexandria, VA: Proceedings of the Section on Survey Research Methods.

Sabol, W. J., Couture, H., & Harrison, P. M. (2007). *Prisoners in 2006* (NCJ-219416). Washington, DC: Office of Justice Programs, Bureau of Justice Statistics.

Stephan, J. J. & Karberg, J. C. (2003). *Census of state and federal correctional facilities, 2000* (NCJ-198272). Washington, DC: Office of Justice Programs, Bureau of Justice Statistics.

Substance Abuse and Mental Health Services Administration. (2006a). *Substance abuse treatment in adult and juvenile correctional facilities*. Rockville, MD: Office of Applied Studies.

Substance Abuse and Mental Health Services Administration. (2006b). *Results from the 2004 national survey on drug use and health: National findings*. Rockville, MD: Office of Applied Studies (SMA 05-4062).

Taxman, F. S., & Belenko, S. (2012). *Implementing evidence-based practices in community corrections and addiction treatment*. New York: Springer.

Taxman, F. S., & Bouffard, J. (2000). The importance of systems in improving offender outcomes: New frontiers in treatment integrity. *Justice Research and Policy, 2*(2), 37–58.

Taxman, F. S., & Bouffard, J. (2002). Assessing therapeutic integrity in modified therapeutic communities for drug-involved offenders. *Prison Journal, 82*(2), 189–212.

Taxman, F. S., Henderson, C. E., & Belenko, S. S. (2009). Organizational context, systems change, and adopting treatment delivery systems in the criminal justice system. *Drug and Alcohol Dependence, 103*(S1), S1–S6.

Taxman, F. S., & Marlowe, D. M. (2006). Risk, needs, responsivity: In action or inaction. *Crime and Delinquency, 52*(1), 3–6.

Taxman, F. S., & Perdoni, M. L. (2009). Drug treatment in drug courts: Results from a national survey. In J. M. Miller (Ed.), *21st Century criminology: A reference handbook*. Newbury Park, CA: Sage.

Taxman, F. S., Perdoni, M. L., & Harrison, L. D. (2007). Drug treatment services for adult offenders: The state of the state. *Journal of Substance Abuse Treatment, 32*(3), 239–254.

Taxman, F. S., Young, D. W., Wiersema, B., Rhodes, A., & Mitchell, S. (2007). National criminal justice treatment practices survey: Methods and procedures. *Journal of Substance Abuse Treatment, 32*(3), 225–238.

Taylor, B., Fitzgerald, N., Hunt, D., Reardon, J. A., & Brownstein, H. (2001). *ADAM preliminary 2000 findings on drug use and drug markets: Adult male arrestees*. Washington, DC: Office of National Drug Control Policy.

The National Center on Addiction and Substance Abuse. (2009). *Shoveling it up II: The impact of substance abuse on federal, state, and local budgets*. New York, NY: Columbia University.

The Pew Center on the States. (2007). *Public safety, public spending: Forecasting America's prison population.* Washington, DC: The Pew Charitable Trusts, Public Safety Performance Project.

Tucker, J. A., & Roth, D. L. (2006). Extending the evidence hierarchy to enhance evidence-based practice for substance use disorders. *Addiction, 101*(7), 918–932.

Volkow, N. D., Wang, G. J., Telang, F., Fowler, J. S., Logan, J., Childress, A. R., et al. (2006). Cocaine cues and dopamine in dorsal striatum: Mechanism of craving in cocaine addiction. *Journal of Neuroscience, 26*(24), 6583–6588.

Weinbaum, C. M., Sabin, K. M., & Santibanez, S. S. (2005). Hepatitis B, hepatitis C, and HIV in correctional populations: A review of epidemiology and prevention. *AIDS, 19*(S3), S41–S46.

Wilson, D. B., Gallagher, C. A., & MacKenzie, D. L. (2000). A meta-analysis of corrections-based education, vocation, and work programs for adult offenders. *Journal of Research in Crime and Delinquency, 37*(4), 347–368.

Wilson, D. B., Mitchell, O., & MacKenzie, D. L. (2006). A systematic review of drug court effects on recidivism. *Journal of Experimental Criminology, 2*(4), 459–487.

Chapter 3
The Simulation Modelling Process

Andrew Greasley

Introduction

This chapter outlines the steps required in undertaking a simulation project. In order to use simulation successfully, a structured process must be followed. This chapter aims to show that simulation is more than just the purchase and use of a software package. It requires a range of skills that include project management, client liaison, statistical skills, modelling skills and the ability to understand and map out organisational processes. Particular challenges are presented when modelling scenarios that incorporate people, such as criminal justice systems. This approach calls for methods that model human behaviour.

Simulation modelling is a flexible tool and is capable of analysing most aspects of an organisation. Therefore, to ensure that maximum value is gained from using the technique, it is necessary to define the areas of the organisation that are key to overall performance and select feasible options for the technique in these areas. Another aspect to consider is the nature of the simulation model that is to be developed. In order to assist the decision-making process, it is not always necessary to undertake all the stages of a simulation study. For instance, the development of the process map may be used to help understanding of a problem. The level of usage of simulation is discussed in this chapter. There follows a description of project management concepts, a discussion of methods of modelling human behaviour and an outline of the steps in the simulation modelling process.

A. Greasley, Ph.D. (✉)
Aston Business School, Birmingham, UK
e-mail: a.greasley@aston.ac.uk

F.S. Taxman and A. Pattavina (eds.), *Simulation Strategies to Reduce Recidivism:*   41
Risk Need Responsivity (RNR) Modeling for the Criminal Justice System,
DOI 10.1007/978-1-4614-6188-3_3, © Springer Science+Business Media New York 2013

Determining the Level of Usage of the Simulation Model

An important aspect in the process of building a simulation model is to recognise that there are many possible ways of modelling a system. Choices have to be made regarding the level of detail to use in modelling processes and even whether a particular process should be modelled at all. The way to make these choices is to recognise that before the model is built, the objectives of the study must be defined clearly. It may even be preferable to build different versions of the model to answer different questions about the system, rather than build a single 'flexible' model that attempts to provide multiple perspectives on a problem. This is because two relatively simple models will be easier to validate, and thus there will be a higher level of confidence in their results than a single complex model.

The objective of the simulation technique is to aid decision-making by providing a forum for problem definition and providing information on which decisions can be made. Thus a simulation project does not necessarily require a completed computer model to be a success. At an early stage in the project proposal process, the analyst and other interested parties must decide the role of the model building process within the decision-making process itself. Thus in certain circumstances the building of a computer model may not be necessary. However for many complex, interacting systems (i.e. most business systems), the model will be able to provide useful information (not only in the form of performance measures, but indications of cause and effect linkages between variables) which will aid the decision-making process. The focus of the simulation project implementation will be dependent on the intended usage of the model as a decision-making tool (Table 3.1).

The level of usage categories is defined as follows:

Problem Definition

One of the reasons for using the simulation method is that its approach provides a detailed and systematic way of analysing a problem in order to provide information

Table 3.1 Levels of usage of a simulation model

| | Level of usage | | | |
	Problem definition	Demonstration	Scenarios	Ongoing decision support
Level of development	Process map	Animation	Experimentation	Decision-support system
Level of interaction	None	None Simple menu	Menu	Extended menu
Level of integration	None	Stand-alone	Stand-alone	Stand-alone database
			Database	Real-time data

on which a decision can be made. It is often the case that ambiguities and inconsistencies are apparent in the understanding of a problem during the project proposal formulation stage. It may be that the process of defining the problem may provide the decision-makers with sufficient information on which a decision can be made. In this case model building and quantitative analysis of output from the simulation model may not be required. The outcome from this approach will be a definition of the problem and possibly a process map of the system.

Demonstration

Although the decision-makers may have an understanding of system behaviour, it may be that they wish to demonstrate that understanding to other interested parties. This could be to internal personnel for training purposes or to external personnel to demonstrate capability to perform to an agreed specification. The development of an animated model provides a powerful tool in communicating the behaviour of a complex system over time.

Scenarios

The next level of usage involves the development of a model and experimentation in order to assess system behaviour over a number of scenarios. The model is used to solve a number of pre-defined problems but is not intended for future use. For this reason a simple menu system allowing change of key variables is appropriate. The simulation may use internal data files or limited the use of external databases.

Ongoing Decision Support

The most fully developed simulation model must be capable of providing decision support for a number of problems over time. This requires that the model be adapted to provide assistance to new scenarios as they arise. The menu system will need to provide the ability to change a wider range of variables for ongoing use. The level of data integration may require links to company databases to ensure the model is using the latest version of data over time. Links may also be required to real-time data systems to provide ongoing information on process performance. Animation facilities should be developed to assist in understanding cause and effect relationships and the effect of random events.

If it is envisaged that the client will perform modifications to the simulation model after delivery, then the issue of model reuse should be addressed. Reuse issues include ensuring detailed model code documentation is supplied and detailed operating procedures are provided. Training may also be required in model development and statistical methods. Another reason for developing a model with ongoing decision-support capabilities is to increase model confidence and acceptance particularly among non-simulation experts (Muller, 1996).

Managing the Simulation Project

An important aspect of the project management process is identifying and gaining the support of personnel who have an interest in the modelling process. In addition to the technical skills required to build and analyse the results from a model must be able to communicate effectively with people in the client organisation in order to collect relevant data and communicate model results. Roles within the project team include the following:

- *Client*: Sponsor of the simulation project—usually a manager who can authorise the time and expenditure required.
- *Model user*: Person who is expected to use the model after completion by the modeller. The role of the model user will depend on the planned level of usage of the model. A model user will not exist for a problem definition exercise but will require extended contact with the developer if the model is to be used for ongoing decision support to ensure all options (e.g. menu option facilities) have been incorporated into the design before handover.
- *Data provider*: Often the main contact for information regarding the model may not be directly involved in the modelling outcomes. The client must ensure that the data provider feels fully engaged with the project and has allocated time for liaison and data collection tasks. In addition the modeller must be sensitive to using the data provider's time as productively as possible.

The project report should contain the simulation study objectives and a detailed description of how each stage in the simulation modelling process will be undertaken. This requires a definition of both the methods to be used and any resource requirements for the project. It is important to take a structured approach to the management of the project as there are many reasons why a project could fail. These include:

- The simulation model does not achieve the objectives stated in the project plan through a faulty model design or coding.
- Failure to collect sufficient and relevant data means that the simulation results are not valid.
- The system coding or user interface design does not permit the flexible use of the model to explore scenarios defined in the project.
- The information provided by the simulation does not meet the needs of the relevant decision-makers.

These diverse problems can derive from a lack of communication leading to failure to meet business needs to technical failures, such as a lack of knowledge of statistical issues in experimentation, leading to invalid model results. For this reason the simulation project manager must have an understanding of both the business and technical issues of the project. The project management process can be classified into the four areas of Estimation, Scheduling/Planning, Monitoring and Control and Documentation.

Estimation

This entails breaking down the project into the main simulation project stages (data collection, modelling input data, etc.) and allocating resources to each stage. Estimates are needed of the time required and skill type of people required along with the need for access to resources such as simulation software. These estimates will allow a comparison between project needs and project resources available. If there are insufficient resources available to undertake the project, then a decision must be made regarding the nature of the constraints on the project. A resource-constrained project is limited by resources (i.e. people/software) availability. A time-constrained project is limited by the project deadline. If the project deadline is immovable, then additional resources will need to be requested in the form of additional personnel (internal or external), overtime or additional software licences. If the deadline can be changed, then additional resources may not be required as a smaller project team may undertake the project over a longer time period.

Once a feasible plan has been determined a more detailed plan of when activities should occur can be developed. The plan should take into account the difference between effort time (how long someone would normally be expected to take to complete a task) and elapse time which takes into account availability (actual time allocated to a project and the number of people undertaking the task) and work rate (skill level) of people involved. In addition a time and cost specification should be presented for the main simulation project stages. A timescale for the presentation of an interim report may also be specified for a larger project. Costings should include the cost of the analyst's time and software/hardware costs. If the organisation has access to the appropriate hardware, then there is the choice between 'run-time' licences (if available) providing use of the model but not the ability to develop new models. A full licence is appropriate if the organisation wishes to undertake development work in-house. Although an accurate estimate of the timescale for project completion is required, the analyst or simulation client needs to be aware of several factors that may delay the project completion date.

The most important factor in the success of a simulation project is to ensure that appropriate members of the organisation are involved in the simulation development. The simulation provides information on which decisions are made within an organisational context, so involvement is necessary of interested parties to ensure confidence and implementation of model results. The need for clear objectives is essential to ensure the correct systems components are modelled at a suitable level of detail. Information must also be supplied for the model build from appropriate personnel to ensure important detail is not missing and false assumptions regarding model behaviour are not made. It is likely that during the simulation process, problems with the system design become apparent that require additional modelling and/or analysis. Both analyst and client need to separate between work for the original design and additional activity. The project specification should cover the number of experimental runs that are envisaged for the analysis. Often the client may require additional scenarios tested, which again should be agreed at a required additional time/cost.

Scheduling/Planning

Scheduling involves determining when activities should occur. Steps given in the simulation study are sequential, but in reality they will overlap—the next stage starts before the last one is finished—and are iterative, e.g. validate part of the model, go back and collect more data, model build and validate again. This iterative process of building more detail into the model gradually is the recommended approach but can make judging project progress difficult.

Monitoring and Control

A network plan is useful for scheduling overall project progress and ensuring on-time completion, but the reality of iterative development may make it difficult to judge actual progress.

Documentation

Interim progress reports are issued to ensure the project is meeting time and cost targets. Documents may also be needed to record any changes to the specification agreed by the project team. Documentation provides traceability. For example, data collection sources and content should be available for inspection by users in future in order to ensure validation. Documentation is also needed of all aspects of the model such as coding and the results of the simulation analysis.

Methods of Modelling Human Behaviour

As stated in the introduction to this chapter, a particular challenge presents itself when modelling scenarios that incorporate people, such as criminal justice systems, and this section outlines methods of modelling human behaviour. These methods can be identified and classified by the level of detail (termed abstraction) required to model human behaviour (see Fig. 3.1). Each approach is given a *method name* and *method description* listed in order of the level of abstraction used to model human behaviour.

The methods are classified into those that are undertaken *outside the model* (i.e. elements of human behaviour are considered in the simulation study but not incorporated in the simulation model) and those that incorporate human behaviour within the simulation model, termed *inside the model*. Methods inside the model are classified in terms of a *world view*. Pegden, Shannon, and Sadowski (1995) describe a world view as giving a framework for defining the components of the system in sufficient detail to allow the model to execute and simulate the system. *Model abstraction* is categorised as macro, meso or micro in order to clarify the different levels of abstraction for models 'inside the model'.

The framework then provides a suggested *simulation approach* for each of the levels of abstraction identified from the literature. The simulation approaches identified are continuous simulation which may be in the form of a system dynamics

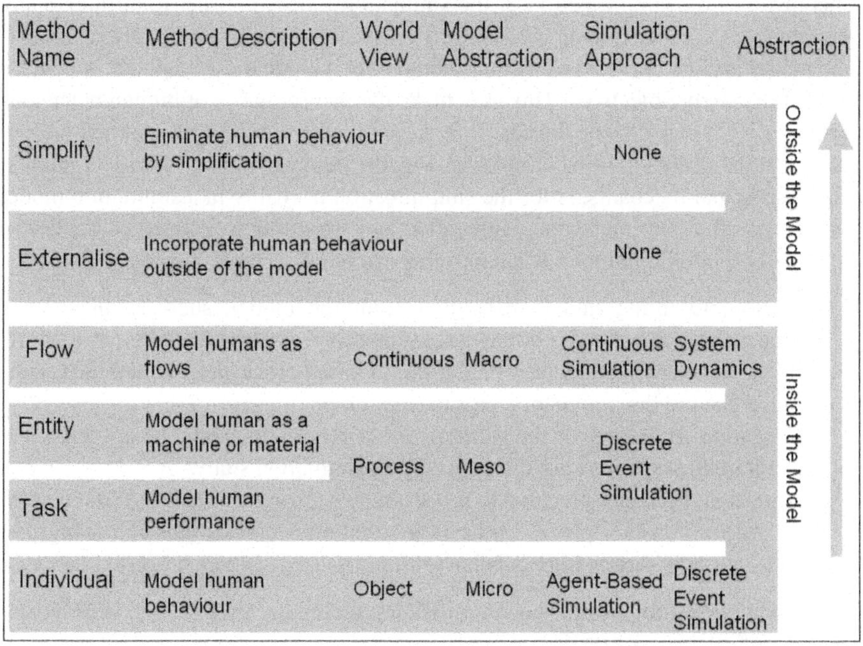

Method Name	Method Description	World View	Model Abstraction	Simulation Approach	Abstraction
Simplify	Eliminate human behaviour by simplification			None	Outside the Model
Externalise	Incorporate human behaviour outside of the model			None	
Flow	Model humans as flows	Continuous	Macro	Continuous Simulation System Dynamics	Inside the Model
Entity	Model human as a machine or material	Process	Meso	Discrete Event Simulation	
Task	Model human performance				
Individual	Model human behaviour	Object	Micro	Agent-Based Simulation Discrete Event Simulation	

Fig. 3.1 Methods of modelling human behaviour in a simulation study

model and discrete-event simulation which may be in the form of an agent-based model. Continuous simulation relates to when the state of the system changes continuously over time and systems are described by differential equations. The implementation of the continuous world view is usually associated with the use of the system dynamics technique (Forrester, 1961). For a discrete-event simulation, a system consists of a number of objects (entity) which flow from point to point in a system while competing with each other for the use of scarce resources (resource). The attributes of an entity may be used to determine future actions taken by the entities. Agent-based simulation is based on the discrete-event mechanism and is an increasingly popular tool for modelling human behaviour (Macal & North, 2006). Agents can be defined as an entity with autonomy (it can undertake a set of local operations) and interactivity (it can interact with other agents to accomplish its own tasks and goals) (Hayes, 1999). A particular class of agent-based systems termed multi-agent simulations are concerned with modelling both individual agents (with autonomy and interactivity) and also the emergent system behaviour that is a consequence of the agent's collective actions and interactions (Shaw & Pritchett, 2005).

The methods of modelling human behaviour shown in Fig. 3.1 are now described in more detail.

Simplify (Eliminate Human Behaviour by Simplification)

This involves the simplification of the simulation model in order to eliminate any requirement to codify human behaviour. This strategy is relevant because a simulation model is not a copy of reality and should only include those elements necessary to meet the study objectives. This may make the incorporation of human behaviour unnecessary. It may also be the case that the simulation user can utilise their knowledge of human behaviour in conjunction with the model results to provide a suitable analysis. Actual mechanisms for the simplification of reality in a simulation model can be classified into omission, aggregation and substitution (Pegden et al., 1995). In terms of modelling human behaviour this can relate to the following:

- Omission: Omitting human behaviour from the model, such as unexpected absences through sickness. It may be assumed in the model that alternative staffing is allocated by managers. Often machine-based processes are modelled without reference to the human operator they employ.
- Aggregation: Processes or the work of whole departments may be aggregated if their internal working is not the focus of the simulation study.
- Substitution: For example, human processes may be substituted by a 'delay' element with a constant process time in a simulation model, thus removing any complicating factors of human behaviour.

An example of the use of the simplification technique is described in Johnson, Fowler, and Mackulak (2005).

Externalise (Incorporate Human Behaviour Outside of the Model)

This approach involves incorporating aspects of human behaviour in the simulation study but externalising them from the simulation model itself. The 'externalise' approach to representing human decision-making is to elicit the decision rules from the people involved in the relevant decisions and so avoid the simplification inherent when codifying complex behaviour. The three approaches to externalising human behaviour are:

- Convert decision points and other aspects of the model into parameters which require human input. Most likely in this context, the model will operate in a gaming mode in order to combine the benefits of real performance while retaining experimenter control and keeping costs low (Warren, Diller, Leung, & Ferguson, 2005).
- Represent decisions in an expert system linked to the simulation model (Robinson, Edwards, & Yongfa, 2003).
- Use the simulation model as a recording tool to build up a set of examples of human behaviour at a decision point. This data set is then used by a neural network to represent human behaviour (Robinson, Alifantis, Edwards, Ladbrook, & Waller, 2005).

Flow (Model Humans as Flows)

At the highest level of abstraction inside the model, humans can be modelled as a group which behaves like a flow in a pipe. In the case of the flow method of modelling human behaviour, the world view is termed continuous and the model abstraction is termed macro. The type of simulation used in the implementation of the flow method is either a continuous simulation approach or the system dynamics technique. Hanisch, Tolujew, Richter, and Schulze (2003) present a continuous model of the behaviour of pedestrian flows in public buildings. Cavana, Davies, Robson, and Wilson (1999) provide systems dynamics analysis of the drivers of quality in health services. Sterman, Repenning, and Kofman (1997) developed a system dynamics model to analyse the behavioural responses of people to the introduction of a Total Quality Management (TQM) initiative.

Entity (Model Human as a Machine or Material)

This relates to a mesoscopic (meso) simulation in which elements are modelled as a number of discrete particles whose behaviour is governed by predefined rules. One way of modelling human behaviour in this way would mean that a human would be represented by a resource, such as equipment that is either 'busy' or 'idle'.

Alternatively, modelling a human as an entity would mean that they would undertake a number of predetermined steps, such as the flow of material in a manufacturing plant. This approach can be related to the process world view which models the flow of entities through a series of process steps. Greasley and Barlow (1998) present a discrete-event simulation of the arrest process in the UK Police service. Here the arrested person is the 'customer' and is represented by an entity object. The police personnel, for example, a Police Constable, is represented by a resource object in the simulation. The police personnel either are engaged in an activity (e.g. interviewing an arrested person) or are 'idle'. This method permits people modelled as resource objects to be monitored for factors such as utilisation in the same way a machine might be.

Task (Model Human Performance)

This method models the action of humans in response to a pre-defined sequence of tasks and is often associated with the term human performance modelling. Human performance modelling relates to the simulation of purposeful actions of a human as generated by well-understood psychological phenomenon, rather than modelling in detail all aspects of human behaviour not driven by purpose (Shaw and Pritchett, 2005). The task approach can be related to the process world view and mesoscopic (meso) modelling abstraction level which models the flow of entities, in this case people, through a series of process steps. The task approach is implemented using a discrete-event simulation which incorporates into the rules governing the behaviour of the simulation attributes of human behaviour. These attributes may relate to factors such as skill level, task attributes such as length of task, and organisational factors such as perceived value of the task to the organisation. Bernhard and Schilling (1997) outline a technique for dynamically allocating people to processes depending on their qualification to undertake the task. When a material is ready to be processed at a work station, the model checks for and allocates the requisite number of qualified workers necessary for the task. The approach is particularly suitable for modelling group work, and the paper investigates the relative overall throughput time for a manufacturing process with different worker skill sets.

Individual (Model Human Behaviour)

This method involves modelling how humans actually behave based on individual attributes such as perception and attention. The approach is associated with an object world view where objects are self-contained units combining data and functions but are also able to interact with one another. The modelling approach can be termed microscopic (micro) and utilises either the discrete-event or agent-based simulation types.

Brailsford and Schmidt (2003) provide a case study of using discrete-event simulation to model the attendance of patients for screening of a sight-threatening complication of diabetes. The model takes into account the physical state, emotions, cognitions and social status of the persons involved. This approach is intended to provide a more accurate method of modelling patients' attendance behaviour compared with the standard approach used in simulation studies of using a simple random sampling of patients. The approach uses the PECS (Schmidt, 2005) architecture for modelling human behaviour at an individual level. This was implemented by assigning numerical attributes, representing various psychological characteristics, to the patient entities. These characteristics included patient anxiety, perceived susceptibility, knowledge of disease, belief about disease prevention, health motivation and educational level. Each characteristic was defined as low, medium or high. These characteristics are then used to calculate the four PECS components of physics, emotion, cognition and status. These in turn were used to calculate the compliance or likelihood that the patient would attend. The form of these calculations and the parameters within them were derived from trial and error in finding a plausible range of values for the compliance compared with known estimates of population compliance derived from the literature.

Another example of the use of discrete-event simulation to model individual human behaviour is the development of a Micro Saint model by Keller (2002) which uses the visual, auditory, cognitive and psychomotor (VCAP) resource components (Wickens, 1984) to estimate the total workload on a person driving a car while talking on a mobile phone. Prichett, Lee, and Goldsman (2001) use agent-based modelling to investigate the behaviour of air traffic controllers. Lam (2007) demonstrates the use of an agent-based simulation to explore decision-making policies for service policy decisions.

Choosing a Method to Simulate Human Behaviour

In terms of choosing an approach the *simplify* approach ignores the role of humans in the process and is appropriate when it is not necessary to model the role of human behaviour to meet the study objectives. Also in practical terms it may take too long to model every aspect of a system even if it was feasible, which it may not be in most cases due to a lack of data, for example. The potential problem with the strategy of simplification is that the resulting model may be too far removed from reality for the client to have confidence in model results. The job of validation is to ensure the 'right' model has been built, and a 'social' role of the simulation developer is to ensure that the simulation client is assured about simulation validity if they are to have confidence in the simulation results.

The *externalise* approach attempts to incorporate human behaviour in the study but not within the simulation model itself. The area of gaming simulation represents a specialist area of simulation when the model is being used in effect to collect data from a human in real time and react to this information. Alternative techniques such as expert systems and neural networks can be interfaced with the simulation and be used to provide a suitable repository for human behaviour. There will, however, be most likely a large overhead in terms of integrating these systems with simulation software.

The *flow* approach models humans at the highest level of abstraction using differential equations. The level of abstraction, however, means that this approach does not possess the ability to carry information about each entity (person) through the system being modelled and is not able to show queuing behaviour of people derived from demand and supply (Stahl, 1995). Thus the simulation of human behaviour in customer processing applications, for example, may not be feasible using this approach.

The *entity* approach models human behaviour using the process world view to represent people by either simulated machines (resources) or simulated materials (entities). This allows the availability of staff to be monitored in the case of resources and the flow characteristics of people, such as customers, to be monitored in the case of entities. Staff may be categorised with different skill sets and variability in process durations can be estimated using sampling techniques. This will provide useful information in many instances but does not reflect the way people actually work. For instance the approach to work of individual staff may be different, particularly in a service context where their day-to-day schedule may be a matter of personal preference.

The *task* approach attempts to model how humans act without the complexity of modelling the cognitive and other variables that lead to that behaviour. The rationale behind the approach is described by Shaw and Pritchett (2005), 'In this approach models are described as modelling performance rather than behaviour because of their scope—the current state of the art is better at capturing purposeful actions of a human as generated by well-understood psychological phenomenon, than it is at modelling in detail all aspects of human behaviour not driven by purpose'. Elliman, Eatock, and Spencer (2005) use task and environmental variables, rather than individual characteristics to model individual behaviour. Bernhard and Schilling (1997) model people using the entity method but separate material flow from people flow. No individual differences are taken into account and the approach uses a centralised mechanism/database to control workers.

The *individual* approach attempts to model the internal cognitive processes that lead to human behaviour. A number of architectures that model human cognition, such as PECS (Schmidt, 2000), Adaptive Control of Thought-Rational (ACT-R) (Anderson & Lebiere, 1998), Soar (Newell, 1990) and Theory of Planned Behavior (TPB) (Ajzen, 1991), have been developed. These have the aim of being able to handle a range of situations in which the person has discretion on what to do next and are more realistic with respect to internal perceptual and cognitive processes for which the external environment constraint is less useful (Pew, 2008). However the difficulty of implementation of the results of studies on human behaviour by behavioural and cognitive researchers into a simulation remains a significant barrier. Silverman, Cornwell and O'Brien (2003) state 'there are well over one million pages of peer-reviewed, published studies on human behaviour and performance as a function of demographics, personality differences, cognitive style, situational and emotive variables, task elements, group and organisational dynamics and culture' but go on to state 'unfortunately, almost none of the existing literature addresses how to interpret and translate reported findings as principles and methods suitable for implementation or synthetic agent development'. Another barrier is the issue of the context of the behaviour represented in the simulation. Silverman (1991) states 'many first principle models from the behavioural science literature have been derived within a particular setting, whereas simulation developers may wish to deploy these models in different contexts'. Further issues include the difficulty of use of these architectures (Pew, 2008) and the difficulty of validation of multiple factors of human behaviour when the research literature is largely limited to the study of the independent rather than the interactive effects of these factors.

The Steps in a Simulation Model Project

The following steps should be undertaken when undertaking a simulation modelling project assignment:

1. Simulation study proposal
2. Data collection
3. Process mapping
4. Modelling input data
5. Building a model
6. Validation and verification
7. Experimentation and analysis
8. Presentation of results
9. Implementation

These steps will now be described in more detail.

Simulation Study Proposal

The requirements for each section of the simulation project proposal are now given.

Study Objectives: A number of specific study objectives should be derived which will provide a guide to the data needs of the model, set the boundaries of the study (scope) and the level of modelling detail and define the experimentation analysis required. It is necessary to refine the study objectives until specific scenarios defined by input variables and measures that can be defined by output variables can be specified. General improvement areas for a project include aspects such as the following:

- Changes in process design: Changes to routing, decision points and layout
- Changes in resource availability: Shift patterns and equipment failure
- Changes in demand: Forecast pattern of demand on the process

Many projects will study a combination of the above, but it is important to study each area in turn to establish potential subjects for investigation at the project pro-posal stage. The next step is to define more specifically the objectives of the study.

Once the objectives and experiments have been defined the scope and level of detail can be ascertained. The model scope is the definition of the boundary between what is to be included in the model and what is considered external to the specification. Once the scope has been determined it is necessary to determine the level of detail in which to model elements within the model scope. In order to keep the model complexity low, only the minimum model scope and level of detail should be used. Regarding model scope there can be a tendency for the model user to want to include every aspect of a process. However this may entail building such a complex model that the build time and the complexity of interpreting model results may lead to a failed study. Regarding model detail, judgement is required in deciding what elements of the system should be eliminated or simplified to minimise unnecessary detail. An iterative process of model validation and addition of model detail should be followed.

Strategies for minimising model detail include:

- Modelling a group of processes by a single process: Often the study requires no knowledge of the internal mechanisms within a process, and only the process time delay is relevant to overall performance.
- Assuming continuous resource availability: The modelling of shift patterns of personnel or maintenance patterns for machinery may not always be necessary if their effect on performance is small.
- Infrequent events such as personnel absence through sickness or machine breakdown may occur so infrequently that they are not necessary to model.

What is important is that any major assumptions made by the developer at the chosen level of detail are stated explicitly in the simulation report, so that the user is aware of them.

Data Collection

The collection of data is one of the most important and challenging aspects of the simulation modelling process. A model which accurately represents a process will not provide accurate output data unless the input data has been collected and analysed in an appropriate manner. Data requirements for the model can be grouped into two areas. In order to construct the process map which describes the logic of the model (i.e. how the process elements are connected), the process routing and decision points are required as follows:

Logic data required for the process map include:

- *Process routing*: All possible routes of people/components/data through the system.
- *Decision points*: Decision points can be modelled by conditional (if … . then x, else y) or probability (with 0.1, x; with 0.5, y; else z) methods.

In order to undertake the model building stage, further data is required in terms of the process durations, resource availability schedules, demand patterns and the process layout.

Additional data required for the simulation model:

- *Process timing*: Durations for all relevant processes (e.g. length of prison stay). Can be a data sample from which a probability distribution is derived
- *Resource availability*: Resource availability schedules for all relevant resources
- *Demand pattern*: A schedule of demand which 'drives' the model (e.g. court commitments to prison)
- *Process layout*: Diagram/schematic of the process which can be used to develop the simulation animation display

Be sure to distinguish between input data which is what should be collected and output data which is dependent on the input data values. For example, customer arrival times would usually be input data, while customer queue time is output data, dependent on input values such as customer arrival rate. However, although we would not enter the data collected on queue times into our model, we could compare these times to the model results to validate the model.

The required data may not be available in a suitable format, in which case the analyst must either collect the data or find a way of working around the problem. In order to amass the data required, it is necessary to use a variety of data sources categorised here as historical records, observations, interviews and process owner/vendor estimates:

Historical records: A mass of data may be available within the organisation regarding the system to be modelled. This data may be in a variety of formats including paper and electronic (e.g. held on a database). However this data may not be in the right format, be incomplete or not be relevant for the study in progress. The statistical validity of the data may also be in doubt.

Observations: A walkthrough of the process by the analyst is an excellent way of gaining an understanding of the process flow. Time studies can also be used to estimate process parameters when current data is not available.

Interviews: An interview with the process owner can assist in the analysis of system behaviour which may not always be documented.

Process owner/vendor estimate: Process owner and vendor estimates are used most often when the system to be modelled does not exist, and thus no historical data or observation is possible. This approach has the disadvantage of relying on the ability of the process owner (e.g. treatment provider) in remembering past performance. If possible a questionnaire can be used to gather estimates from a number of process owners and the data statistically analysed.

As with other stages of a simulation project, data collection is an iterative process with further data collected as the project progresses. For instance, statistical tests during the modelling of input data or experimentation phases of development may suggest a need to collect further data in order to improve the accuracy of results.

Also the validation process may expose inaccuracies in the model which require further data collection activities. Thus it should be expected that data collection activities will be ongoing throughout the project as the model is refined.

Process Mapping

A process map (also called a conceptual model) should be formulated in line with the scope and level of detail defined within the project specification. An essential component of this activity is to construct a diagrammatic representation of the process in order to provide a basis for understanding between the simulation developer and process owner. Two diagramming methods used in discrete-event simulation are activity cycle diagrams and process maps. Activity cycle diagrams can be used to represent any form of simulation system. Process maps are most suited to representing a process-interaction view that follows the life cycle of an entity (e.g. client) through a system comprising a number of activities with queuing at each process (e.g. waiting for service). Most simulation applications are of this type and the clear form of the process map makes it the most suitable method in these instances.

Two main problems associated with data are that little useful data is available (e.g. when modelling a system that does not yet exist) or that the data is not in the correct format. If no data exist, you are reliant on estimates from vendors or other parties, rather than samples of actual performance, so this needs to be emphasised during the presentation of any results. An example of data in the wrong format is a customer service time calculated from entering the service queue to completion of service. This data could not be used to approximate the customer service time in the simulation model as you require the service time only. The queuing time will be generated by the model as a consequence of the arrival rate and service time parameters. In this case the client may assume that your data requirements have been met and will specify the time and cost of the simulation project around that. Thus it is important to establish as soon as possible the actual format of the data and its suitability for your needs to avoid misunderstandings later.

A number of factors will impact on how the data collection process is undertaken including the time and cost within which project must be conducted. Compromises will have to be made on the scale of the data collection activity and so it is important to focus effort on areas where accuracy is important for simulation results and to make clear assumptions made when reporting simulation results. If it has not been possible to collect detailed data in certain areas of the process, it is not sensible to then model in detail that area. Thus there is a close relationship between simulation objectives, model detail and data collection needs. If the impact of the level of data collection on results is not clear, then it is possible to use sensitivity analysis (i.e. trying different data values) to ascertain how much model results are affected by the data accuracy. It may be then necessary to either undertake further data collection or quote results over a wide range.

Activity Cycle Diagrams: Activity cycle diagrams can be used to construct a conceptual model of a simulation which uses the event, activity or process orientation. The diagram aims to show the life cycles of the components in the system. Each component is shown in either of two states, the *dead* state is represented by a circle and the *active* state is represented by a rectangle. Each component can be shown moving through a number of dead and active states in a sequence that must form a loop. The dead state relates to a conditional ('C') event where the component is waiting for something to happen such as the commencement of a service, for example. The active state relates to a bound ('B') event or a service process, for example. The duration of the active state is thus known in advance while the duration of the dead state cannot be known, because it is dependent on the behaviour of the whole system.

Process Maps: The construction of a process flow diagram is a useful way of understanding and documenting any business process and showing the interrelationships between activities in a process. These diagrams have become widely used in Business Process Reengineering (BPR) projects, and the use of process mapping in this context is evaluated in Peppard and Rowland (1995). For larger projects it may be necessary to represent a given process at several levels of detail. Thus a single activity may be shown as a series of sub-activities on a separate diagram. In simulation projects this diagram is often referred to as the simulation conceptual model, and the method is particularly suitable when using process-oriented simulation languages and visual interactive modelling systems.

Modelling Input Data

It is important to model randomness in such areas as arrival times and process durations. Taking an average value will not give the same behaviour. Queues are often a function of the *variability* of arrival and process times and not simply a consequence of the relationship between arrival interval and process time. The method of modelling randomness used in the simulation will be dependent on the amount of data collected on a particular item. For less than 20 data points, a mean value or theoretical distribution must be estimated. Larger samples allow the user to fit the data to a theoretical distribution or to construct an empirical distribution. Theoretical and empirical distributions are classified as either continuous or discrete. Continuous distributions can return any real value quantity and are used to model arrival times and process durations. Discrete distributions return only whole number or integer values and are used to model decision choices or batch sizes. Guidance on possible methods for modelling randomness with increasing levels of data is now provided.

Less than 20 Data Points: *Estimation*—If it is proposed to build a model of a system that has not been built or there is no time for data collection, then an estimate must be made. This can be achieved by questioning interested parties such as the process owner or the equipment vendor. A sample size of below 20 is probably too small to fit a theoretical distribution with any statistical confidence although it may be appropriate to construct a histogram to assist in finding a representative distribution.

The simplest approach is to simply use a fixed value to represent the data representing an estimate of the mean. Otherwise a theoretical distribution may be chosen based on knowledge and statistical theory. Statistical theory suggests that if the mean value is not very large, interarrival times can be simulated using the exponential distribution. Service times can be simulated using a uniform or symmetric triangular distribution with the minimum and maximum values at a percentage variability from the mean. For example, a mean of 100 with a variability of +/− 20% would give values for a triangular distribution of 80 for minimum, 100 for mode and 120 for maximum. The normal distribution may be used when an unbounded (i.e. the lower and upper levels are not specified) shape is required. The normal distribution requires mean and standard deviation parameters. When only the minimum and maximum values are known and the behaviour between those values is not known, a uniform distribution generates all values with an equal likelihood.

20+ Data Points: *Deriving a Theoretical Distribution*—For 20+ data points, a theoretical distribution can be derived. The standard procedure to match a sample distribution to a theoretical distribution is to construct a histogram of the data and compare the shape of the histogram with a range of theoretical distributions. Once a potential candidate is found it is necessary to estimate the parameters of the distribution which provides the closest fit. The relative 'goodness of fit' can be determined by using an appropriate statistical method.

200+ Data Points: *Constructing an Empirical Distribution*—For more than 200 data points, the option of constructing a user-defined distribution is available. An empirical or user-defined distribution is a distribution that has been obtained directly from the sample data. An empirical distribution is usually chosen if a reasonable fit cannot be made with the data and a theoretical distribution. It is usually necessary to have in excess of 200 data points to form an empirical distribution. In order to convert the sample data into an empirical distribution, the data is converted into a cumulative probability distribution using the following steps:

1. Sort values into ascending order.
2. Group identical values (discrete) or group into classes (continuous).
3. Compute the relative frequency of each class.
4. Compute the cumulative probability distribution of each class.

Historical Data Points: A simulation driven by historical data is termed a 'trace-driven' simulation. An example would be using actual arrival times of customers in a bank directly in the simulation model. The major drawback of this approach is that it prevents the simulation from being used in the 'what-if' mode as only the historical data is modelled. It also does not take account of the fact that in the future the system will most likely encounter conditions out of the range of the sample data used. This approach can be useful, however, in validating model performance when the behaviour of the model can be compared to the real system with identical data.

Building the Model

This involves using computer software to translate the process map into a computer simulation model which can be 'run' to generate model results. This will entail the use of simulation software such as ARENA, WITNESS or SIMUL8. Kelton, Sadowski, and Sturrock (2007) cover the use of ARENA and Greasley (2004) covers all three of the above software packages.

Validation and Verification

Before experimental analysis of the simulation model can begin, it is necessary to ensure that the model constructed provides a valid representation of the system we are studying. This process consists of verification and validation of the simulation model. Verification refers to ensuring that the computer model built using the simulation software is a correct representation of the process map of the system under investigation. Validation concerns ensuring that the assumptions made in the process map about the real-world system are acceptable in the context of the simulation study. Both topics will now be discussed in more detail.

Verification: Verification is analogous to the practice of 'debugging' a computer program. Thus many of the following techniques will be familiar to programmers of general-purpose computer languages.

Model Design: The task of verification is likely to become greater with an increase in model size. This is because a large complex program is both more likely to contain errors and these errors are less likely to be found. Due to this behaviour, most practitioners advise on an approach of building a small simple model, ensuring that this works correctly and then gradually adding embellishments over time. This approach is intended to help limit the area of search for errors at any one time. It is also important to ensure that unnecessary complexity is not incorporated in the model design. The design should incorporate only enough detail to ensure the study objectives and not attempt to be an exact replica of the real-life system.

Structured Walkthrough: This enables the modeller to incorporate the perspective of someone outside the immediate task of model construction. The walkthrough procedure involves talking through the program code with another individual or team. The process may bring fresh insight from others, but the act of explaining the coding can also help the person who has developed the code discover their own errors. In discrete-event simulation, the code is executed nonsequentially and different coding blocks are executed simultaneously. This means that the walkthrough may best be conducted by following the 'life history' of an entity through the simulation coding, rather than a sequential examination of coding blocks.

Test Runs: Test runs of a simulation model can be made during program development to check model behaviour. This is a useful way of checking model behaviour as a defective model will usually report results (e.g. machine utilisation, customer wait times) which do not conform to expectations, either based on the real system performance or common-sense deductions. It may be necessary to add performance measures to the model (e.g. costs) for verification purposes, even though they may not be required for reporting purposes. One approach is to use historical (fixed) data, so model behaviour can be isolated from behaviour caused by the use of random variates in the model. It is also important to test model behaviour under a number of scenarios, particularly boundary conditions that are likely to uncover erratic behaviour. Boundary conditions could include minimum and maximum arrival rates, minimum and maximum service times and minimum and maximum rate of infrequent events (e.g. machine breakdowns).

Trace Analysis: Due to the nature of discrete-event simulation, it may be difficult to locate the source of a coding error. Most simulation packages incorporate an entity trace facility that is useful in providing a detailed record of the life history of a particular entity. The trace facility can show the events occurring for a particular entity or all events occurring during a particular time frame. The trace analysis facility can produce a large amount of output so it is most often used for detailed verification.

The animation facilities of simulation software packages provide a powerful tool in aiding understanding of model behaviour. The animation enables the model developer to see many of the model components and their behaviour simultaneously. A 'rough-cut' animated drawing should be sufficient at the testing stage for verification purposes. To aid understanding, model components can be animated which may not appear in the final layout presented to a client. The usefulness of the animation technique will be maximised if the animation software facilities permit reliable and quick production of the animation effects.

It is important to document all elements in the simulation to aid verification by other personnel or at a later date. Any general-purpose or simulation coding should have comments attached to each line of code. Each object within a model produced on a visual interactive modelling system requires comments regarding its purpose and details of parameters and other elements.

Validation: A verified model is a model which operates as intended by the modeller. However this does not necessarily mean that it is a satisfactory representation of the real system for the purposes of the study. Validation is about ensuring that the model behaviour is close enough to the real-world system for the purposes of the simulation study. Unlike verification, the question of validation is one of judgement. Ideally the model should provide enough accuracy for the measures required while limiting the amount of effort required to achieve this. For most systems of any complexity, this aim can be achieved in a number of ways, and a key skill of the simulation developer is finding the most efficient way of achieving this goal. Pegden et al. (1995) outline three aspects of validation:

- Conceptual validation: Does the model adequately represent the real-world system?
- Operational validity: Are the model generated behavioural data characteristic of the real-world system behavioural data?

- Believability: Does the simulation model's ultimate user have confidence in the model's results?

Conceptual Validity: Conceptual validation involves ensuring that the model structure and elements are correctly chosen and configured in order to adequately represent the real-world system. As we know that the model is a simplification of the real world, then there is a need for a consensus around the form of the conceptual model between the model builder and the user. To ensure a credible model is produced, the model builder should discuss and obtain information from people familiar with the real-world system including operating personnel, industrial engineers, management, vendors and documentation. They should also observe system behaviour over time and compare with model behaviour and communicate with project sponsors throughout the model build to increase credibility.

Operational Validity: This involves ensuring that the results obtained from the model are consistent with real-world performance. A common way of ensuring operational validity is to use the technique of sensitivity analysis to test the behaviour of the model under various scenarios and compare results with real-world behaviour. The technique of common random numbers (CRN) can be used to isolate changes due to random variation. The techniques of experimental design can also be employed to conduct sensitivity analysis over two or more factors. Note that for validation purposes, these tests are comparing simulation performance with real-world performance while in the context of experimentation they are used to compare simulation behaviour under different scenarios.

Sensitivity analysis can be used to validate a model, but it is particularly appropriate if a model has been built of a system which does not exist as the data has been estimated and cannot be validated against a real system. In this case the main task is to determine the effect of variation in this data on model results. If there is little variation in output as a consequence of a change in input, then we can be reasonably confident in the results. It should also be noted that an option may be to conduct sensitivity analysis on subsystems of the overall system being modelled which do exist. This emphasises the point that the model should be robust enough to provide a prediction of what would happen in the real system under a range of possible input data. The construction and validation of the model should be for a particular range of input values defined in the simulation project objectives. If the simulation is then used outside of this predefined range, the model must be revalidated to ensure additional aspects of the real system are incorporated to ensure valid results.

An alternative to comparing the output of the simulation to a real system output is to use actual historical data in the model, rather than derive a probability distribution. Data collected could be used for elements such as customer arrival times and service delays. By comparing output measures across identical time periods, it should be possible to validate the model. Thus the structure or flow of the model could be validated and then probability distributions entered for random elements. Thus any error in system performance could be identified as either a logic error or from an inaccurate distribution. The disadvantage of this method is that for a model of any size the amount of historical data needed will be substantial. It is also necessary to read this data, either from a file or array, requiring additional coding effort.

Sensitivity analysis should be undertaken by observing the output measure of interest with data set to levels above and below the initial set level for the data. A graph may be used to show model results for a range of data values if detailed analysis is required (e.g. a non-linear relationship between a data value and output measure is apparent). If the model output does not show a significant change in value in response to the sensitivity analysis, then we can judge that the accuracy of the estimated value will not have a significant effect on the result.

If the model output is sensitive to the data value, then preferably we would want to increase the accuracy of the data value estimate. This may be undertaken by further interviews or data collection. In any event the simulation analysis will need to show the effect of model output on a range of data values. Thus for an estimated value we can observe the likely behaviour of the system over a range of data values within which the true value should be located. Further sensitivity analysis may be required on each of these values to separate changes in output values from random variation.

When it is found that more than one data value has an effect on an output measure, then the effects of the individual and combined data values should be assessed. This will require 3^k replications to measure the minimum, initial and maximum values for k variables. Fractional factorial design techniques (Law & Kelton, 2000) may be used to reduce the number of replications required.

Believability: In order to ensure implementation of actions recommended as a result of simulation experiments requires that the model output is seen as credible from the simulation user's point of view. This credibility will be enhanced by close co-operation between model user and client throughout the simulation study. This involves agreeing clear project objectives explaining the capabilities of the technique to the client and agreeing assumptions made in the process map. Regular meetings of interested parties, using the simulation animation display to provide a discussion forum, can increase confidence in model results. Believability emphasises how there is no one answer to achieving model validity and the perspective of both users and developers need to be satisfied that a model is valid.

Experimentation and Analysis

The stochastic nature of simulation means that when a simulation is run, the performance measures generated are a sample from a random distribution. Thus each simulation 'run' will generate a different result, derived from the randomness which has been modelled. In order to interpret the results (i.e. separate the random changes in output from changes in performance), statistical procedures are outlined in this chapter which are used for the analysis of the results of runs of a simulation. There are two types of simulation system that need to be defined, each requiring different methods of data analysis.

Terminating systems run between pre-defined states and times where the end state matches the initial state of the simulation. For example, a simulation of a retail shop from opening to closing time. Nonterminating systems do not reach pre-defined states or times. In particular the initial state is not returned to, for example, a manufacturing facility. Most service organisations tend to be terminating systems which close at the end of each day with no in-process inventory (i.e. people waiting for service) and thus return to the 'empty' or 'idle' state they had at the start of that day. Most manufacturing organisations are nonterminating with inventory in the system that is awaiting a process. Thus even if the facility shuts down temporarily, it will start again in a different state to the previous start state (i.e. the inventory levels define different starting conditions). However the same system may be classified as terminating or nonterminating depending on the objectives of the study. Before a nonterminating system is analysed, the bias introduced by the non-representative starting conditions must be eliminated to obtain what are termed steady-state conditions from which a representative statistical analysis can be undertaken.

Statistical Analysis for Terminating Systems: This section will provide statistical tools to analyse either terminating systems or the steady-state phase of nonterminating systems. The statistics relevant to both the analysis of a single model and comparing between different models will now be outlined in turn. The output measure of a simulation model is a random variable, and so we need to conduct multiple runs (replications) of the model to provide us with a sample of its value. When a number of replications have been undertaken, the sample mean can be calculated by averaging the measure of interest (e.g. time in queue) over the number of replications made. Each replication will use different set of random numbers and so this procedure is called the method of independent replications.

Establishing a Confidence Interval: To assess the precision of our results, we can compute a confidence interval or range around the sample mean that will include, to a certain level of confidence, the true mean value of the variable we are measuring. Thus confidence intervals provide a point estimate of the expected average (average over infinite number of replications) and an idea of how precise this estimate is. The confidence interval will fall as replications increase. Thus a confidence interval does not mean that say 95% of values fall within this interval but that we are 95% sure that the interval contains the expected average.

For large samples (replications) of over around 50, the normal distribution can be used for the computations. However the sample size for a simulation experiment will normally be less than this with ten replications of a simulation being common. In this case provided the population is approximately normally distributed, the sampling distribution follows a *t*-distribution.

Both the confidence interval analysis and the *t*-tests presented later for comparison analysis assume the data measured is normally distributed. This assumption is usually acceptable if measuring an average value for each replication as the output variable is made from many measurements and the central limit theorem applies. However the central limit theorem applies for a large sample size, and the definition of what constitutes a large sample depends partly on how close the actual distribution of the output variable is to the normal distribution. A histogram can be used to observe how close the actual distribution is to the normal distribution curve.

Comparing Alternatives: When comparing between alternative configurations of a simulation model, we need to test whether differences in output measures are statistically significant or if differences could be within the bounds of random variation. Alternative configurations which require this analysis include:

- Changing input parameters (e.g. changing arrival rate)
- Changing system rules (e.g. changing priority at a decision point)
- Changing system configuration (comparing manual vs. automated system)

Whatever the scale of the differences between alternative configurations, there is a need to undertake statistical tests. The tests will be considered for comparing between two alternatives and then between more than two alternatives.

The following assumptions are made when undertaking the tests:

- The data collected *within* a given alternative are independent observations of a random variable. This can be obtained by each replication using a different set of random numbers (method of independent replications).
- The data collected *between* alternatives are independent observations of a random variable. This can be obtained by using a separate number stream for each alternative. This can be implemented by changing the seeds of the random number generator between runs. Note, however, that certain tests use the ability to use CRN for each simulation run in their analysis (see paired *t*-test using CRN).

Hypothesis Testing: When comparing simulation scenarios, we want to know if the results of the simulation for each scenario are different because of random variability or because of an actual change in performance. In statistical terms we can do this using a hypothesis test to see if the sample means of each scenario differ.

A hypothesis test makes an assumption or hypothesis (termed the null hypothesis, H_0) and tries to disprove it. Acceptance of the null hypothesis implies that there is insufficient evidence to reject it (it does not prove that it is true). Rejection of the null hypothesis, however, means that the alternative hypothesis (H_1) is accepted. The null hypothesis is tested using a test statistic (based on an appropriate sampling distribution) at a particular significance level which relates to the area called the critical region in the tail of the distribution being used. If the test statistic (which we calculate) lies in the critical region, the result is unlikely to have occurred by chance and so the null hypothesis would be rejected. The boundaries of the critical region, called the critical values, depend on whether the test is two-tailed (we have no reason to believe that a rejection of the null hypothesis implies that the test statistic is either greater or less than some assumed value) or one-tailed (we have reason to believe that a rejection of the null hypothesis implies that the test statistic is either greater or less than some assumed value).

We must also consider the fact that the decision to reject or not reject the null hypothesis is based on a probability. Thus at a 5% significance level there is a 5% chance that H_0 will be rejected when it is in fact true. In statistical terminology this is called a type I error. The converse of this is accepting the null hypothesis when it is in fact false, called a type II error. Usually values of 0.05 (5%) or 0.01 (1%) are used. An alternative to testing at a particular significance level is to calculate the p-value which is the lowest level of significance at which the observed value of the test statistic is significant. Thus a p-value of 0.045 (indicating a type I error occurring 45 times out of 1000) would show that the null hypothesis would be rejected at 0.05 but only by a small amount.

Paired t-Test: The test calculates the difference between the two alternatives for each replication. It tests the hypothesis that if the data from both models is from the same distribution, then the mean of the differences will be 0.

Paired t-Test Using CRN: The idea of using CRN is to ensure that alternative configurations of a model differ only due to those configurations and not due to the different random number sets used to drive the random variables within the model. It is important that synchronisation of random variables occurs across the model configurations, so the use of a dedicated random number stream for each random variate is recommended. Again as with other variance reduction techniques (VRT) the success of the method will be dependent on the model structure and there is no certainty that variance will be actually reduced. Another important point is that by driving the alternative model configuration with the same random numbers, we are assuming that they will behave in a similar manner to large or small values of the random variables driving the models. In general it is advisable to conduct a pilot study to ensure that the CRN technique is in fact reducing the variance between alternatives.

Because the output from a simulation model is a random variable, the variance of that output will determine the precision of the results obtained from it. Statistical techniques to reduce that variance may be used to either obtain smaller confidence intervals for a fixed amount of simulating time or achieve a desired confidence interval with a smaller amount of simulating.

A variety of VRT are discussed in Law and Kelton (2000). The use of CRN in conjunction with the paired t-test is described for comparing alternative system configurations. The paired-t approach assumes the data is normally distributed (see 'testing for normality' section to check if the data is normally distributed) but does not assume all observations from the two alternatives are independent of each other, as does the two-sample-t approach.

One-Way ANOVA: One-way analysis of variance (ANOVA) is used to compare the means of several alternative systems. Several replications are performed of each alternative and the test attempts to determine whether the variation in output performance is due to differences *between* the alternatives or due to inherent randomness *within* the alternatives themselves. This is undertaken by comparing the ratio of the two variations with a test statistic. The test makes the following assumptions:

- Independent data both within and between the data sets.
- Observations from each alternative are drawn from a normal distribution.
- The normal distributions have the same variance.

The first assumption implies the collection of data using independent runs or the batch means technique but precludes the use of VRT (e.g. CRN). The second assumption implies that each output measure is the mean of a large number of observations. This assumption is usually valid but can be tested with the chi-square or Kolmogorov–Smirnov test if required. The third assumption may require an increase in replication run length to decrease the variances of mean performance. The F-test can be used to test this assumption if required. The test finds if a significant difference between means is apparent but does not indicate if all the means are different or if the difference is between particular means. To identify where the differences occur, then tests such as Tukey's HSD test may be used. Alternatively confidence intervals between each combination can provide an indication (Law & Kelton, 2000).

Statistical Analysis for Nonterminating Systems: The previous section considered statistical analysis for terminating systems. This section provides details of techniques for analysing steady-state systems in which the start conditions of the model are not returned to. These techniques involve more complex analysis than for a terminating system and so consideration should be given to treating the model as a terminating system if at all possible.

A non-terminating system generally goes through an initial transient phase and then enters a steady-state phase when its condition is independent of the simulation starting conditions. This behaviour could relate to a manufacturing system starting from an empty ('no-inventory') state and then after a period of time moving to a stabilised behaviour pattern. A simulation analysis will be directed towards measuring performance during the steady-state phase and avoiding measurements during the initial transient phase. The following methods of achieving this are discussed:

Setting Starting Conditions: This approach involves specifying start conditions for the simulation which will provide a quick transition to steady-state conditions. Most simulations are started in an empty state for convenience, but by using knowledge of steady-state conditions (e.g. stock levels) it is possible to reduce the initial bias phase substantially. The disadvantage with this approach is the effort in initialising simulation variables, of which there may be many, and when a suitable initial value may not be known. Also it is unlikely that the initial transient phase will be eliminated entirely. For these reasons the warm-up period method is often used.

Using a Warm-Up Period: Instead of manually entering starting conditions, this approach uses the model to initialise the system and thus provide starting conditions automatically. This approach discards all measurements collected on a performance variable before a preset time in order to ensure that no data is collected during the initial phase. The point at which data is discarded must be set late enough to ensure that the simulation has entered the steady-state phase but not so late that insufficient data points can be collected for a reasonably precise statistical analysis. A popular method of choosing the discard point is to visually inspect the simulation output behaviour of the variable over time. Welch (1983) suggests a procedure using the moving average value in order to smooth (i.e. separate the long-term trend values from short-term fluctuations) the output response. It is important to ensure the model is inspected over a time period which allows infrequent events (e.g. machine breakdown) to occur a reasonable number of times.

In order to determine the behaviour of the system over time and in particular to identify steady-state behaviour, a performance measure must be chosen. A value such as work in progress (WIP) provides a useful measure of overall system behaviour. In a manufacturing setting this could relate to the amount of material within the system at any one time. In a service setting (as is the case with the bank clerk model) the WIP measure represents the number of customers in the system. While this measure will vary over time in a steady-state system, the long-term level of WIP should remain constant.

Using an Extended Run Length: This approach simply consists of running the simulation for an extended run length, so reducing the bias introduced on output variables in the initial transient phase. This approach is best applied in combination with one or both of the previous approaches.

Batch Means Analysis: To avoid repeatedly discarding data during the initial transient phase for each run, an alternative approach allows all data to be collected during one long run. The batch means method consists of making one very long run of the simulation and collecting data at intervals during the run. Each interval between data collection is termed a batch. Each batch is treated as a separate run of the simulation for analysis. The batch means method is suited to systems that have very long warm-up periods and so avoidance of multiple replications is desirable. However with the increase in computing power available this advantage has diminished with run lengths needing to be extremely long in order to slow down analysis considerably. The batch means method also requires the use of statistical analysis methods which are beyond the scope of this book (see Law & Kelton, 2000).

Presentation of Results

For each simulation study the simulation model should be accompanied by a project report, outlining the project objectives and providing the results of experimentation. Discussion of results and recommendations for action should also be included. Finally a further work section will communicate to the client any possible developments and subsequent results it is felt could be obtained from the model. If there are a number of results to report, an appendix can be used to document detailed statistical work, for example. This enables the main report to focus on the key business results derived from the simulation analysis. A separate technical document may also be prepared which may incorporate a model and/or model details such as key variables and a documented coding listing. Screenshots of the model display can also be used to show model features. If the client is expected to need to develop the code in-house, then a detailed explanation of model coding line by line will be required. The report structure should contain the following elements:

- Introduction
- Description of the problem area
- Model specification
- Simulation experimentation
- Results
- Conclusions and recommendations
- Further studies
- Appendices: Process logic, data files and model coding

A good way of 'closing' a simulation project is to organise a meeting of interested parties and present a summary of the project objectives and results. Project documentation can also be distributed at this point. This enables discussion of the outcomes of the project with the client and provides an opportunity to discuss further analysis. This could be in the form of further developments of the current model ('updates' or 'new phase') or a decision to prepare a specification for a new project.

Implementation

It is useful to both the simulation developer and client if an implementation plan is formed to undertake recommendations from the simulation study. Changes in the system studied may also necessitate model modification. The level of support at this time from the developer may range from a telephone 'hotline' to further personal involvement specified in the project report. Results from a simulation project will only lead to implementation of changes if the credibility of the simulation method is assured. This is achieved by ensuring each stage of the simulation project is undertaken correctly.

Organisational Context of Implementation

A simulation modelling project can use extensive resources both in terms of time and money. Although the use of simulation in the analysis of a one-off decision, such as investment appraisal, can make these costs low in terms of making the correct decision, the benefits of simulation can often be maximised by extending the use of the model over a period of time. It is thus important that during the project proposal stage, elements are incorporated into the model and into the implementation plan that assist in enabling the model to provide ongoing decision support. Aspects include the following:

- Ensure that simulation users are aware at the project proposal stage that the simulation is to be used for ongoing decision support and will not be put to one side once the immediate objectives are met.

- Ensure technical skills are transferred from simulation analysts to simulation users. This ensures understanding of how the simulation arrives at results and its potential for further use in related applications.
- Ensure communication and knowledge transfer from simulation consultants and industrial engineers to business managers and operational personnel.

The needs of managerial and operational personnel are now discussed in more detail.

Managerial Involvement: The cost associated with a simulation project means that the decision of when and where to use the technique will usually be taken by senior management. Thus an understanding of the potential and limitations of the technique is required if correct implementation decisions are to be made. The Simulation Study Group (1991) found 'there is a fear among UK managers of computerisation and this fear becomes even more pronounced when techniques that aid decision making are involved'. This is combined with the fact that even those who do know how to use simulation become 'experts' within a technically oriented environment. This means that those running the business do not fully understand the technique which could impact on their decision to use the results of the study.

Operational Involvement: Personnel involved in the day-to-day operation of the decision area need to be involved in the simulation project for a number of reasons. They usually have a close knowledge of the operation of the process and thus can provide information regarding process logic, decision points and activity durations. Their involvement in validating the model is crucial in that any deviations from operational activities seen from a managerial view to the actual situation can be indicated. The use of process maps and a computer-animated simulation display both provide a means of providing a visual method of communication of how the whole process works (as opposed to the part certain personnel are involved in) and facilitates a team approach to problem solving by providing a forum for discussion.

Simulation can be used to develop involvement from the operational personnel in a number of areas. It can present an ideal opportunity to change from a top-down management culture and move to greater involvement from operational personnel in change projects. Simulation can also be a strong facilitator of communicating ideas up and down an organisation. Engineers, for example, can use simulation to communicate reasons for taking certain decisions to operational personnel who might suggest improvements. The use of simulation as a tool for employee involvement in the improvement process can be a vital part of an overall change strategy. The process orientation of simulation provides a tool for analysis of processes from a cross-functional as opposed to a departmental perspective. This is important because powerful political forces may need to be overcome in ensuring departmental power does not prevent change from a process perspective.

The choice of simulation software should also take into consideration ongoing use of the technique by personnel outside of the simulation technicians. For ongoing use, software tools need to provide less complex model building tools. This suggests the use of visual interactive modelling tools which incorporate iconic model building and menu facilities making this type of simulation more accessible.

There is also a need for training in statistical techniques for valid experimentation analysis. In summary the following needs are indicated:

- Knowledge transfer from technical personnel to managerial and operational staff of the potential application of simulation.
- Training at managerial/operational levels in statistical techniques from companies and universities.
- Training at managerial/operational levels in model building techniques from companies and universities.
- Use of simulation as a communication tool between stakeholders in a change program. The use of animation is useful.
- Use of suitable software, such as a visual interactive modelling system, to provide a platform for use by non-technical users.

Summary

This chapter has described the main activities in the design and implementation of a simulation model. These activities include the collection of data, the construction of a process map and the statistical analysis of both input data and output statistics. An important element of a simulation project is a project report which documents the simulation model, presents the results of the analysis and suggests further studies. The organisational context of the use of simulation is discussed in terms of managerial and operational involvement in the simulation study process.

References

Ajzen, A. (1991). The theory of planned behavior. *Organizational Behavior and Human Decision Processes, 50*, 179–211.
Anderson, J. R., & Lebiere, C. (1998). *The atomic components of thought*. Mahwah, NJ: Erlbaum.
Bernhard, W. & Schilling, A. (1997). *Simulation of group work processes in manufacturing*. In Proceedings of the 1997 Winter Simulation Conference, SCS, Atlanta, GA, pp. 888–891.
Brailsford, S., & Schmidt, B. (2003). Towards incorporating human behavior in models of health care systems: An approach using discrete-event simulation. *European Journal of Operational Research, 150*, 19–31.
Cavana, R. Y., Davies, P. K., Robson, R. M., & Wilson, K. J. (1999). Drivers of quality in health services: Different worldviews of clinicians and policy managers revealed. *System Dynamics Review, 15*(3), 331–340.
Elliman, T., Eatock, J., & Spencer, N. (2005). Modelling knowledge worker behavior in business process studies. *Journal of Enterprise Information Management, 18*(1), 79–94.
Forrester, J. (1961). *Industrial dynamics*. MA: Productivity Press, Cambridge.
Greasley, A. (2004). *Simulation modelling for business*. Hants: Ashgate.
Greasley, A., & Barlow, S. (1998). Using simulation modelling for BPR: Resource allocation in a police custody process. *International Journal of Operations and Production Management, 18*, 978–988.

Hanisch, A., Tolujew, J., Richter, K. & Schulze, T. (2003). *Online simulation of pedestrian flow in public buildings*. In Proceedings of the 2003 Winter Simulation Conference, SCS, New Orleans, LA, pp. 1635–1641.

Hayes, C. C. (1999). Agents in a nutshell—A very brief introduction. *IEEE Transactions on Knowledge and Data Engineering., 11*(1), 127–132.

Johnson, R.T., Fowler, J.W. & Mackulak, G.T. (2005). *A discrete event simulation model simplification technique*. In *Proceedings of the 2005 Winter Simulation Conference*, SCS, Orlando, FL, pp. 2172–2176.

Keller, J. (2002). *Human performance modelling for discrete-event simulation: Workload*. In Proceedings of the 2002 Winter Simulation Conference, San Diego, CA, pp. 157–162.

Kelton, W. D., Sadowski, R. P., & Sturrock, D. T. (2007). *Simulation with arena* (4th ed.). New York, NY: McGraw-Hill.

Lam, R. B. (2007). *Agent-based simulations of service policy decisions*. In Proceedings of the 2007 Winter Simulation Conference, SCS, Washington, DC, pp. 2241–2246.

Law, A. M., & Kelton, W. D. (2000). *Simulation modeling and analysis* (3rd ed.). Singapore: McGraw-Hill.

Macal, C.M. & North, M.J. (2006). *Tutorial on agent-based modeling and simulation Part 2: How to model with agents*. In Proceedings of the 2006 Winter Simulation Conference, SCS, Monterey, CA, pp. 73–83.

Muller, D. J. (1996). Simulation: "What to do with the model afterward". In J. M. Charnes, D. J. Morrice, D. T. Brunner, & J. J. Swain (Eds.), *Proceedings of the 1996 Winter Simulation Conference*. San Diego, CA: Society for Computer Simulation.

Newell, A. (1990). *Unified theories of cognition*. Cambridge, MA: Harvard University Press.

Pegden, C. D., Shannon, R. E., & Sadowski, R. P. (1995). *Introduction to simulation using SIMAN* (2nd ed.). Singapore: McGraw-Hill.

Peppard, J., & Rowland, P. (1995). *The essence of business process Re-engineering*. Hemel Hempstead: Prentice Hall.

Pew, R. W. (2008). More than 50 years of history and accomplishments in human performance model development. *Human Factors, 50*(3), 489–496.

Prichett, A. R., Lee, S. M., & Goldsman, D. (2001). Hybrid-system simulation for national airspace safety systems analysis. *AIAA Journal of Aircraft, 38*(5), 835–840.

Robinson, S., Edwards, J. S., & Yongfa, W. (2003). Linking the witness simulation software to an expert system to represent a decision-making process. *Journal of Computing and Information Technology, 11*(2), 123–133.

Robinson, S., Alifantis, T., Edwards, J. S., Ladbrook, J., & Waller, T. (2005). Knowledge based improvement: Simulation and artificial intelligence for identifying and improving human decision-making in an operations system. *Journal of the Operational Research Society, 56*(8), 912–921.

Schmidt, B. (2000). *The modelling of human behavior*. Erlangen: SCS.

Schmidt, B. (2005). *Human factors in complex systems: The modeling of human behavior*. In Proceedings 19th European Conference of Modelling and Simulation, ECMS, Riga, Latvia.

Shaw, A. P., & Pritchett, R. (2005). Agent-based modeling and simulation of socio-technical systems. In W. B. Rouse & K. R. Boff (Eds.), *Organizational simulation*. New York, NY: Wiley.

Silverman, B. G., Cornwell, J., & O'Brien, K. (2003). Human performance simulation. In J. W. Ness, D. R. Rizer, & V. Tepe (Eds.), *Metrics and methods in human performance research toward individual and small unit simulation*. Washington DC: Human Systems Information Analysis Centre.

Silverman, B. G. (1991). Expert critics: Operationalizing the judgement/decision making literature as a theory of "bugs" and repair strategies. *Knowledge Acquisition, 3*, 175–214.

Simulation Study Group. (1991). *Simulation in U.K. manufacturing industry*. In R. Horrocks (ed.). Coventry: The Management Consulting Group, University of Warwick Science Park.

Stahl, I. (1995). *New product development: When discrete simulation is preferable to system dynamics*. In Proceedings of the 1995 EUROSIM conference. Amsterdam: Elsevier Science B.V.

Sterman, J. D., Repenning, N. P., & Kofman, F. (1997). Unanticipated side effects of successful quality programs: Exploring a paradox of organizational improvement. *Management Science, 43*(4), 503–521.

Warren, R., Diller, D.E., Leung, A., & Ferguson, W. (2005), *Simulating scenarios for research on culture & cognition using a commercial role-play game.* In Proceedings of the 2005 Winter Simulation Conference, SCS, Orlando, FL, pp. 1109–1117.

Welch, P. D. (1983). The statistical analysis of simulation results. In S. S. Lavenberg (Ed.), *The computer performance modeling handbook* (Vol. 9, No. 2, pp. 111–116). New York, NY: Academic.

Wickens, C. D. (1984). *Engineering psychology and human performance.* New York, NY: Harper Collins.

Chapter 4
The Empirical Basis for the RNR Model with an Updated RNR Conceptual Framework

Faye S. Taxman, April Pattavina, Michael S. Caudy, James Byrne, and Joseph Durso

Introduction

Since the early 1990s, the risk-need-responsivity (RNR) model for correctional programming has served as a framework to promote the use of evidence-based correctional strategies. The model emphasizes the importance of both classification (by risk level and treatment need) and rehabilitative correctional programming (including quantity and quality) for achieving three critically important correctional goals of (1) recidivism reduction, (2) least restrictive sanctioning, and (3) cost-effectiveness. In the RNR model, offenders are matched to appropriate controls, supervision levels, and treatment-related services based on their static risk level and dynamic criminogenic needs. The framework emerges from an empirical body of research demonstrating that providing appropriate treatment services (e.g., cognitive behavioral therapy (CBT), drug treatment courts, therapeutic communities) should result in reductions in recidivism.

The RNR framework, as shown below in Fig. 4.1, is conceptually clear with three core principles: (1) identify the risk level of the individual, (2) identify the dynamic risk factors (needs) that are associated with offending behavior and that affect psychosocial functioning, and (3) identify appropriate correctional interventions that are suitable to address the risk–need interaction. Adherence to all three of these principles promises a greater impact on the individual which can also impact the overall performance of the corrections system. The RNR model has implications

F.S. Taxman (✉) • M.S. Caudy • J. Durso
Department of Criminology, Law and Society, George Mason University,
10900 University Boulevard, Fairfax, VA 20110, USA
e-mail: ftaxman@gmu.edu

A. Pattavina • J. Byrne
School of Criminology and Justice Studies, University of Massachusetts Lowell,
University Avenue 1, Lowell, MA 01854, USA
e-mail: april_pattavina@uml.edu

F.S. Taxman and A. Pattavina (eds.), *Simulation Strategies to Reduce Recidivism:*
Risk Need Responsivity (RNR) Modeling for the Criminal Justice System,
DOI 10.1007/978-1-4614-6188-3_4, © Springer Science+Business Media New York 2013

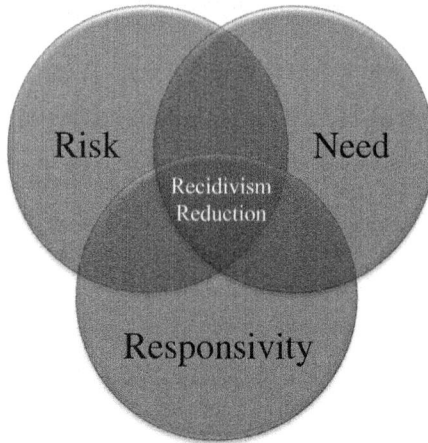

Fig. 4.1 Risk-need-responsivity model

for the management of offenders, but more importantly it dictates a set of principles regarding organizational and programmatic factors that affect the operations of supervision and correctional agencies.

With recidivism rates hovering around 67% (Langan & Levin, 2002, namely, rearrests within 3-year post-release from prison), depending on the distribution of risk and need profiles in an offender population and the quality and type of programming available in the justice system, using the RNR principles can have a dramatic impact (see Chap. 7 for a discussion of the potential impact). Recidivism reduction is more likely to occur for higher-risk/higher-need offenders placed in appropriate programs[1] than for moderate-risk offenders with fewer criminogenic needs or low-risk offenders without criminogenic needs. Since the distribution of risk and need profiles varies across jurisdictions, as does the availability of effective sanctions and treatment programming, the potential for recidivism reduction also varies. In line with the RNR model, expanding the capacity for human service programming in the justice system can have a large impact on recidivism rates.

In this chapter we provide an overview of the literature on (1) risk assessment, (2) needs assessment, and (3) responsivity. The summary of literature offered here does not provide the details contained in the classic *The Psychology of Criminal Conduct* (Andrews & Bonta, 2010). In their 672-page treatises on the topic, Andrews and Bonta (2010), outline empirical evidence to support their position. We provide an updated review of studies pertinent to the basic premise of the model and offer a critique of measurement and definitions of key constructs. Additionally, we hope to clarify some of the misconceptions regarding how the responsivity principle can be implemented into practice based on the available literature.

---

[1] The term programs will be used to refer to supervision strategies, treatment programs, services, and control techniques.

The Risk-Need-Responsivity Model

Andrews and Bonta (2010) state explicitly that the RNR framework covers a "normative and organizational context" (p. 45) that affects justice and correctional agencies. That is, while the RNR model relies upon a cognitive social learning framework for programming grounded in general personality theories, the effectiveness of the model is based on the delivery of clinical, social, and human services to both individuals and groups. Therefore, the framers of the RNR model expect it to be delivered in environments supportive of human services. In fact, Andrews and Bonta specify in their core RNR principles (principle 4) that justice agencies should "introduce human service: introduce human service into the justice context. Do not rely on the sanction to bring about reduced offending. Do not rely on deterrence, restoration, or other principles of justice" (p. 46). Without justice agencies shifting toward being human service providers, it is unlikely that the RNR framework can be effective in improving offender outcomes.

The core constructs of the RNR model (Andrews & Bonta, 2010, p. 46) are:

- *Risk*: Match the intensity of service to the offender's risk level. Work with moderate- and higher-risk cases. Generally, avoid creating interactions of low-risk cases with higher-risk cases.
- *Need*: Target the individual level factors that affect engagement in criminal conduct. Needs are those "drivers" or factors that affect involvement in criminal behavior. Define criminogenic needs both in terms of deficits and in terms of strengths to improve overall psychosocial and social functioning.
- *General responsivity*: Employ behavioral, social learning, and cognitive behavioral treatment strategies.
- *Specific responsivity*: Adapt the style and model of service according to the setting of service and to relevant characteristics of individual offenders, such as their strengths, motivations, preferences, personality, age, gender, ethnicity, cultural identifications, and other factors.

What Do We Know About Recidivism?

To improve recidivism outcomes by identifying targets for correctional interventions, we first need to better understand the relationship between key individual-level characteristics and recidivism. In this section, we briefly review the literature on recidivism with a focus on the impact of demographics, criminal history risk, and dynamic risk (need) factors on recidivism. While risk and needs constitute key components of the RNR model, demographic characteristics are generally neglected in the conceptual framework. Throughout this discussion, suggestions are made for improving the RNR model by considering key demographic characteristics that are associated with offending and redefining some constructs within the existing conceptualization.

Recidivism Defined

When we think about the risk posed by an individual offender, we are typically considering whether this individual will commit another criminal offense at some point in the future. Accordingly, recidivism can be defined as "the act of engaging in criminal offending despite having been punished" (Pew Center on the States, 2011, p. 7). We are often imprecise on how far into the future (1, 2, 3, 5, 10, 20 years) we are projecting these outcomes since the general concern in the criminal justice system is preventing *any* future criminal behavior.

One challenge posed by the extant recidivism literature is that any new offense or technical violation can be considered a continuation of offending behavior. Consequently, an "effective" intervention is expected to impact all types and patterns of offending for all types of offenders. This is a tall order that assumes that all offending behavior is the same and therefore will respond to the same intervention (or punishment). This perspective places an unrealistic expectation on correctional programs/interventions/punishment and violates one of the central tenets of the RNR model—targeting dynamic needs that are related to a certain type of offending behavior. We can not expect one approach to affect all future behavior.

Under the RNR model, correctional interventions are matched to a risk-need profile with the expectation that there are many different profiles and therefore a need for differing types of interventions. This is not the case in practice, however, because the expectation is that all criminal behavior is the same and will respond to the same patterns of response. If a person is placed in a program, it is expected that the program will affect their criminal thinking, substance use, employability, and an array of other need factors. A parallel example comes from education: if you have a student who has below grade-level reading skills and you provide tutoring, the expectation is that student will achieve grade-appropriate reading levels. It is not expected that the student will become a fan of science or excel in math, merely that the reading level will be addressed.

Recidivism Among Offenders Under Correctional Control

Much of what we currently "know" about recidivism of offenders in our prison system is based on a single, large Bureau of Justice Statistics (BJS) study conducted nearly 20 years ago (Langan & Levin, 2002).[2] This study tracked the recidivism of 272,111 former inmates released in 1994 from prisons across 15 states. This cohort of state prisoners was followed for 3 years after release from prison. A total of 67.5% of these offenders were rearrested for a new offense within 3 years of release from prison (Langan & Levin, 2002). The 183,675 prisoners rearrested during this period were charged with a total of 800,240 new crimes during the 3-year follow-up period. Within this subgroup of recidivists, there were a small

---

[2] BJS is currently conducting a new study on a 2005 cohort, but these results are not yet available.

number of high-rate offenders who accounted for a significant proportion of all crimes (as measured by arrests) committed during the follow-up period. According to Langan and Levin (2002), a small proportion of all releases (12%) account for a significant proportion (34.4%) of all crimes committed by the release cohort. We estimate that about half of all higher-risk offenders are also higher-rate offenders. Accordingly, targeting this group of higher-rate offenders will have a significant impact on recidivism rates. Targeting effective risk reduction strategies and resources toward this subgroup of offenders will have a greater overall crime reduction effect, even though high-risk/high-rate offenders are usually a smaller proportion (generally 10–20%) of any offender cohort.

A recent report published by the Pew Center on the States (2011) revealed that little has changed in terms of recidivism rates since the 1994 BJS study. Defining recidivism as any return to incarceration within 3 years of release, the PEW study found that about 44% of releases recidivated during the observation period. They noted that recidivism rates have remained stable at about 40% since 1994. The clear conclusion from these national studies is that recidivism rates are high, especially for a small group of higher-risk offenders, and that correctional interventions have done little to alter these rates of offending over the past 20 years.

Recidivism and Offender Characteristics

Criminal History Risk and Recidivism

Nearly every major review of the research on recidivism conducted over the past two decades has found that the strongest predictor of future criminal behavior is past criminal behavior (Gottfredson & Gottfredson, 1986; NRC, 2007). The concept of risk assessment (to be discussed below) is based on the notion that one's past criminal justice involvement can predict their future offending behavior. In the BJS study, about 48% of offenders with 3 or fewer prior arrests (about 22% of the total cohort) were rearrested within 3 years of release from prison. By comparison, over 80% of the offenders with more than 10 prior arrests (34.2% of the cohort) were rearrested during this same review period. The BJS study did not use a risk assessment tool (a measure of criminal history and risk of reoffending) but, like other studies, distinguished higher-risk offenders from moderate- and low-risk offenders based on prior arrests alone (criminal history).[3] Consistent with this empirical reality, prioritizing treatment for high-risk offenders characterized by chronic criminal histories is a central component of the RNR model.

One question that invariably is raised is what other factors, besides historical criminal justice factors, are related to recidivism. Answering this question requires us to assess whether recidivism rates differ by demographic factors such as gender, race/ethnicity, and age. It is assumed that risk models will predict recidivism for one

---

[3] Prior arrest history is just one ingredient of a risk assessment tool.

group (e.g., males) compared to another (e.g., females). But is this assumption supported by empirical research? Empirical evidence suggests that, even after controlling for criminal justice history, demographics of the offender population have an impact on recidivism. Andrews and Bonta (2010) nested these demographics into responsivity while others argue that gender and age should be considered within the framework of risk and criminogenic needs.

Gender and Recidivism

Gender differences in patterns of offending—and reoffending—have been identified across a broad range of studies, using a variety of data sources, methodologies, and outcome measures. Women are more likely to score lower on risk assessment tools and some factors that predict risk for recidivism may be different for women than for men. The notion that there are gendered pathways to offending has implications for the RNR model in that risk assessment, needs assessment, and intervention strategies may be very different for women than for men. Van Voorhis, Bauman, Wright, and Salisbury (2009) and Van Voorhis, Wright, Salisbury, and Bauman (2010), examined gender-responsive risk and need factors and interventions, and generally found that the set of measures that are important for women are also important for men. They are following trends: (a) assessments within probation samples should include factors related to parental stress, family support, self-efficacy, educational assets, housing safety, anger/hostility, and current mental health factors; (b) assessments within prisoner samples should include measures of child abuse, anger/hostility, relationship dysfunction, family support, and current mental health factors; and (c) assessments for returning citizens should include factors of adult victimization, anger/hostility, educational assets, and family support. They also find similar criminogenic needs, with an emphasis on factors that affect psychosocial functioning, and note that there is a greater need for gender-responsive treatments. There is some disagreement in the field with Jennings et al. (2010) who no empirical support for gendered pathways or risk factors.

In the 1994 BJS recidivism study, 8.7% of the release cohort ($n=272,111$) were women. The study findings reported that men were more likely than the women to be rearrested (68.4% vs. 57.6%), reconvicted (47.6% vs. 39.9%), resentenced to prison for a new crime (26.2% vs. 17.3%), and returned to prison with or without a new prison sentence (53.0% vs. 39.4%). Given these differences, it makes sense to consider that recidivism reduction models should be tailored to both common criminogenic needs and the gender-specific needs of the offender.

Race/Ethnicity and Recidivism

A second area of inquiry involves racial/ethnic differences in official rates of offending and reoffending. We know from previous research that racial/ethnic variations in offending rates, and risk of reoffending, have been identified (Hawkins, Laub, &

Lauritsen, 1998; Hindelang, 1978; LaFree, 1995; Morenoff, 2005; NRC, 2007; Piquero & Brame, 2008; Sampson & Lauritsen, 1997). One key assumption of the RNR model is that the factors related to criminality do not vary as much across demographic subgroups as by risk and need areas. Unfortunately, the empirical research supporting this assumption is weak. While some report that the variables used in risk classification instruments may be biased, resulting in the overclassification of minority offenders as higher-risk offenders (Clear, 1988), there is scant research on this important topic available for review.

In the 1994 BJS recidivism study, 50.4% of the release cohort was classified as white, 48.5% as black, and 1.1% as other. Separate classification by ethnicity identified 24.5% of the cohort as Hispanic. Since Hispanics are included in both the white and black categories, it is impossible to distinguish racial/ethnic differences in reoffending in this cohort. Nonetheless, Langan and Levin (2002) reported that blacks were more likely than whites to be rearrested (72.9% vs. 62.7%), reconvicted (51.1% vs. 43.3%), returned to prison with a new prison sentence (28.5% vs. 22.6%), and returned to prison with or without a new prison sentence (54.2% vs. 49.9%).[4] Conversely, non-Hispanics were more likely than Hispanics to be rearrested (71.4% vs. 64.6%), reconvicted (50.7% vs. 43.9%), and returned to prison with or without a new prison sentence (57.3% vs. 51.9%). Hispanics (24.7%) and non-Hispanics (26.8%) did not differ significantly in terms of likelihood of being returned to prison with a new prison sentence. The prima fascia evidence suggests a need for more detailed analyses of the risk of recidivism among specific offender subgroups.

Age and Recidivism

A third demographic characteristic that has been linked to offending/reoffending rates is age. A strong empirical relationship between age and offending has been established as shown by the well-known age-crime curve (see, e.g., Cohen, Piquero, & Jennings, 2010; Farrington, 1986; Greenberg, 1985; Hirschi & Gottfredson, 1983; Moffitt, 1993; NRC, 2007; Quetelet, 1831/1984; Thornberry, 1997). It is generally accepted that for most types of offending, rates peak in late teens and then drop significantly in mid-20s. According to the recent review by the National Research Council: "Perhaps the most obvious and simplest pathway to desistance from crime is aging: offending declines with age for all offenses" (2007, p. 26). In the BJS recidivism cohort, the younger the prisoner was when released, the higher the rate of recidivism (Langan & Levin, 2002). Consider the following age-specific rearrest levels: 82.1% of those under age 18 were rearrested, 75.4% of those 18–24, 70.5% of those 25–29, 68.8% of those 30–34, 66.2% of those 35–39, 58.4% of those 40–44, and 45.3% of those 45 or older. It should be noted that most incarcerated offenders have peaked in their offending careers prior to their first incarceration, and therefore the deterrent effect of incarceration on overall crime rates is minimal (see Nagin, Cullen, & Jonson, 2009). In the BJS recidivism study, only 21.3% of the release cohort was under 24 and 33.2% were 35 or older at the time of their release.

---

[4] These numbers do not control for criminal justice history.

Risk of Recidivism and Location: The Community Context of Failure

A number of recent research studies have emphasized the importance of examining the influence of both individual and community risk factors (see Byrne, 2009). We know much more about individual risk factors than we know about community-level risk factors that can be directly linked to recidivism. This paucity of empirical research on the community context of recidivism is a major impediment to efforts to improve individual level outcomes and correctional outcomes. From the research conducted over the past 2 decades available for review (see Byrne & Pattavina, 2006; Gottfredson & Taylor, 1985; Hipp, Petersilia, & Turner, 2010; Kubrin & Stewart, 2006), it appears that an offender's risk of recidivism is influenced by both individual- and community-level risk factors. The location where the offender resides makes a difference in the likelihood of recidivism.

The notion that offenders with similar individual risk profiles (based on such factors as prior offense history, prior incarceration, history of substance abuse, and employment/education deficits) are more likely to fail if they are released to a small number of identifiable high-risk communities is a factor that has emerged or reside in the literature in recent years. To the extent that high-risk communities are also resource-poor communities, it seems logical to suggest that as a general principle, you cannot change offenders unless you also change the communities in which offenders reside (Byrne, 2009). The influence of community resource limitations and programming capacity is generally ignored within the RNR framework; however, the framework can be used to guide resource allocation efforts to improve access to services in disadvantaged communities.

Summary of Knowledge About Recidivism

The purpose of this review was to examine extant correlates that are related to recidivism but not typically included in standard measures of risk and needs of offenders. A long history in the classification and assessment literature in criminal justice has emphasized the importance of tools being "demographic neutral" to avoid imposing systematic biases that might occur particularly related to age, gender, ethnic, or racial categories. Yet, as shown above, in any model that examines the impact of programming on recidivism, we cannot ignore key demographic factors that are related to recidivism such as age and gender. The literature suggests that intervention designs, as well as simulation tools, need to consider these factors, even if they are not front and center in the RNR model. The literature reviewed above also stresses the importance of considering community-level context in simulation models and using the RNR principles to help demographic-related factors and build up the capacity for effective treatment in resource-depleted communities.

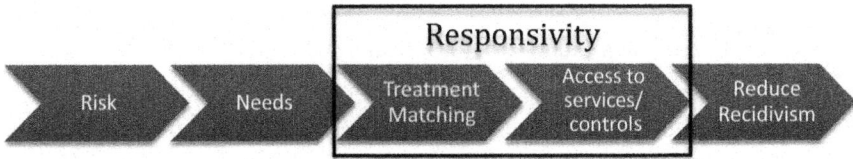

Fig. 4.2 Expanded RNR framework

Criminal Justice Risk: An Overview

In the RNR model, the first ingredient is static risk level (see Fig. 4.2). Risk in this context refers specifically to an offender's likelihood of reoffending and is operationalized by indicators of prior criminal involvement. Since the 1920s, there has been a growing awareness of the factors that predict the likelihood of reoffending. Risk assessment tools are designed to measure the degree to which the individual is likely to have negative outcomes (e.g., more recidivism) during or after experience with the justice system. There are a number of standard risk assessment tools such as the Wisconsin Risk and Needs Instrument, the Level of Service Inventory-Revisited (LSI-R), the Correctional Offender Management Profiling for Alternative Sanctions (COMPAS), the Ohio Risk Assessment System (ORAS), and a myriad of other tools, some in the public domain and others are proprietary. Each of these tools has some standard reference to historical involvement with the justice system (i.e., takes into account the criminal history). These assessments tap into static factors (Box 4.1) such as the number of prior arrests, the number of prior convictions, the number of prior probation experiences, prior escapes, and prior probation violations.

The Design and Development of Risk Assessment Instruments

Risk assessment is not a new concept in the field of criminology. The use of interviews and assessments that collect data on individuals has been a commonplace within the criminal justice system for many years. Screening and assessment instruments have been developed for each decision point in the system such as pretrial release, probation supervision level, parole release, and parole decision-making. The most recent discussion of the evolution of risk assessment instruments in criminology is offered by Andrews, Bonta, and Wormith (2006) who point out how assessment tools support the RNR framework. The development of risk assessments is generally discussed in terms of the "generation" of the approach with techniques

Box 4.1 Definition of Static Risk for Recidivism

Static risk of recidivism refers to the historical involvement in the justice system that is used to characterize the person's likelihood of future involvement. These static factors cannot be changed except for being increased. Risk refers to objective factors such as:

- Age of first arrest
- Number of prior arrests
- Number of prior convictions
- Number of escapes or infractions in prison
- Number of probation violations
- Number of incarcerations

Static risk factors do not include the following typical proxies used in the research literature:

- Seriousness of offense (stakes involved with the offense)
- Type of offender such as drug, property, violent, and sex (stakes of an offender or the degree to which the offender presents serious harm)
- Rate of offending
- Age of offender (demographic)
- Socioeconomic status

recently emerging that are designed to guide the responsivity process. The four generations of risk assessment are briefly reviewed here:

- *First generation*: *Clinical assessment*. First-generation risk assessments rely solely on professional judgment. Depending on the skill of the interviewer, the nature of the decision, and the degree to which the offender is forthcoming in the interview, the clinical interview can generate valid information about risk. But, the interview process has been critiqued due to the potential for bias related to subjective factors (i.e., age, race, gender, and other demographics of the offender) that may affect the decision of the interviewer and/or system.[5] This criticism has led to concerns about introducing systematic bias into the risk assessment process. These assessments are based on criminal justice history (generally referred to arrest history or rap sheet) along with unspecified characteristics of interest of interviewers.

---

[5] While extralegal factors such as demographics may be empirically linked to recidivism outcomes, they are not generally included in risk assessments due to the potential for bias. Our conceptualization of the RNR model considers age and gender as key components of responsivity rather than risk.

Table 4.1 Indicators of criminal history risk in third-generation risk instruments

Items in common risk instruments			
Item	LSI-R[a]	ORAS[b]	Wisconsin[c]
Prior arrests (convictions)	✓	✓	✓
Age of first arrest (conviction)	✓		✓
Prior incarceration	✓	✓	
History of escape (or attempt) from correctional facility	✓		
History of institutional misconduct	✓	✓	
Number of prior periods of probation/parole supervision		✓	✓
Probation/parole revocations	✓	✓	✓
History of assault or violence	✓		✓
Juvenile conviction for burglary, theft, autotheft, robbery, or forgery			✓

[a]Level of Service Inventory-Revised
[b]Ohio Risk Assessment System
[c]Wisconsin Risk-Need Assessment

- *Second generation*: *Actuarial risk assessments*. The second stage in the development of risk assessment techniques involves a risk-only statistical calculation. Beginning with parole-hearing officers in the 1920s, there was a growing realization that prior involvement with the justice system can be a strong predictor of future involvement with the justice system (recidivism). Key factors such as the number of prior arrests, the number of incarcerations, attempts to escape, behavior in the institution, prior probation violations, and age of first arrest have been found over the past 90 years to be consistent predictors of continued involvement in the justice system (see Gottfredson, 1987; Hoffman & Beck, 1974). These are generally referred to as static risk factors (Table 4.1) that can be used to create a statistical (or actuarial) assessment of the likelihood of failure. The statistical assessment wards against the problems associated with interviewer or systematic bias from the justice system by ensuring that only the offenders' criminal justice history influences recommendations and decisions. Interviewer or system basis regarding the demographic characteristics of the offender is minimized in this approach. Since the statistical assessment is generally demographically blind—it does not include gender, age, or race factors—it minimizes the potential that justice workers will allow bias regarding these factors to impact decisions. Actuarial risk assessment has also been on the forefront of advancing the use of risk information to improve resource allocation. For example, O'Leary and Clear (1984) proposed that risk assessment tools could be used to identify risk level and that higher-risk offenders should be placed in more intensive and more control-oriented supervision programs. This resource allocation model represents a practical use of the risk assessments tool and makes up a central component of the RNR model. Risk assessments can have an impact on recidivism through the use of tools to assign higher-risk offenders to more intensive programming (responsivity).

Table 4.2 Dynamic risk factors in third-generation instruments

Item	LSI-R[a]	ORAS[b]	COMPAS[c]	Wisconsin[d]
Education/employment	✓	✓	✓	✓
Financial	✓	✓	✓	✓
Family/marital	✓	✓	✓	✓
Accommodation	✓		✓	
Leisure/recreation	✓		✓	
Companions/associates	✓	✓	✓	✓
Substance use	✓	✓	✓	✓
Emotional/personal	✓		✓	✓
Attitudes/orientation	✓	✓	✓	
Neighborhood		✓	✓	
Mental health				✓
Health/wellness				✓
Sexual behavior				✓

[a]Level of Service Inventory-Revised
[b]Ohio Risk Assessment System (community supervision tool)
[c]Correctional Offender Management Profiling for Alternative Sanctions
[d]Wisconsin Risk-Needs Assessment

- *Third generation*: *Actuarial risk and dynamic risk* (*needs*). The third generation of risk assessment tools added in dynamic risk factors to improve the alignment between programming and offender needs. Dynamic risk factors are offender characteristics that are amenable to change. When these dynamic risk factors are directly related to recidivism, they are labeled criminogenic needs (Andrews & Bonta, 2010). Beginning with the Wisconsin Risk and Needs Assessment Tool, this era of instruments added needs or psychosocial factors into the equation of risk assessment. Depending upon the instrument, the added factors may include attitudes and orientation, employment, substance abuse, living arrangements, mental health status, leisure time activities, criminal peers, and other areas that have a hypothesized correlation with offending (Table 4.2). The third generation has two means by which to assess outcomes: historical risk factors (static risk) and dynamic offender needs (dynamic risk). In some risk-need instruments, risk and needs are added together to calculate a total score; in others they are considered separate scores with a matrix guiding the user as to the proper placement for the offender being assessed. Third-generation instruments represent an important advancement because they provide more structured information about the offender that can be used to identify factors that drive criminal behavior and identify targets to be addressed through correctional interventions.
- *Fourth generation*: *Case management through risk and needs assessment*. Fourth-generation risk and needs assessment tools are an extension of third-generation tools with a focus on treatment matching/case management. In these instruments, risk and need factors are used to identify appropriate services (treatments and controls) to reduce the risk of recidivism. These tools focus on core components of responsivity by creating service matching algorithms based

on the risk and need profile of the offender. Fourth-generation tools are designed to improve the quality of assessments and to make them more applicable to offender case management and resource allocation by identifying individual-specific behavioral targets for intervention.

Regardless of the generation of the tool being used, risk assessment tools create an opportunity to classify the level of risk of one offender relative to other offenders in a population. Based on an actuarial risk assessment, risk scores can be divided into levels (e.g., high, moderate, and low) that reflect the probability that an individual offender will recidivate. Accordingly, one can then assess the distribution of risk across the population and use this information to guide decision-making. Given that risk is defined and measured in different ways across jurisdictions and tools, it is difficult to compare how the tools can be used to guide populations or resources. The use of actuarial risk and needs assessments is a core component of evidence-based decision-making and has been identified as a best practice in the field of corrections.

Risk and Need Factors in Various Tools

The most common risk-need classification instrument used today is the LSI-R (Level of Service Inventory-Revised) which is a third-generation tool that includes both risk and need factors (Taxman, Perdoni, & Harrison, 2007). The LSI-R instrument requires the classification of each offender across 10 unique subscales, using a total of 54 items, including the following: criminal history (10 items), education and employment (10 items), financial (2 items), family and marital (4 items), accommodations (3 items), leisure and recreation (2 items), companions (5 items), alcohol and drugs (9 items), emotional and personal (5 items), and attitude and orientation (4 items). Latessa and colleagues (2010) have updated the style of LSI-R in their Ohio Risk Assessment System (ORAS) for community supervision[6] which has seven domains including criminal history (static risk factors), family and social support, substance use, criminal attitudes and behavioral patterns, education/employment/financial factors, neighborhood problems, and peer associations (Table 4.2).

Table 4.1 compares the type of static risk information that is included in different risk assessment tools. We can see the similarity of the items included in four of the most commonly used risk-needs assessments (LSI-R, ORAS, COMPAS, and Wisconsin Risk and Needs) in the justice system. Four need domains (Table 4.2) are common to all four instruments: educational/employment need, family/marital need, antisocial companions/associates, financial and substance use. Antisocial

---

[6]There are versions of the ORAS for pretrial, prison, and reentry that have slightly fewer domains; see Latessa and Lovins (2010).

attitudes is measured in three of the four instruments. A primary limitation of these tools that is discussed in more detail later in this chapter is their limited coverage of substance use and mental health disorders. While all four instruments assess substance use, they typically do not provide enough diagnostic information to guide treatment matching strategies. Mental health needs are only directly measured in one instrument. Noticeably absent is the type of offense or offending behaviors since prior research has concluded that specific crimes are not risk factors in the same manner as the history of involvement with the justice system.

A major debate in the literature concerns the scoring of third-generation tools. Baird (2009) notes that third-generation tools that include risk and need factors do not classify offenders as well as second-generation tools that focus only on risk factors. In analyzing the discriminate and predictive validity of the LSI-R, it has been found that static risk factors are generally more predictive of recidivism than the LSI-R score that includes all domains. In the LSI-R, 10 of the 54 risk items concern criminal history, and they have the same predictive validity as the full instrument (Austin, 2006). Latessa, Smith, Lemke, Makarios, and Lowenkamp (2009) report the same pattern with the criminal history subscale of the ORAS. They find that criminal history predicts risk for recidivism more so than other domains, and it has more discriminant validity. The question that remains is whether risk levels should include only risk factors or risk and need factors.

Summary of Key Risk-Related Assumptions

Recidivism is a product of, at a minimum, risk level, age, gender, and location. We can treat all conviction offenses alike, given that the risk level will control for any offense-specific links to recidivism. In this section, we have defined recidivism, described the overall recidivism patterns of offenders released from prison, and identified significant variations in recidivism patterns by criminal history, age, gender, race/ethnicity, and location. In addition, we have reviewed the content of risk instruments used to classify offenders into risk levels and have discussed the importance of using actuarial risk and need assessments to guide resource allocation and facilitate treatment matching. We note that recidivism risk is more complex than just a high rate of offending, the seriousness of instant offense, or the length of an offenders' criminal career. Risk, in this sense, is a predictor of future criminal offending based on prior involvement in the criminal justice system, not on any offense-specific information. This operationalization of risk is one of the three core principles of the RNR model. The Andrews and Bonta model stresses that offenders with more extensive criminal histories (i.e., high-risk offenders) should be prioritized for more intensive programming and controls. To achieve the risk principle, it is required that all offenders are assessed with validated risk assessment instruments and that risk information is used in assigning offenders to levels of programming.

Dynamic Risk (Criminogenic Needs): An Overview

The second ingredient in the RNR model is criminogenic need or dynamic risk. In the third and fourth generation of assessment tools, there is a recognition that an individual's current situation—dynamic factors—also influences his or her involvement in criminal behavior. For the most part, these factors such as substance abuse, mental health, employment retention, pro-social values, friends and families, and criminal thinking are amenable to change. The question is: which of these needs are criminogenic (needs that are directly related to recidivism), and which are more related to psychosocial functioning? The constructs—criminogenic needs and psychosocial functioning needs—are different in terms of their relationship to recidivism. Criminogenic needs have a direct link to recidivism whereas other types of psychosocial functioning needs have a spurious or indirect. In other words, criminogenic needs can predict recidivism while psychosocial functioning cannot (once controls for static risk and demographics have been included in multivariate models).

There are a number of challenges that exist in distinguishing different types of need factors. Criminogenic needs should be seen as "crime-producing factors that are strongly correlated with risk" for recidivism (Latessa & Lovins, 2010, p. 209). In reality, criminogenic needs can be described as the subset of dynamic risk factors that are correlated—typically in a direct and robust manner—with risk of recidivism. These are factors that can be improved over time through correctional responses. Static risk factors (e.g., age of first arrest, number of prior arrests, number of prior probation violations) and demographics (e.g., gender, race/ethnicity) are not amenable to change and therefore should be included in the treatment placement decisions but after an emphasis on the primary criminogenic need. In contrast, dynamic needs such as substance abuse, criminal thinking, and peer associations can be altered to reduce the risk of recidivism.

Andrews and Bonta (2010) identified seven key dynamic needs (they have eight central needs but one is static risk factor). The first four needs listed below—antisocial history, antisocial attitudes, antisocial peers, and criminal personality—are considered to be more robust and predictive of recidivism outcomes than the remaining factors. The latter four are considered to be of lesser importance in that they have weaker correlations with recidivism and are often not directly related to recidivism outcomes. The "Central Eight" (see Andrews et al., 2006) are:

1. A history of antisocial behaviors (static risk)
2. Antisocial attitudes, values, and beliefs
3. Antisocial/pro-criminal associates
4. Antisocial personality pattern
5. Family/marital factors, such as lack of social support, as well as neglect and abuse
6. Low levels of educational, vocational, or financial achievement

7. A lack of pro-social leisure activities
8. Abuse of drugs and alcohol

In this list of potential intervention target areas, Andrews and Bonta (2010) identified factors that are not necessarily criminogenic but affect the overall functioning of the offender. These additional dynamic factors include non-criminogenic factors such as educational deficits or literacy problems, poor employment history, unstable housing situation, mental health disorders, and lack of pro-social leisure time activities. These factors affect the ability of a person to lead a productive life but do not necessarily correlate with continued criminal behavior; the factors by themselves are not "crime producing." For example, unemployment is not just a factor associated with offender subpopulations; it is a societal problem that affects many people, many of whom are not involved in offending. While not directly predictive of criminal conduct, non-criminogenic need factors are relevant because they affect how well the person will function in society and in correctional programming (e.g., mental health status can affect how well the person participates in treatment programs or responds in prison).

The following sections briefly reviews the literature on each dynamic need. The goal is to distinguish between criminogenic and non-criminogenic in a human service-driven model of recidivism reduction. The measurement of these factors may affect the relationship between a factor and offending. This discussion extends the RNR model beyond Andrews and Bonta's original conceptualization and takes into account empirical evidence regarding the link between dynamic risk factors and recidivism outcomes. A primary focus in this discussion regards the handling of substance use in the RNR model. As discussed above, substance use is not considered a priority criminogenic need in the Andrews and Bonta model.

Substance Abuse

Empirical research consistently demonstrates that justice-involved individuals have significantly higher rates of substance use disorders (SUDs) than the general population (Lurigio, Cho, Swartz, Graf, & Pickup, 2003; Mumola & Bonczar, 1998; Staton-Tindall, Havens, Oser, & Burnett, 2011; Taxman, Perdoni & Harrison, 2007; Taxman, Cropsey, Young, & Wexler, 2007). A national survey of adults on probation conducted in 1995 by the BJS indicates that over two thirds (70%) of all probationers reported lifetime drug use, one third reported using illegal drugs in the month prior to their offense, and 14% reported being under the influence of illegal drugs at the time of their offense (Mumola & Bonczar, 1998). Based on findings from the ongoing *Arrestee Drug Abuse Monitoring* (ADAM) project, drug testing of arrestees routinely finds that 60% of the arrestees test positive for some substance at the time of their arrest (ONDCP, 2011; Taylor et al., 2001).

Another key source of data on offender substance abuse prevalence is the BJS special report on drug use among federal and state inmates (Mumola, 1999). The BJS report defined "regular drug use" as "using at least once a week for at least a month" (Mumola, 1999, p. 2) regardless of the drug of choice. The definition of a substance user in an RNR model has important implications for responsivity. The RNR model generally neglects the fact that there are different types of drug users, and the linkage to criminal conduct varies considerably by substance use disorder problem severity and by drug of choice. Some substance abuse behaviors drive criminal behavior while other use behaviors have negligible or no impact on offending behavior. A difference exists among types of drug use: it is important to distinguish between lifetime use (ever used), regular use (abuse), and use that impacts decision-making and daily life (dependence). From a recidivism reduction perspective, it is important to separate the drug user from those engaged in drug dependence and the abuse of more criminogenic drugs.

Various Definitions of Substance Abuse

There are several methods available to assess and determine severity of substance abuse need for justice populations. Most criminal justice risk assessments include substance abuse-related questions (see Table 4.2). For example, in the Wisconsin Risk and Needs Instrument and the Level of Service Inventory-Revised (among others), the substance abuse questions generally ask whether the individual has ever had a problem with substance abuse (generally with three response categories: no, some, severe). In the new Ohio Risk Assessment System (ORAS), Latessa and Lovins (2010) measure drug use with items that reflect lifetime use, longest period of abstinence, and employment-related problems stemming from drug use. There is a weak correlation between the substance use subscale of the ORAS and rearrest ($r=0.14$) (Latessa & Lovins, 2010). While offender risk assessment instruments generally include items related to substance use needs, they typically do not provide enough information to adequately facilitate treatment matching strategies.

Validated substance use screening and assessment tools, such as the Addiction Severity Index, the TCU Drug Screen, and others (see Taxman, Cropsey, et al., 2007), tend to garner a better understanding of the offender's pattern of drug use and misuse. These instruments include more detailed information about lifetime use of drugs, drug of choice, mode of delivery (e.g., smoking, injection), longest periods of abstinence, and whether drug use affects employment or is related to legal problems. In general, the risk-need screening instruments used in the justice system tend to cast a large net to describe an array of substance use behaviors, some of which are pertinent to predicting recidivism and some of which are not. The instruments are useful to screen for substance use disorders, but do not adequately examine how these disorders impact criminal behavior. While researchers tend to find a correlation between lifetime substance abuse and rearrest (see White & Gorman, 2000), others find a more consistent relationship when the emphasis is on current or recent drug use patterns. Accordingly, it is important to consider how drug use is measured and to differentiate the population of abusers based on the pattern of use, frequency of current use, and types of drugs used. All of these factors are important in terms of responsivity—to determine the appropriate treatment program to impact recidivism rates.

Clinical Standards

The clinical diagnostic standard for SUDs identifies four levels of use: dependent (chronic, drug-seeking behavior), abuse (frequent use which interferes with daily life), recreational, and no use (DSM-V, 2000). Clinical standards focus on identifying drug use that interferes with functioning (e.g., employment, family responsibilities, legal problems, and health status). The emphasis on functioning derives from the effort to discern whether the use pattern warrants treatment attention. Some of the frequently used instruments in assessing substance abuse disorders (e.g., the Addiction Severity Index) examine the type of drug used, the method of using the drug, and the time frame for use (e.g., monthly, weekly). This information is then used to assess the severity of the substance use disorder. It is important to distinguish between lifetime use and current use in identifying substance use treatment needs. While lifetime use may be related to offender outcomes, current use is much more important from a responsivity perspective. Clinical dependence is seldom measured in the justice system, and therefore it is often difficult for criminological studies to differentiate substance-dependent offenders from offenders with other or no SUDs.

Type of Drug Used and Criminal Behavior

The drug-crime nexus is complicated by a number of qualifying issues regarding whether drug use affects criminal conduct (White & Gorman, 2000). For instance, the literature on opiate and cocaine use illustrates a closer connection between drug use and involvement in criminal behavior (Nurco, Hanlon, Kinlock, & Slaght, 1984; Taylor et al., 2001). Conversely, the literature thus far has not shown that marijuana use affects involvement in criminal behavior (White & Gorman, 2000), even though a number of offenders may use marijuana as part of their lifestyle (Taylor et al., 2001). The drug-crime nexus theory suggests that involvement in criminal behavior will be disrupted by providing substance abuse treatment in the justice system (see Chandler, Fletcher, & Volkow, 2009; Hubbard et al., 1988) especially for those users for whom drug use directly affects criminal behavior. Most scholars report that when drug users are in treatment, the number of criminally active days declines significantly, contributing to the perspective that participation in drug treatment will reduce aggregate rates of criminal behavior (Ball, Shaffer, & Nurco, 1983; Nurco et al., 1984; Sampson & Lauritsen, 1997).

Many explanations and elaborations of the drug-crime nexus have been offered (see, e.g., Goldstein, 1985; White & Gorman, 2000). Recent work by Bennett, Holloway, and Farrington (2008) provides strong meta-analytic support for the relationship between drug use and offending. Additionally, this research suggests the importance of distinguishing between drugs of choice. Based on their meta-analysis of over 30 primary studies, Bennett and colleagues find that the odds of offending are 3–4 times higher for drug users than nondrug users. Additionally, their work shows that drug of choice matters. The odds of offending are about 6 times greater for crack users (OR = 6.09) relative to non-crack users, about 3 times greater for heroin users (OR = 3.08), about 2.5 times greater for cocaine users (OR = 2.56) and about 1.5 times greater for marijuana users (OR = 1.46) relative to non-marijuana users (Bennett et al., 2008). These findings confirm that certain drug use patterns are more strongly correlated with offending, and therefore it is important to consider the nuances of the drug-crime relationship when targeting recidivism reduction strategies for substance-involved offenders.

Assumptions for the RNR Model: Substance Abuse

Given that the measurement of substance abuse in most third- and fourth-generation risk assessment tools is often inconsistent with a DSM-V clinical definition of a substance use disorder (SUD), there is a need to refine how substance abuse is measured in justice settings. The commonly cited 70–80% of offenders with substance use issues usually refer to the percentage of offenders who report any use of illicit substances during their lifetime (or the lifetime prevalence), and biological tests at the time of arrest (the ADAM program) report that nearly 60% of the offenders have some type of substance in their system at the time of the drug test (although the majority of drug use is marijuana) (ONDCP, 2011). The problematic tendency in the criminal justice-related instruments is to assess whether the individual has a substance abuse problem based on their lifetime use of substances. The use of lifetime prevalence measures limits the usability of these tools for guiding treatment strategies and often results in offenders without current SUDs being placed into substance abuse treatment programs, a waste of resources that may also jeopardize program effectiveness.

More refined measures of substance abuse behaviors vary considerably across the offender population, as shown in Table 4.3 below. Using the 2000 ADAM data, Taylor et al. (2001), report that 30% of the arrestees are drug-dependent, 19% are substance abusers, 19% are users, and 30% are not users of drugs or alcohol. In terms of responsivity, this means that substance abuse treatment-only interventions should focus on the 30% of offenders who are drug-dependent. In similar analyses, Belenko and Peugh (2005) found that nearly 30% of the prisoners were drug-dependent and used hard drugs (i.e., cocaine, opiates, amphetamines) on a frequent basis. This group is more likely to be responsive to treatment focused on substance abuse, and addressing their more severe SUD will have a potentially greater impact on their likelihood of recidivism.

Based on our review of the literature, substance abuse as a criminogenic need is more likely to be related to recidivism for offenders with dependency disorders and those who abuse hard drugs. Based on the extant empirical research reviewed above,

Table 4.3 Drug use distributions

Clinical categories[a]	Percent
General use of drugs	
Never used drugs	30
Use	19
Abuse	19
Dependent	30
Use of hard drugs[b]	
Never used hard drugs or marijuana	30
Used marijuana, but no hard drugs within 30 days of arrest	9
Used hard drugs but not within 30 days of arrest	25
Used a single hard drug weekly or monthly or used more frequently	35

[a]Taylor et al. (2001)
[b]Belenko and Peugh (2005):275

we propose the following refinements to the consideration of substance use in an RNR model:

1. Some types of drug use do not impact criminal behavior. The literature illustrates that opiates and cocaine use are more likely than other drugs to influence criminal conduct through the need to acquire money and to participate in the drug trade to support one's own habit and the decisions made by an individual in the inebriated state. There is increasing evidence that methamphetamine use falls into this category as well. For alcohol use, involvement in drunk driving is the major linkage between drinking behaviors and criminal conduct.
2. Marijuana use does not appear to follow the same trend as other "hard drugs" (e.g., cocaine, opiates, amphetamines), and therefore marijuana abuse should not be considered a criminogenic need. Marijuana users with criminal thinking patterns should be prioritized for interventions that can address their antisocial thinking needs rather than their substance use issues.
3. The offense should not drive a determination of a substance abuse treatment need since many offenders are arrested on charges that are not related to substance abuse, and many times substance abuse is a secondary driver of criminal behavior. Additionally, many offenders arrested for drug-related offenses do not have SUDs that necessitate intensive treatments and therefore are unlikely to benefit from substance abuse treatment programs.
4. Offenders should be divided into four categories regarding substance abuse: (a) nonusers, (b) used drugs in the past with current use of soft drugs, (c) used hard drugs in the past, and (d) used hard drugs prior to the arrest. This classification is similar to the recommendation of Belenko and Peugh (2005). This classification is suggested because most studies and assessment tools examine these issues, but they do not include factors that are used in clinical assessments (such as the DSM-V). This pattern is also more similar to proposed changes in the DSM-V which only focuses on substance use disorders.

This suggested revision to the RNR framework increases the transportability of the model and aligns it better with the literature regarding the relationship between substance dependence, substance abuse treatment needs, and offending. This alteration also improves the potential for responsivity and can improve the resource allocation function of the model. Triaging offenders based on the severity of their substance abuse treatment needs will improve cost-effectiveness and recidivism reduction potential by increasing the likelihood that the people being assigned to treatment are the ones who need it and for whom the treatment will be most effective.

Criminogenic Needs (All but Substance Abuse)

In their initial book (1996), Andrews and Bonta report that factors involved in a subculture of offending behavior (e.g., antisocial cognitions, antisocial peers, and criminal personality) are the strongest predictors of recidivism. They describe four primary criminogenic needs including history of antisocial behavior, antisocial values and opinions, antisocial peers, and criminal personality factors[7]. These factors have since been labeled as the "Big Four" criminogenic needs and the argument can be made that these items represent one underlying construct of a "criminal lifestyle." The literature on criminogenic needs recognizes the interrelated nature of criminal thinking and criminal behavior and this "criminal lifestyle" is the most important target for correctional intervention in the RNR model. Criminal lifestyle is a global construct that encompasses the decision-making skills of the offender, the attitudes and values that affect involvement in criminal conduct, and the supports (or lack of supports) that either encourage or constrain criminal behavior.

Third- and fourth-generation risk tools include items related to antisocial thinking and offending lifestyles. For instance, the Wisconsin Risk and Needs Tool, the LSI-R, and COMPAS each have items devoted to various facets of criminal lifestyle; the instruments have different domains and varying measurement schemes regarding lifestyle factors, as shown in Table 4.2. These definitions vary considerably. Researchers have not developed psychometrically sound subscales of the criminal lifestyle construct. Essentially, each tool screens for involvement in a criminal lifestyle without having predictive or construct validity which means that these may not affect outcomes because we may not be measuring the constructs accurately. Unlike substance abuse, where the DSM-V has accepted criteria for assessing a substance use disorder (including different levels of use), no one has developed a clinical definition of criminal lifestyle. Risk assessment tools do not have well-defined, valid measures of criminal lifestyle. This leaves us in a quandary in that we

---

[7]The discussion of criminogenic needs excludes a history of antisocial behavior and substance abuse. History of antisocial behavior is equivalent to criminal history risk. Both risk and substance use disorders are covered in detail above.

acknowledge that these lifestyle-related factors are criminogenic needs that are related to continued offending but tools and measurement strategies have not been sufficiently developed.

Antisocial Cognitions/Thinking

Antisocial cognitions are the means by which people can rationalize their deviant behavior or neutralize or reduce negative consequences resulting from offending. These thinking patterns have been identified as a Big Four criminogenic need (Andrews & Bonta, 2010; Andrews et al., 2006). In their seminal work in this area, Yochelson and Samenow (1976) identified 36 thinking errors of offenders. The typical thinking errors include dominance, entitlement, self-justification, displacing blame, optimistic perceptions of realities, and "victim stance" (e.g., blaming society because they are considered outcasts). Offenders tend to exhibit more of these thinking errors than members of the general population (see Walters, 2003a, 2003b). Lipsey and Landenberg (2006) note that "distorted thinking may misperceive benign situations as threats (e.g., predisposed to perceive harmless remarks as disrespectful or deliberately provocative), demand instant gratification, and confuse want with needs" (p. 57). The individual offender is interested in their own needs, and their behaviors are focused on fulfilling these needs instead of considering consequences or the needs of others. Criminal thinking or antisocial cognitions are included in many criminological theories including subcultural, anomie, differential association, control, labeling, and self-control theory. More importantly, antisocial cognitions contribute to continued involvement in criminal behavior.

Over the last 2 decades, a number of instruments have been developed to measure criminal thinking. These instruments include the PICTS (Psychological Inventory of Criminal Thinking Styles) (Walters, 2002), the Criminal Sentiments Scale (Shields & Simourd, 1991), the Measure of Offender Thinking Styles (Mandracchia, Morgan, Garos, & Garland, 2007), the TCU Criminal Thinking Scales (Knight, Garner, Simpson, Morey, & Flynn, 2006), the Criminogenic Thinking Profile (Mitchell & Tafrate, 2012), and the Criminal Cognitions Scale (Tangney et al., 2012). The instruments vary in length (from 80 items for the PICTS to 25 items for the CCS) and number of subscales (from 8 to 4). Some of the scales (MOT, CSS, TCU CTS) do not have predictive validity studies documenting that higher criminal thinking scores are directly related to recidivism while the PICTS (Walters, 2012) and CCS (Tangney et al., 2012) report a positive relationship between criminal thinking and criminal history. The PICTS has also been useful in predicting recidivism although it has a small mean effect size (r) of 0.20 (Walters, 2012). While considerable strides have been made in this area further research is needed to establish the reliability and validity of these instruments and clarify the empirical relationship between criminal thinking and recidivism risk.

Mandracchia and Morgan (2012) recently explored the impact of demographic characteristics of offenders on criminal cognitions. Using different scales (PICTS,

MOT, CSS) of criminal cognitions, they found young offenders are more likely to have criminal cognitions than older offenders. And, their study found that race, gender, and educational attainment are not related to criminal cognitions, which is a finding that is different from the work of other researchers. Some of the reliability and validity studies of the criminal cognition scales mentioned above also found that age was negatively correlated with criminal cognition scores. The significance of this work is that it is unclear whether criminal cognitions are a function of age or developmental issues, a factor that should be researched further. The correlation with age is an important finding because it suggests that demographic differences are important to consider in the research process.

Antisocial Personality Pattern

Andrews and Bonta (2010) also identify an antisocial personality pattern as one of the Big Four criminogenic needs. This criminogenic need factor is generally characterized by proneness for adventurous pleasure-seeking behavior, low self-control, and/or aggressiveness (Andrews et al., 2006). The DSM-V criteria for an antisocial personality disorder (ASPD) is characterized by a callous disregard for the feelings of others, gross or persistent attitude of irresponsibility; disregard for social norms, rules, or obligations; incapacity to maintain enduring relationships, low tolerance, frustration, and use of aggression or violence; incapacity to experience guilt or to profit from experience; or marked proneness to blame others for the behavior that the person exhibits. DSM-V criteria for ASPD offer one possible operationalization of this antisocial personality criminogenic need factor. As with any personality trait, it is important to recognize that all human beings exhibit some aspect of this but the criminal offender tends to have sustained characteristics. ASPD is the diagnostic criteria closet to criminal personality (as well as psychopathy). This conceptualization recognizes that such behaviors are common and that the issue is whether the individual consistently displays the traits.

Another possible indicator of an antisocial personality pattern that has been operationalized is the construct of low self-control. Impulsive and risk taking behavior is a dynamic characteristic often noted among offenders. The general premise is that low self-control does not define criminal behavior; instead it provides a context for criminal acts depending on opportunities and other motivating factors. A person's decision to become engaged in criminal acts is affected by other factors such as natural constraints, attachments to parents, school, and employment (Gottfredson & Hirschi, 1990, pp. 95–97). Low self-control is often due to the offender being easily persuaded by situational and environmental factors; unless the individual has a support system to constrain the individual, the person engaged in the behavior. Empirical support for low self-control as a predictor of offending, recidivism, and other antisocial behaviors has been well established in the field of criminology (see Pratt & Cullen, 2000 for a review). Based on a meta-analysis of 126 independent effect sizes from 21 different studies, Pratt and Cullen (2000) conclude that low self-control is a strong predictor of offending whether it is measured behaviorally or attitudinally. Meta-analysis reveals a mean effect size of low self-control on crime of $r = 0.26$ ($k = 82$) when attitudinal measures are used and a mean effect size of $r = 0.28$ ($k = 12$) when behavioral measures are used (Pratt & Cullen, 2000). These findings suggest a potential direct relationship between low self-control and recidivism outcomes.

While empirical studies indicate a relationship between different measures of an antisocial personality pattern and recidivism outcomes, a question that still needs to be addressed is whether this criminogenic need can be successfully addressed through correctional intervention. If these personality traits are stable by early adulthood, as has been argued by Gottfredson and Hirschi (1990), then interventions targeting this need for adults involved in the justice system will have little impact on recidivism. Andrews and Bonta's model holds that an antisocial personality pattern is in fact amenable to change and can be addressed through cognitive restructuring interventions (e.g., CBT).

Antisocial Associates and Negative Social Supports

Consistent with its grounding in social learning theory, the RNR model stresses antisocial associates (family and peers) as a criminogenic need. At the same time, the model identifies pro-social family and peers as potential protective factors against continued involvement in offending. Empirical evidence shows that strong family relations are important for reducing offending behaviors (Berg & Huebner, 2011; Cobbina, Huebner, & Berg, 2012; Laub & Sampson, 2003; Sampson & Laub, 1993). Family ties can provide emotional support and facilitate the offender change process. Cobbina et al. (2012) found that strong, pro-social family ties contributed to a decline in criminal behavior. But, histories of family involvement in crime and prior incarcerations are usually indicators or precursors to learned behaviors regarding criminal behavior and drug use. The issues regarding family are complex in that the household may allow and tolerate certain behaviors, including substance use or criminal behavior, and this may reinforce negative behaviors. This would then have a negative impact on the individual.

Relations with others, including antisocial peers and families, can affect three aspects of future offending involvement. As noted by Huebner and Berg (2011), three mechanisms through which family and peer associations (referred to as social ties) may foster desistance are as follows: (1) ties with pro-social individuals control offender behavior, (2) ties can provide emotional support, and (3) ties can help offenders facilitate a transformation in identity from offender to citizen or reinforce an offender identity. While adult pro-social bonds to family and spouse can serve as a catalyst for desistance (Laub & Sampson, 2003; Sampson & Laub, 1993), a strong empirical link has been established between associations with antisocial peers and offending (see, e.g., Warr, 1998).

For offenders, having close relationships with criminally involved peers and little interaction with those who are not justice-involved (referred to as pro-social peers) has a direct effect on drug use and criminal involvement (Haynie, 2003).

While peer affiliations can change over time in terms of the number of criminally involved peers, a steady relationship with people involved in criminal behavior reinforces further involvement in criminal behavior and has been empirically linked to recidivism (Giordano, Cernkovich, & Holland, 2003; Hawkins & Fraser, 1987; Warr, 1998; Wright & Cullen, 2004; Yahner & Visher, 2008). From a dynamic criminogenic need perspective, the offender's social network is important because it indicates access and opportunities to others involved in offending. Alternatively, offenders who have pro-social networks are less at risk for continued involvement in offending. Interventions designed to improve recidivism outcomes by addressing this criminogenic need generally focus on reducing associations with antisocial peers and fostering relationships with pro-social family and community institutions (Andrews et al., 2006).

Other Dynamic Offender Needs

In addition to the Big Four criminogenic needs, Andrews and Bonta's model identifies several additional dynamic needs that generally have weaker direct relationships with recidivism outcomes. While these needs, and others not included in Andrews and Bonta's conceptual model (e.g., housing and mental illness), are less directly linked to recidivism, they are still important for offender functioning and should be addressed through correctional programming when appropriate. The non-criminogenic needs reviewed here include employment and education, housing, and mental health issues.[8] Distinguishing between criminogenic and non-criminogenic needs is important in that it helps justice agencies identify targets for programs and can help facilitate treatment prioritization efforts.

Employment and Educational Attainment

Offenders tend to have lower educational attainment than the general population. This includes a higher percentage of offenders that do not graduate from high school or receive a general educational development degree (GED) (Greenberg, Dunleavy, Kutner, & White, 2007; Harlow, 2003). However, not completing high school is not criminogenic in that it does not "cause" a person to commit crimes. Not graduating high school however may affect verbal intelligence or literacy levels, which may affect the ability to obtain employment. Huebner and Berg (2011) found that men who did not graduate high school were more likely to recidivate than those who graduated high school. Much of the impact of education on recidivism outcomes may operate through its impact on employment which has generally received more empirical support as a correlate of recidivism outcomes.

---

[8] As discussed above, we suggest that substance dependence on hard drugs is a criminogenic need but substance abuse is a non-criminogenic need. Housing and mental health status are not included as dynamic needs in Andrews and Bonta's model.

Offenders tend to have lower steady employment than non-justice-involved individuals (Petersilia, 2001, 2003, 2005; Western, 2006). Again, employment-related issues are non-criminogenic needs given that there is no causal linkage between employment and recidivism. But, employment has an indirect relationship with recidivism and therefore may be targeted through correctional programming when other more direct needs are not present. Employment after release from prison is shown to foster better outcomes, but the overall pattern is unclear regarding the relationship between employment and recidivism outcomes (see Berg & Huebner, 2011; Huebner & Berg, 2011; Makarios, Steiner, & Travis, 2010; Redcross, Millenky, Rudd, & Levshin, 2012; Tripodi, Kim & Bender, 2010; Zweig, Yahner, & Redcross, 2010). Further research is needed to (1) clarify the mechanisms through which employment and education affect recidivism outcomes and (2) clarify the recidivism reduction potential of educational and employment-related treatment programs.

Mental Health Status

Mental health functioning is a complex need given the prevalence and diversity of mental health conditions in the justice system and their varying impact on offender functioning. Recent research by Steadman, Osher, Robbins, Case, and Samuels (2009) finds that about 15% of male and 31% of female jail inmates have experienced a recent serious mental illness (Steadman et al., 2009). The prevalence of serious mental illness is also elevated among state and federal prisoners, probationers, and parolees (Feucht & Gfroerer, 2011; James & Glaze, 2006). An additional concern is the high-rate of co-occurring substance use and mental health disorders in the justice system. It is well observed that about half the offenders with substance use disorders also have mental health issues (Abram & Teplin, 1991; Abram, Teplin, & McClelland, 2003; National GAINS Center, 2004).

Despite the high prevalence of mental health and co-occurring disorders (CODs) in the justice system, few empirical studies have found that the presence of a mental health condition is a direct predictor of criminal conduct (see Lovell, Gagliardi, & Peterson, 2002 for a discussion). Some research does suggest that offenders with CODs are at an increased risk for recidivism relative to offenders with mental illness alone (Baillargeon et al., 2010; Hartwell, 2004). While it does not appear that mental health status is directly related to recidivism (see also Skeem & Louden, 2006; Yahner & Visher, 2008), mental health conditions may negatively impact the performance of offenders in programming and can increase the risk for technical violations related to failure to complete mandated treatment and rearrests related to increased monitoring (Skeem & Louden, 2006). While there is no direct link between mental health issues and recidivism, failure to provide mental health services to those with mental illness can negatively impact the outcomes of offenders and lead to increased recidivism rates (Baillargeon et al., 2010; Hoge, 2007; Osher, Steadman, & Barr, 2002, 2003; Peters & Hills, 1997).

Housing

Being homeless has not been shown to be predictive of future criminal behavior (Broner, Lang, & Behler, 2009). Although housing status does not predict recidivism outcomes, instability in housing indirectly affects outcomes because it impacts the overall stability of the offender in the community. Broner et al. (2009) found that housing instability did not predict recidivism outcomes, but it did predict successful completion of mental health court. DeLisi (2000) found that homeless jail inmates were not significantly more likely to reoffend than jail inmates with a place to live. Homelessness, however, is fundamentally different than housing instability. While the impact of housing stability varies across studies, some emerging research has supported housing instability as a correlate of recidivism. For example, Makarios et al. (2010) found that number of post-release residence changes was significantly related to recidivism. Other research has shown that stability of housing can serve as a protective factor against recidivism (see, e.g., Yahner & Visher, 2008). While the relationship between housing stability and recidivism remains empirically unclear, addressing housing issues can provide stability in an offender's life and potentially have an indirect impact on recidivism outcomes.

Identifying Targets for Programming

In the RNR model, Andrews and Bonta (2010) consider criminal subculture and personality traits as independent factors. In various meta-analyses Andrews and Bonta report that each Big Four factor has a correlation coefficient with a wide range from 0.16 to 0.48 with recidivism. The other four factors (i.e., family/marital problems, substance abuse, education/employment, leisure recreation) tend to have weaker correlations with recidivism that range from 0.06 to 0.43 (Andrews & Bonta, 2010). For this reason, Andrews and Bonta separate the Big Four from the other needs that are a lesser priority. While they advance the argument that most of the potential influence of non-criminogenic needs on recidivism operates indirectly through the Big Four (Andrews et al., 2006), the application of the model can be advanced by identifying the main dynamic factors (that are directly related to recidivism) in a treatment matching scheme. Many individuals have more than one of the Big Four or Central Eight needs. By examining the number and type of criminogenic needs an offender presents, it is possible to identify offenders who are more ingrained in a criminal lifestyle and therefore should be targeted for more intensive services designed to affect criminal thinking and potentially reduce recidivism. Additionally, assessing needs provides guidance regarding which needs are present and which needs should be prioritized in interventions.

Given the framework for the RNR model, the goal is to identify dynamic factors—amendable to change—that contribute either directly or indirectly to criminal behavior. Identifying dynamic needs that affect both the opportunity to participate in and the desire to commit criminal acts is important. Similarly, it is important to give attention to factors that either can serve to protect the individual from criminal involvement (stabilizers) or those that may contribute to the opportunity or desire to commit crimes (destabilizers). Included in this framework are factors that tend to have an indirect relationship with recidivism including mental health status, educational attainment, employment history and options, housing, location of residence, and other factors frequently discussed in the realm of dynamic offender needs. Clinically relevant factors are built on the notion that these non-criminogenic factors impact the decisions and choices made by offenders and therefore need to be included in decisions regarding treatment matching and amenability to change. To some degree, these factors may also indicate severity of problem behavior warranting attention.

Summary of Key Need-Related Assumptions

Given that criminogenic needs and other dynamic risk factors are important to the concept of responsivity, a review of the existing literature about factors that affect recidivism can be used to develop a number of key assumptions about the priority of these factors in treatment matching strategies. While Andrews and Bonta outlined their Central Eight, a reexamination of the literature leads us to slightly different prioritization of offender needs based on consideration of direct impact on recidivism and the clinical relevance of different non-criminogenic factors. The following assumptions are grounded in the linkage between recidivism and dynamic offender needs. The following conceptualization of dynamic need factors converts an individual need approach to a spectrum of needs. This is done to augment the Andrews and Bonta RNR model and foster more efficient resource allocation:

1. Primary criminogenic needs fall into three categories: substance dependence (on a criminogenic drug), criminal lifestyle (made up of criminal thinking and other Big Four criminogenic needs), and specific offender type (e.g., sex offender, domestic violence offender, or drunk driver). This priority is based on the magnitude of the relationship to recidivism and knowledge of effective interventions and the availability of programming to target these specific criminogenic factors.
2. Offenders who have three or more criminogenic needs should be prioritized for treatment. Research has shown that the number of needs affects recidivism (see Andrews & Bonta, 2010), and therefore offenders who present with higher scores on criminal attitudes and values, antisocial peers, family criminal networks, and history of antisocial behaviors need to be prioritized for more intensive and structured programming.
3. Core demographics are important in guiding assignments to programming. Gender, age, and location of residence warrant consideration in the treatment matching process.
4. Substance abuse and mental illness, although not directly related to recidivism, should be prioritized as intervention targets over other non-criminogenic needs

because they are clinically relevant and treatment programs are available that target these specific offender needs.

5. Given their generally weaker correlations with recidivism, the remaining dynamic need factors will be considered as either stabilizers (protective factors) or destabilizers (negative indicators). This distinction provides for a greater opportunity to consider factors that should be used to adjust programming given the number and type of stabilizers and destabilizers in a person's life.

Responsivity: An Overview

Responsivity is the third component of the RNR model (Box 4.2). Andrews and Bonta (2010) originally conceptualized responsivity in terms of the personal characteristics of the offender "that should be assessed, since these factors can affect their engagement in treatment. They should include areas like mental and emotional problems cognitive functioning and level of motivation and readiness to change" (Latessa and Lovens, 2010, p. 210). The term has evolved into more of a general principle for matching the appropriate type of programming based on the risk and need factors of offenders—recognizing that these risk-need factors should drive the level and type of programming (see Fig. 4.2). This conceptualization mirrors some other efforts to develop treatment placement criterion (e.g., American Society of Addiction Medicine) but recognizes that the setting and the type of intervention are important in affecting outcomes. Embedded in this conceptualization of responsivity are a number of factors that affect the quality and potential impact of the programming, namely, the dosage, the content, the fidelity or adherence to the program model, and the quality of program resources. It is beyond this book to review all of these components (see Taxman & Belenko, 2012 for a full discussion on implementation). But, consistent with the evidence-based practice literature, we are aware that:

- Dosage or the number of hours that a person is involved in the correctional program should vary by risk level and that higher-risk offenders should be involved in larger quantities of programming (see Bourgon & Armstrong, 2005). As discussed in Chap. 6, dosage refers to the frequency and length of the program.
- Content or the theoretical orientation of the intervention should include cognitive behavioral therapy, therapeutic communities, or integrated service models like drug treatment court. The nature of the treatment programming has an impact on results.
- Fidelity, adherence to core correctional practices, or the degree to which the program is implemented as planned is directly related to program effectiveness. This construct includes the dosage, content, type of staff, curriculum, and other key functional components of the program and is essential for bringing about offender change.
- Program resources include facilities, resources, and staffing available to the program. These should be aligned with the intended goals of the intervention. Program resources are important to ensure the clinical or responsivity components are adequately resourced to be effective.

Box 4.2 Responsivity Defined

Responsivity refers to the appropriate treatment and correctional programming given the risk and need level of the offender. The goal is to place offenders in appropriate criminal justice settings and correctional control designed to reduce the risk of recidivism.

Table 4.4 Evidence-based interventions to reduce recidivism

Intervention	% Reduction	# of Studies
General interventions		
CBT for general population prisoners	8.2	25
CBT for low-risk sex offenders on probation	31.2	6
CBT for sex offenders in prison	14.9	5
ISP with treatment	21.9	10
Basic adult education in prison	5.0	7
Vocational education for general population in prison	12.6	3
Employment training and job assistance in the community	4.8	16
Correctional industry for general population prisoners	7.8	4
Drug treatment		
CBT drug treatment for general population prisoners	6.8	8
Adult drug courts	10.7	56
Drug treatment for offenders in the community	12.4	5
TC drug treatment in prison with aftercare	6.9	6
TC drug treatment in prison without aftercare	5.3	7
Drug treatment in jail	6.0	9

Source: Aos et al. (2006)

Correctional programming is complicated. It can occur in different settings—prison, jail, probation/parole offices, community treatment provider, halfway house, and so on. Very little is known about these settings and the potential impact of program setting on offender-level recidivism, other than the therapeutic community literature that reports the importance of providing treatment in areas separate from the general population (see Simpson, Wexler, & Inciardi, 1999). A number of unanswered questions exist about offering different types of programs in different types of settings. It should also be noted that oftentimes similar programs are offered by prison or community correctional agencies, and except for a few studies that report better findings in community-based programming (Aos, Miller, & Drake, 2006; Lipsey & Landenberg, 2005), we do not know setting may affect the program operations, fidelity, or adherence to the theoretical model. As discussed in Chap. 6 and new tools used by agencies to assess the quality of the interventions (Lipsey, 2009), it is important to understand the quality of the programming in order to realize appreciable impacts on treatment outcomes.

The research literature has identified a number of evidence-based interventions, including cognitive behavioral therapy, drug treatment courts, and drug treatment programs that use a therapeutic community (with aftercare) model, as shown in Table 4.4. This list has not changed drastically in the last few years, but more research is available about the type of offenders and risk-need profiles that might do better in a given program. For example, Landenberg and Lipsey (2005) detail some of these issues and generally find that the following implementation variables affect more positive recidivism outcomes: having a multifaceted treatment intervention that includes cognitive restructuring and/or anger therapy, researcher involvement in program, planning, longer and more intensive duration of services, and serving higher-risk offenders.

The review of the literature in this chapter has helped to crystalize some new parameters including the criteria to support responsivity decisions. That is, a review of the literature has helped us to better understand the factors that affect outcomes and therefore should be used to inform the responsivity decisions. While the original RNR framework stands, the review of the literature has assisted in reconfiguring some of the decision criteria to include:

1. Static risk level should drive dosage and type of programming. Since static risk accounts for the majority of the variance in explaining recidivism, it appears that placing moderate- to high-risk offenders, with various criminogenic needs, into the more intensive programming should be the priority.
2. The concept of problem severity should be integrated into the RNR framework. While the original RNR model is separated by priority (Big Four) and lesser priority (non-criminogenic) needs, it appears that this distinction does not separate out those that have more severe behavioral issues from those with less severe issues. The responsivity model should consider all dynamic offender needs, including the host of clinically relevant, non-criminogenic factors in identifying problem severity.
3. Substance use dependency and criminal lifestyle (a composite measure of problem severity) should be prioritized for more structured, intensive programming. These severe behavioral issues generally require more intensive services, particularly for moderate- to high-risk offenders.
4. Age and gender are important responsivity factors that should adjust the content of the program/services.
5. Programming content should be categorized into four levels: cognitive and behavioral, interpersonal and social skills, lifestyle skills, and punishment (no programming). Cognitive and behavioral programming should include interpersonal and social skills and lifestyle skills (see Chap. 6 for an explanation) (Table 4.5).

Conclusion

The original RNR framework has had a positive impact on the growth of evidence-based programming and treatments in the field of corrections. This model identified the importance of individual-level factors in making determinations about the appropriate level of programming that is needed. However, a review of the literature

Table 4.5 RNR model responsivity matrix

Risk	Need	Stabilizers	Destabilizers	Correctional programming
High Moderate	Substance dependence	N/A	N/A	Therapeutic community with aftercare (CBT) Drug court (CBT)
Low				
High Moderate	Criminal thinking ("Big Four")	N/A	N/A	Therapeutic community with aftercare (CBT) Problem-solving court (CBT) RNR supervision
High Moderate	1 Criminogenic need	Social ties/ network Employment Education	N/A	RNR supervision CBT therapy
High Moderate	1 Criminogenic need	N/A	Mental health Employment Educational Housing	RNR supervision with supportive living CBT therapy
Moderate	1 Criminogenic need	Social ties Employment Education	N/A	RNR supervision
Moderate	1 Criminogenic need	N/A	Mental health Employment Educational Housing	RNR supervision with CBT
Low	1 Criminogenic need	N/A	Mental health Employment Educational Housing	RNR supervision
Low	1 Criminogenic need	Social ties Education Employment	N/A	Probation

on recidivism found that the original positioning of various criminogenic needs in the RNR framework may need to be altered. This is especially true when examining how these individual factors affect the results from participation in a different type of programming or the type of programming that is likely to reduce recidivism for specific profiles of offenders. The quagmire of recidivism reduction is that (1) there is great variability across studies regarding the size and (sometimes) the direction of the effect between an individual-level factor and recidivism; (2) often-times an individual-level factor is indirectly related to recidivism, usually when other individual-level variables are present; and (3) program-level effects may vary from merely examining the impact of various individual factors on offender out-comes. That being said, these inconsistencies in the literature do not challenge the RNR framework but merely suggest a slightly different ordering of importance of different factors and the inclusion of clinically relevant factors such as mental health, housing stability, and substance abuse (not dependence). And, it means that there should be certain drivers of who needs more intensive programming—namely, individuals with moderate to higher risk or individuals with more severe constella-tions of dynamic needs. But overall the RNR framework, as configured in the RNR Simulation Tools, offers great promise to improve the delivery of outcomes by hav-ing sound criteria to determine whether a particular risk-need profile is appropriate for a specific level of programming.

We would be remiss if we did not discuss a few caveats. First, the size of the potential impact of expanding correctional treatment programs is debated in the lit-erature. Some researchers contend that the impact will be rather insignificant given that the "evidence" points to "small effect sizes" (i.e., under 20%) for cor-rectional interventions (see Austin, 2009 for a discussion). These effect sizes are based on experimental studies testing new programs and services that are typically in researcher-controlled settings. And, they do not consider the uptake of the inter-vention into the organizational culture that may further dilute the effect and reduce the size of the recidivism reduction (see Taxman & Belenko, 2012 for a discussion). This "small effect" finding may be a function of (1) the failure of the current cor-rectional systems to use evidence-based criteria to assign offenders to appropriate treatment, supervision, services, or controls; (2) the lack of specific criteria to assess whether the current programs and services are delivered with fidelity or adherence to the main ingredients to bring about offender change; and (3) whether the risk-need configuration of the offender is adequately dealt with by the program or services.

The RNR model outlines a system that can integrate these core components into routine strategies for managing offenders in the correctional system. This is necessitated by the insufficient quantity of programming currently available in the system. As revealed by national surveys of prisons, jails, and community correc-tions agencies, less than 10% of offenders on any given day are involved in pro-grams (see Taxman, Perdoni, et al., 2007), and for the most part, existing programs do not embrace evidence-based practices (Friedmann, Taxman, & Henderson, 2007). Small effect sizes can therefore be a product of a system that is not consis-tently using the RNR principles and that is not routinely implementing evidence-based correctional programming (Lipsey & Cullen, 2007). To have a larger impact on recidivism reduction, we will need to expand the availability of effective ser-vices. In fact, Caudy and colleagues (see Chap. 7) illustrate that by expanding the availability of RNR-based correctional interventions, we can increase the potential for recidivism reduction at a system level. To illustrate the potential impact of moving to a human service-focused justice system, they point to different impacts from various scenarios: using RNR methods to match offenders to treatment will prevent 1 recidivism event for each four people treated; using treatment programming (with voluntary placement by the individual), 1 recidivism event will be prevented by treating eight people; and finally using a punishment only approach, one recidivism event can be prevented by punishing 33 people. Therefore it is not only the availability of treatment but also the type of treatment programming that matters. The RNR framework challenges the system to employ these decision criteria to improve the overall outcomes form the system but also to have a more evidence-based justification and rationalization for placing different risk-need profiles into different programming.

References

Abram, K. M., & Teplin, L. A. (1991). Co-occurring disorders among mentally ill jail detainees: Implications for public policy. *American Psychologist, 46*, 1036–1045.

Abram, K. M., Teplin, L. A., & McClelland, G. M. (2003). Comorbidity of severe psychiatric disorders and substance use disorders among women in jail. *The American Journal of Psychiatry, 160*, 1007–1010.

American Psychiatric Association. (2000). *Diagnostic and statistical manual of mental disorders* (Revised 4th ed.). Washington, DC.

Andrews, D. A., & Bonta, J. (2010). *The psychology of criminal conduct* (5th ed.). Cincinnati, OH: Anderson.

Andrews, D. A., Bonta, J., & Wormith, J. S. (2006). The recent past and near future of risk and/or need assessment. *Crime and Delinquency, 52*(1), 7–27.

Aos, S., Miller, M., & Drake, E. (2006). *Evidence-based public policy options to reduce future prison construction, criminal justice costs and crime rates*. Olympia, WA: Washington State Institute for Public Policy.

Austin, J. (2006). How much risk can we take? The misuse of risk assessment in corrections. *Federal Probation, 70*(2), 58–63.

Austin, J. (2009). The limits of prison based treatment. *Victims & Offenders, 4*, 311–320.

Baillargeon, J., Penn, J. V., Knight, K., Harzke, A. J., Baillargeon, G., & Becker, E. A. (2010). Risk of reincarceration among prisoners with co-occurring severe mental illness and substance use disorders. *Administration and Policy in Mental Health, 37*(4), 367–374.

Baird, C. (2009). *A question of evidence: A critique of risk assessment models used in the justice system*. Madison, WI: National Council on Crime and Delinquency (Special Report).

Ball, J. C., Shaffer, J. W., & Nurco, D. N. (1983). The day-to-day criminality of Heroin addicts in Baltimore: A study in the continuity of offense rates. *Drug and Alcohol Dependence, 12*, 119–142.

Belenko, S., & Peugh, J. (2005). Estimating drug treatment needs among state prison inmates. *Drug and Alcohol Dependence, 77*, 269–281.

Bennett, T. H., Holloway, K., & Farrington, D. P. (2008). The statistical association between drug misuse and crime: A meta-analysis. *Aggression and Violent Behavior, 13*(2), 107–118.

Berg, M., & Huebner, B. M. (2011). Reentry and the ties that bind: An examination of social ties, employment, and recidivism. *Justice Quarterly, 28*(1), 382–410.

Bourgon, G., & Armstrong, B. (2005). Transferring the principles of effective treatment into a "real world" prison setting. *Criminal Justice and Behavior, 32*(1), 3–25.

Broner, N., Lang, M., & Behler, S. A. (2009). The effect of homelessness, housing type, functioning, and community reintegration supports on mental health court completion and recidivism. *Journal of Dual Diagnosis, 5*(3–4), 323–356.

Byrne, J. (2009). *Maximum impact: Targeting supervision on higher risk people, places, and times*. Washington, DC: Public Safety Performance Project, the PEW Center on the States.

Byrne, J., & Pattavina, A. (2006). Assessing the role of clinical and actuarial risk assessment in an evidence-based community corrections system: Issues to consider. *Federal Probation, 70*(2), 64–67.

Chandler, R. K., Fletcher, B. W., & Volkow, N. D. (2009). Treating drug abuse and addiction in the criminal justice system: Improving public health and safety. *Journal of the American Medical Association, 301*(2), 183–190.

Clear, T. (1988). *Prediction methods in corrections*. Washington, DC: Research in Corrections. National Institute of Corrections.

Cobbina, J. E., Huebner, B. M., & Berg, M. T. (2012). Men, women, and postrelease offending an examination of the nature of the link between relational ties and recidivism. *Crime & Delinquency, 58*(3), 331–361.

Cohen, M. A., Piquero, A. R., & Jennings, W. G. (2010). Monetary costs of gender and ethnicity disaggregated group-based offending. *American Journal of Criminal Justice, 35*, 159–172.

Committee on Community Supervision and Desistance from Crime, National Research Council. (2007). *Parole, desistance from crime, and community integration*. Washington, DC: The National Academies Press.

DeLisi, M. (2000). Who is more dangerous? Comparing the criminality of adult homeless and domiciled jail inmates: A research note. *International Journal of Offender Therapy and Comparative Criminology, 44*(1), 59–69.

Farrington, D. P. (1986). Age and crime. In M. Tonry & N. Morris (Eds.), *Crime and justice: An annual review of research* (Vol. 7). Chicago, IL: University of Chicago Press.

Feucht, T. E., & Gfroerer, J. (2011). *Mental and substance use disorders among adult men on probation or parole: Some success against a persistent challenge*. Rockville, MD: SAMHSA, Center for Behavioral Health Statistics and Quality.

Friedmann, P. D., Taxman, F. S., & Henderson, C. E. (2007). Evidence-based treatment practices for drug-involved adults in the criminal justice system. *Journal of Substance Abuse Treatment, 32*(3), 267–277.

Giordano, P. C., Cernkovich, S. A., & Holland, D. D. (2003). Changes in friendship relations over the life course: Implications for desistance from crime. *Criminology, 41*(2), 293–328.

Goldstein, P. J. (1985). The drugs/violence nexus: A tripartite conceptual framework. *Journal of Drug Issues, 39*, 143–174.

Gottfredson, D. (1987). Prediction and classification in criminal justice decision making. *Crime and Justice: A review of research, 9*(1).

Gottfredson, S. D., & Gottfredson, D. M. (1986). Accuracy of prediction models. In A. Blumstein (Ed.), *Criminal careers and careers criminals* (Vol. II, pp. 212–290). Washington, DC: National Academy Press.

Gottfredson, M., & Hirschi, T. (1990). *A general theory of crime* (1st ed.). Stanford, CA: Stanford University Press.

Gottfredson, S. D., & Taylor, R. B. (1985). Person-environment interactions in the prediction of recidivism. In J. M. Byrne & R. Sampson (Eds.), *The social psychology of crime*. New York, NY: Springer.

Greenberg, D. F. (1985). Age, crime, and social explanation. *The American Journal of Sociology, 91*(1), 1–21.

Greenberg, E., Dunleavy, E., Kutner, M., & White, S. (2007). *Literacy behind bars: Results from the 2003 National Assessment of Adult Literacy Prison Survey*. Washington, DC: National Center for Education Statistics.

Harlow, C. W. (2003). *Education and correctional populations. Bureau of Justice Statistics special report*. Washington, DC: Bureau of Justice Statistics, U.S. Department of Justice.

Hartwell, S. W. (2004). Comparison of offenders with mental illness only and offenders with dual diagnoses. *Psychiatric Services, 55*(2), 145–150.

Hawkins, J. D., & Fraser, M. W. (1987). The social networks of drug abusers before and after treatment. *International Journal of the Addictions, 22*(4), 343–355.

Hawkins, D. F., Laub, J. H., & Lauritsen, J. L. (1998). Race, ethnicity, and serious juvenile offending. In R. Loeber & D. P. Farrington (Eds.), *Serious & violent juvenile offenders: Risk factors and successful interventions* (pp. 30–46). Thousand Oaks, CA: Sage.

Haynie, D. L. (2003). Contexts of risk? Explaining the link between girls pubertal development and their delinquency involvement. *Social Forces, 82*, 355–397.

Hindelang, M. J. (1978). Race and involvement in common law personal crimes. *American Sociological Review, 43*, 93–109.

Hipp, J., Petersilia, J., & Turner, S. (2010). Parolee recidivism in California: The effect of neighborhood context and social service agency characteristics. *Criminology, 48*(4), 947–979.

Hirschi, T., & Gottfredson, M. R. (1983). Age and the explanation of crime. *The American Journal of Sociology, 89*(3), 552–584.

Hoffman, 1. P. B., & Beck, J. L. (1974). Parole decision-making: A salient factor score. *Journal of Criminal Justice, 2*(3), 195–206.

Hoge, S. K. (2007). Providing transitional and outpatient services to the mentally ill released from correctional institutions. In R. B. Greifinger (Ed.), *Public health behind bars: From prisons to communities* (pp. 461–477). New York, NY: Springer.

Hubbard, R. L., Marsden, M. E., Rachal, J. V., Harwood, H. J., Cavanaugh, E. R., & Ginzburg, H. M. (1988). *Drug abuse treatment: A national study of effectiveness*. Chapel Hill, NC: The University of North Carolina Press.

Huebner, B. M., & Berg, M. (2011). Examining the sources of variation in risk for recidivism. *Justice Quarterly, 28*(1), 146–173.

James, D. J., & Glaze, L. E. (2006). *Mental health problems of prison and jail inmates*. Washington, DC: U.S. Department of Justice (BJS Special Report Document 2).

Jennings, W. G., Maldonado-Molina, M. M., Piquero, A. R., Odgers, C. L., Bird, H., & Canino, G. (2010). Sex differences in trajectories of offending among Puerto Rican youth. *Crime & Delinquency, 56*(3), 327–357.

Knight, K., Garner, B. R., Simpson, D. D., Morey, J. T., & Flynn, P. M. (2006). An assessment for criminal thinking. *Crime & Delinquency, 52*(1), 159–177.

Kubrin, C. E., & Stewart, E. A. (2006). Predicting who reoffends: The neglected role of neighborhood context in recidivism studies. *Criminology, 44*, 165–197.

LaFree, G. (1995). Race and crime trends in the United States, 1946–1990. In D. F. Hawkins (Ed.), *Ethnicity, race, and crime: Perspectives across time and place* (pp. 169–193). Albany, NY: State University of New York Press.

Landenberg, N., & Lipsey, M. (2005). The positive effects of cognitive behavioral programs for offenders: A meta-analysis of factors associated with effective treatment. *Journal of Experimental Criminology, 1*, 451–476.

Langan, P., & Levin, D. (2002). *Recidivism of prisoners released in 1994*. Washington, DC: Bureau of Justice Statistics (BJS Special Report No. NCJ 193427).

Latessa, E. J., & Lovins, B. (2010). The Role of offender risk assessment: A policy maker guide. *Victims & Offenders, 5*(3), 203.

Latessa. E., Smith, P, Lemke, R., Makarios, M., & Lowenkamp, C. (2009). Creation and validation of the Ohio risk assessment system final report. University of Cincinnati. Available at www.uc.edu/criminaljustice.

Laub, J. H., & Sampson, R. J. (2003). *Shared beginnings, divergent lives: Delinquent boys to age 70*. Cambridge, MA: Harvard University Press.

Lipsey, M. W. (2009). The primary factors that characterize effective interventions with juvenile offenders: A meta-analytic overview. *Victims & Offenders, 4*(2), 124.

Lipsey, M. W., & Cullen, F. T. (2007). The effectiveness of correctional rehabilitation: A review of systematic reviews. *Annual Review of Law and Social Science, 3*, 297–320.

Lipsey, M. W., & Landenberg, N. A. (2005). Cognitive-behavioral interventions: A meta-analysis of randomized controlled studies. In B. C. Welsh & D. P. Farrington (Eds.), *Preventing crime: What works for children, offenders, victims, and places*. Berlin: Springer.

Lipsey, M. W., & Landenberg, N. A. (2006). Cognitive-behavioral interventions. In B. C. Welsh & D. P. Farrington (Eds.), *Preventing crime: What works for children, offender, victims, and places*. Great Britain: Springer.

Lovell, D., Gagliardi, G., & Peterson, P. (2002). What happens to mentally ill offenders released from prison? Findings from Washington's community transitions study. *Psychiatric Services, 53*(10), 1290–1296.

Lurigio, A., Cho, Y., Swartz, J., Graf, I., & Pickup, L. (2003). Standardized assessment of substance-related, other psychiatric, and comorbid disorders among probationers. *International Journal of Offender Therapy and Comparative Criminology, 47*, 630–652.

Makarios, M., Steiner, B., & Travis, L. F. (2010). Examining the predictors of recidivism among men and women released from prison in Ohio. *Criminal Justice and Behavior, 37*(12), 1377–1391.

Mandracchia, J. T., & Morgan, R. D. (2012). Predicting offenders' criminogenic cognitions with status variables. *Criminal Justice and Behavior, 39*(1), 5–25.

Mandracchia, J. T., Morgan, R. D., Garos, S., & Garland, J. T. (2007). Inmate thinking patterns: an empirical investigation. *Criminal Justice and Behavior, 34*(8), 1029–1043.

Mitchell, D., & Tafrate, R. C. (2012). Conceptualization and measurement of criminal thinking: Initial validation of the criminogenic thinking profile. *International Journal of Offender Therapy and Comparative Criminology, 56*(7), 1080–1102.

Moffitt, T. E. (1993). Adolescence-limited and life-course-persistent antisocial behavior: A developmental taxonomy. *Psychological Review, 100*(4), 674–701.

Morenoff, J. D. (2005). Racial and ethnic disparities in crime and delinquency in the United States. In M. Rutter & M. Tienda (Eds.), *Ethnicity and causal mechanisms* (pp. 139–173). New York, NY: Cambridge University Press.

Mumola, C. J. (1999, January). *Substance abuse and treatment, state and federal prisoners, 1997.* Washington, DC: U.S. Department of Justice, Office of Justice Programs. (Bureau of Justice Statistics Special Report).

Mumola, C. J., & Bonczar, T. P. (1998). *Substance abuse and treatment of adults on probation, 1995.* Washington, DC: Bureau of Justice Statistics.

Nagin, D., Cullen, F., & Jonson, C. (2009). Imprisonment and reoffending. *Crime and Justice, 38*(115), 1–91.

National GAINS Center for People with Co-Occurring Disorders in the Justice System. (2004). *The prevalence of co-occurring mental illness and substance use disorders in jails* (Fact Sheet Series). Delmar, NY: Author.

Nurco, D. N., Hanlon, T. E., Kinlock, T. W., & Slaght, E. (1984). *Variations in criminal patterns among narcotic addicts in Baltimore and New York City, 1983–1984.* Baltimore, MD: Friends Medical Science Research Center.

O'Leary, V., & Clear, T. R. (1984). *Directions for Community Corrections in the 1990s.* Washington, DC: U.S. Department of Justice, National Institute of Corrections.

Office of National Drug Control Policy (ONDCP). (2011). *National Drug Control Strategy.* Washington, DC: Executive Office of the President.

Osher, F., Steadman, H. J., & Barr, H. (2002). *A best practice approach to community re-entry from jails for inmates with co-occurring disorders: The APIC model.* Delmar, NY: The National GAINS Center.

Osher, F., Steadman, H. J., & Barr, H. (2003). A best practice approach to community re-entry from jails for inmates with co-occurring disorders: The APIC model. *Crime and Delinquency, 49*, 79–96.

Peters, R. H., & Hills, H. A. (1997). *Intervention strategies for offenders with co-occurring disorders: What works.* Delmar, NY: National GAINS Center for People with Co-Occurring Disorders in the Criminal Justice System.

Petersilia, J. (2001). Prisoner reentry: Public safety and reintegration challenges. *Prison Journal, 81*, 360–375.

Petersilia, J. (2003). *When prisoners come home: Parole and prisoner reentry.* New York, NY: Oxford University Press.

Petersilia, J. (2005). Hard time: Ex-offenders returning home after prison. *Corrections Today, 67*, 66–72.

Pew Center on the States. (2011). *State of recidivism: The revolving doors of America's prisons.* Washington, DC: Pew Charitable Trusts.

Piquero, A. R., & Brame, R. (2008). Assessing the race-crime and ethnicity-crime relationship in a sample of serious adolescent delinquents. *Crime & Delinquency, 54*(3), 390–422.

Pratt, T. C., & Cullen, F. T. (2000). The empirical status of Gottfredson and Hirschi's general theory of crime: A meta-analysis. *Criminology, 38*, 931–964.

Quetelet, A. (1984). *Research on the propensity for crime at different ages.* (S. F. Sylvester, Trans.). Cincinnati, OH: Anderson. (Original work published 1831).

Redcross, C., Millenky, M, Rudd, T, & Levshin, V (2012). More than a job final results from the evaluation of the Center for Employment Opportunities (CEO) Transitional Jobs Program. Retrieved September 30, 2012 from http://www.mdrc.org/publications/616/overview.html

Sampson, R. J., & Laub, J. H. (1993). *Crime in the making: Pathways and turning points through life.* Cambridge, MA: Harvard University Press.

Sampson, R. J., & Lauritsen, J. L. (1997). Racial and ethnic disparities in crime and criminal justice in the United States. In M. Tonry (Ed.), *Ethnicity, crime, and immigration: Comparative and cross-national perspectives, crime and justice. An annual review of research* (21, pp. 311–374). Chicago, IL: University of Chicago Press.

Shields, I. W., & Simourd, D. J. (1991). Predicting predatory behavior in a population of young offenders. *Criminal Justice and Behavior, 18*, 180–194.

Simpson, D. D., Wexler, H. K., & Inciardi, J. A. (1999). Introduction. *Prison Journal, 79*, 381–383.

Skeem, J. L., & Louden, J. E. (2006). Toward evidence-based practice for probationers and parolees mandated to mental health treatment. *Psychiatric Services, 57*(3), 333–342.

Staton-Tindall, M., Havens, J. R., Oser, C. B., & Burnett, M. C. (2011). Substance use prevalence in criminal justice settings. In C. Leukefeld, T. P. Gullota, & J. Gregrich (Eds.), *Handbook of evidence-based substance abuse treatment in criminal justice settings* (pp. 81–101). New York, NY: Springer Science + Business Media.

Steadman, H. J., Osher, F. C., Robbins, P. C., Case, B., & Samuels, S. (2009). Prevalence of serious mental illness among jail inmates. *Psychiatric Services, 60*(6), 761–765.

Tangney, J. P., Stuewig, J., Furukawa, E., Kopelovich, S., Meyer, P. J., & Cosby, B. (2012). Reliability, validity, and predictive utility of the 25-item Criminogenic Cognitions Scale (CCS). *Criminal Justice and Behavior, 39*(10), 1340–1360.

Taxman, F. S., & Belenko, S. (2012). *Implementing evidence-based practices in community corrections and addiction treatment.* New York, NY: Springer.

Taxman, F. S., Cropsey, K. L., Young, D., & Wexler, H. (2007). Screening, assessment, and referral practices in adult correctional settings: a national perspective. *Criminal Justice and Behavior, 34*(9), 1216–1234.

Taxman, F. S., Perdoni, M., & Harrison, L. D. (2007). Drug treatment services for adult offenders: The state of the state. *Journal of Substance Abuse Treatment, 32*, 239–254.

Taylor, B. G., et al. (2001, December). *ADAM preliminary 2000 findings on drug use and drug markets—Adult male arrestees.* Washington, DC: U.S. Department of Justice, National Institute of Justice (Research Report, NCJ189101).

Thornberry, T. P. (1997). Introduction: Some advantages of developmental and life-course perspectives for the study of crime and delinquency. In T. P. Thornberry (Ed.), *Developmental theories of crime and delinquency: Advances in criminological theory* (Vol. 7, pp. 1–10). New Brunswick, NJ: Transaction.

Tripodi, S. J., Kim, J. S., & Bender, K. (2010). Is employment associated with reduced recidivism? The complex relationship between employment and crime. *International Journal of Offender Therapy and Comparative Criminology, 54*(5), 706–720.

Van Voorhis, P., Bauman, A., Wright, E., & Salisbury, E. (2009). Implementing the women's risk/needs assessments (WRNAs): Early lessons from the field. *Women, Girls, and Criminal Justice, 10*(6), 81–82. 89–91.

Van Voorhis, P., Wright, E. M., Salisbury, E., & Bauman, A. (2010). Women's risk factors and their contributions to existing risk/needs assessment: The current status of a gender-responsive supplement. *Criminal Justice and Behavior, 37*(3), 261–288.

Walters, G. D. (2002). The psychological inventory of criminal thinking styles (PICTS): A review and meta-analysis. *Assessment, 9*(3), 278–291.

Walters, G. D. (2003a). Changes in criminal thinking and identity in novice and experienced inmates: Prisonization revisited. *Criminal Justice and Behavior, 30*(4), 399–421.

Walters, G. D. (2003b). Predicting criminal justice outcomes with the psychopathy checklist and lifestyle criminality screening form: A meta-analytic comparison. *Behavioral Sciences & the Law, 21*(1), 89.

Walters, G. D. (2012). Substance abuse and criminal thinking: Testing the countervailing, mediation, and specificity hypotheses. *Law and Human Behavior, 36*(6), 506–512.

Warr, M. (1998). Life-course transitions and desistance from crime. *Criminology, 36*(2), 183–216.

Western, B. (2006). *Punishment and inequality in America.* New York, NY: Russell Sage Foundation.

White, H. R., & Gorman, D. M. (2000). Dynamics of the drug-crime relationship. In G. LaFree (Ed.), *Criminal justice 2000 The nature of crime: Continuity and change* (pp. 151–218). Washington, DC: U.S. Department of Justice.

Wright, J. P., & Cullen, F. T. (2004). Employment, peers, and life-course transitions. *Justice Quarterly, 21*(1), 183–205.

Yahner, J., & Visher, C. (2008). *Illinois prisoners' reentry success three years after release.* Washington, DC: Urban Institute.

Yochelson, S., & Samenow, S. E. (1976). *The criminal personality* (A profile for change, Vol. I). New York, NY: Jason Aronson.

Zweig, J., Yahner, J., & Redcross, C. (2010). Recidivism effects of the Center for Employment Opportunities (CEO) program vary by former prisoners' risk of reoffending. Urban Institute/ MDRC.

Part II
Simulation Inputs for an RNR Model

Chapter 5
Creating Simulation Parameter Inputs with Existing Data Sources: Estimating Offender Risks, Needs, and Recidivism

Stephanie A. Ainsworth and Faye S. Taxman

Introduction

There are no single data sources currently available that capture all of the offender information that is needed to measure the key Risk-Need-Responsivity (RNR) concepts. In fact, there is no national estimate of risk or need profiles. Jurisdiction-specific data files are rarely complete enough to develop comprehensive RNR-based offender classification systems. For this reason, the first part of developing a simulation tool involved the creation of offender profiles of risk and need factors combined with estimated recidivism rates for different offenders. This database is critical to the task of trying to simulate outcomes under different scenarios, such as placing offenders in programming that is more responsive to the risk-need profiles. The database is the foundation of both the static and discrete event models (see Chap. 1).

Every simulation model requires input parameters, as discussed in Chap. 3 by Dr. Greasley on the procedures and steps to develop a simulation tool. For the RNR Simulation Tool the input parameters reflect the national correction population across the spectrum of potential correctional placement: pretrial, probation, jail, prison, and parole. Parameters include demographic, static risk, criminogenic need, and recidivism distributions among prison, jail, and community correction populations from a US national perspective—that is, instead of developing jurisdiction- or site (i.e., site can refer to justice setting, type of organizational, location)-specific components, which create untold complexities in converting to a representative data set, we begin with a national perspective. And, then as is discussed in Chap. 8 on the RNR Simulation Tool, the national data set framework can then be converted to a file reflective of a jurisdiction's or specific agency's correctional population. This chapter starts with the process to define the key measures behind the national database. These

S.A. Ainsworth, M.A. (✉) • F.S. Taxman, Ph.D.
Department of Criminology, Law and Society,
George Mason University, 10900 University Boulevard, Fairfax, VA 20110, USA
e-mail: Sainswo1@masonlive.gmu.edu

F.S. Taxman and A. Pattavina (eds.), *Simulation Strategies to Reduce Recidivism:*   115
Risk Need Responsivity (RNR) Modeling for the Criminal Justice System,
DOI 10.1007/978-1-4614-6188-3_5, © Springer Science+Business Media New York 2013

measures and the resulting offender profiles are comprehensive enough to allow for the profiles to be weighted by the distribution of characteristics of offenders in a given area (refer to Chap. 8 for a discussion of the reweighting option).

The creation of a synthetic data set involves merging data files together to create a combined file with the key ingredients measuring risk, needs, demographic characteristics, and recidivism. While the ideal situation would involve having a single, nationally representative data set that contains this information, such a data set does not exist. Therefore, we focus on creating such a data set to provide an "analysts" dream tool—a comprehensive data file of offender profiles throughout the landscape of the correctional system. In this chapter, we describe the sets to create this ideal tool and our efforts to validate each measure and approach.

Creating the RNR Simulation Model Assumptions

The RNR principles provide a theoretical framework for classifying and treating criminal justice populations to maximize the reduction in recidivism. As discussed in prior chapters (Chap. 2 and 3), the risk principle states that the intensity of services and supervision for individual offenders should match each offender's level of criminal justice risk (historical factors that describe prior involvement in the justice system). The need principle states that offenders should be matched to programming and services that target their criminogenic needs—characteristics that are directly related to reoffending. Finally, the responsivity principle consists of two levels: general and specific responsivity. General responsivity states that cognitive-behavioral techniques should be used to address criminogenic needs, as research indicates that cognitive-behavioral therapy (CBT) is effective at reducing recidivism (Landenberger & Lipsey, 2005; Lipsey & Cullen, 2007). The specific responsivity principle states that factors such as gender, ethnicity, age, culture, individual learning styles, and other individual factors need to be taken into account when matching offenders to services and controls. A detailed description of the model is provided in Chap. 3 which outlines the empirical evidence and base for the RNR framework.

Besides the overarching RNR principles, Andrews and Bonta (2010) define (static) risk factors and (criminogenic) need factors based on variables that have a direct correlation with offending behavior. As discussed in Chap. 3, however, the RNR framework also includes other factors that show weak or inconsistent correlations with offending. There are "eight central factors" which are part of this model. The factors with direct correlations to offending behavior are termed the "the Big Four"—history of antisocial behavior, antisocial personality pattern, antisocial cognition, and antisocial associates (Andrews & Bonta, 2010, pp. 58–59). The variable *history of antisocial behavior* is a unidirectional measure that can only increase over time since it reflects the cumulative life experiences of the offender. According to the RNR principles, this history should determine an offender's risk level. Conversely, antisocial personality pattern, antisocial cognition, and antisocial associates are all dynamic risk factors, sometimes referred to as criminogenic needs,

Table 5.1 Reconsideration of risk, need, stabilizer, and destabilizer variables in the RNR Simulation Tool

Static risk factors	Criminogenic need factors
Age at first arrest	Substance dependence[a]
Criminal history	Antisocial values and attitudes
Prior probation/parole experiences	Low self-control
Prior probation/parole violations	Unstable lifestyle[b]
Prior incarceration	
Prior institutional misconduct	
History of violence	
Clinically relevant destabilizers	
Substance abuse[c]	
Mental illness (untreated or improperly managed)	
Lifestyle factors	
Stabilizers	Destabilizers
Stable housing (not homeless within year prior to arrest)	Unstable housing (homeless within year prior to arrest)
Education (at least a high school diploma)	Education (GED or less than a high school diploma)
Full-time employment	Less than full-time employment
Social support (someone they can lean on—prosocial friend or family member)	Lack of social support (no one to lean on—no prosocial friend or family member)
Stable family/marital (married, good relationship with family, prosocial relatives)	Unstable family/marital (divorced, estranged relationship with family, criminal family members)
Stable finances (working, earning a legitimate income)	Unstable finances (not working, earning an illegitimate income)
<3 Antisocial associates	3+ Antisocial associates

[a]Substance dependence is defined as dependence on a substance for which there is a clear drug-crime nexus such as cocaine, crack, heroin, methamphetamine, other amphetamines, and other opiates.
[b]An unstable lifestyle refers to an individual with three or more destabilizing factors.
[c]Substance abuse refers to abuse of a substance for which there is a clear drug-crime nexus, such as cocaine, crack, heroin, methamphetamine, other amphetamines, and other opiates, or dependence on a substance for which there is no clear drug-crime nexus such as marijuana or alcohol.

since these factors are more amenable to change over time. Andrews and Bonta (2010) identify several additional dynamic factors that exhibit a moderate causal link with offending behavior which are considered of lesser importance in the "Central Eight." These factors include substance abuse, family/marital circumstances, school/work, and leisure/recreation (Andrews & Bonta, 2010). As noted in Chap. 3, the "Central 8" are based on correlations between these characteristics and need. A review of these correlations and recent research suggests that this configuration does not take into account the factors that are protective or prevent further involvement. And, the RNR framework does not identify the psychosocial factors that may contribute to difficulties in adjusting in the community.

Table 5.1 reflects a slightly revised conceptualization of the RNR principles based on a review of the literature (described in Chap. 3) regarding direct and indirect relationships to offending behavior (recidivism). In this chapter, we describe how we measured each of these variables. While we recognize the original "Central 8" tenets, the review of the literature finds that (1) the measurement of key variables tend to be global variables that may not reflect the risk and need factors that are more directly related to recidivism; (2) certain psychosocial factors such as mental health disorders and employment history have been found to mediate the outcomes, and therefore it is necessary to reconceptualize the RNR framework to include these factors; and (3) lifestyle behaviors that exist along a continuum from unhealthy to healthy are useful for determining how the person is currently functioning which is better able to reflect outcomes. In the RNR Simulation Tool, we have reordered and redefined the risk and need factors to include a broader range of variables that have been found to be important in affecting recidivism. The rationale for this approach is described in more detail in Chap. 3. In this reconceptualization, the dynamic factors that are "non-criminogenic" (i.e., have an indirect correlation with recidivism) are important in making responsivity decisions since these factors often affect how well the person is likely to do in programming and whether the programming needs to be adjusted to meet the individual's unique needs. We define the term stabilizers to include key factors that are not directly included in the "lesser priority areas" in the Andrews and Bonta's RNR model but are strengths that the person has that may serve to increase their success. That is, stabilizers are the positive extralegal and psychosocial factors that influence the degree to which offenders can be responsive to services and controls. Destabilizers are the extralegal and psychosocial factors that may prevent a person from being successful and often require additional services and controls to minimize the negative influence of these factors.

The RNR model emphasizes that criminal justice risk is an important (and unforgiving) factor. Criminal justice risk refers to an individual's propensity to commit crime. It can be comprised of static (historical) and dynamic (malleable) factors. Static factors are often historical in nature and can only increase over time, while dynamic factors reflect contemporary situations that can be altered by either improving or decompensating. Static risk is also referred to as criminal history risk because it is largely composed of factors from an individual's offense and criminal justice history. These factors include age at first arrest, number of prior offenses, prior terms of probation or parole, prior incarceration, instances of institutional misconduct, and a history of violence (Table 5.1). These items are unidirectional in nature, meaning that they are unable to be reduced, and can only increase as an offender has more criminal justice events that add to their criminal history. Research indicates that offenders should be separated for treatment and supervision based on an offender's level of risk (Andrews & Bonta, 2010; Andrews, Bonta, & Wormith, 2006; Lowenkamp, Latessa, & Holsinger, 2006). That is, by placing high-risk and low-risk offenders in programming together, this may have the contagion effect of increasing the recidivism for low-risk offenders (Lowenkamp & Latessa, 2005; Lowenkamp, Latessa, & Holsinger, 2006).

Dynamic risk is another part of a person's risk for reoffending. This construct includes contemporary factors that are known to be directly related to offending behavior. These factors include substance dependence, antisocial values and attitudes, antisocial peers, and low self-control. Each factor represents an area of need that can be addressed through correctional programming. Targeting these factors can help reduce an offender's overall risk to reoffend (Lowenkamp, Latessa, & Smith, 2006). A failure to target these factors will limit the effectiveness of correctional programming and may in fact increase offending behavior (Andrews & Bonta, 2010).

In the revised RNR framework, clinically relevant factors are included to encompass the factors that affect how well the person may respond to the proposed criminal justice controls (e.g., drug testing, curfews) and treatment. Clinically relevant factors are separated from criminogenic needs because they are not directly related to offending behavior; instead they refer to factors that affect responsivity. Two items—substance abuse (not dependence) and mental illness—have an indirect association with offending, yet they can directly impact how an offender performs under correctional control. While it is important that programming for offenders targets their criminogenic needs, it is also important that treatment providers and supervision/correctional officers recognize that some offenders may need assistance to stabilize these factors that may interfere with their ability to respond to treatment and controls. That is, any potential substance abuse or mental illness can affect receptivity to programming. Addressing these "unmet" needs may help to ensure that offenders are able to get the most out of their treatment and supervision.

Another set of clinically relevant factors are lifestyle stabilizers/destabilizers. Again, these are factors that are not directly related to offending behavior but may impact an offender's receptivity to treatment and supervision. Lifestyle factors include housing, education, employment, social and family supports, finances, and antisocial associates. Some of these factors are often included as requirements for individuals under supervision. For example, one common condition of supervision is that an offender must be employed. Another is that offenders must not associate with other "known criminal associates." A failure to comply with these conditions can lead to a technical violation or a failure to successfully complete the supervision process. These factors can also act as stabilizers in an offender's life in the sense that they may serve to intensify prosocial attributes—access and opportunity to be socially productive—that may positively affect offender outcomes.

In creating the RNR Simulation Tool, it was necessary to assess the degree to which these risk, need, stabilizing, and destabilizing variables exist in national data sets. As shown in Chap. 3, these variables and constructs are measured in different ways. Existing measures of risk and need factors are often not as specific as they are described in the RNR Simulation Tool. It is thus important to not only identify but also improve upon existing measures of risk and need, as these measures often are not specific enough to capture distinctions that are important in responding to offenders' risk and needs.

Preparing the Data Sets for the Simulation Model and Tool

To develop a database to support the RNR Simulation Tool, we assessed the different types of extant data available. We searched data sets available from the Inter-university Consortium for Political and Social Research (ICPSR). The results of the search did not yield any national-level data collection efforts that contained comprehensive information on the variables referenced in Table 5.1 such as offender risk profiles, need profiles, stabilizers/destabilizers, and recidivism. The search was then expanded to include national data sets which might contain relevant information to construct risk and needs associated with different offender profiles. This search resulted in the identification of the *Survey of Inmates in State and Federal Correctional Facilities, 2004* (SISFCF); the *Survey of Inmates in Local Jails, 2002* (SILJ); the *State Court Processing Statistics* (SCPS); and the *Recidivism of Inmates Released in 1994* data sets—all collected by the Bureau of Justice Statistics (BJS), a federal agency with the goal of collecting data to highlight trends in the field of criminal justice (see http://bjs.ojp.usdoj.gov/). Each data set was examined regarding the type of data elements collected with attention to the information available for the model input parameters (demographic, risk, need, and recidivism information). The data was then cleaned, as necessary, to address missing data,[1] construct profiles, and validate measures:

- *Survey of Inmates in State and Federal Correctional Facilities, 2004* (SISFCF, prison sample), data is a nationally representative cross section of prison inmates in state and federal correctional facilities in 14 states. The data consists of extensive information regarding inmates' criminal history, substance use history, mental health history, education, and current offense for 18,185 offenders (BJS, 2004).
- *Survey of Inmates in Local Jails, 2002* (SILJ, jail sample), data contains information from 3,000 local jails across the USA and contains information nearly identical to the SISFCF for 7,750 offenders (BJS, 2002). Given comparable information in these two data sets, they were combined into a single database to create a representative sample of prison and jail inmates.[2]
- *State Court Processing Statistics* (SCPS, community sample) data contains felony court case information from 40 of the most populous counties in the USA (BJS, 2006). This file includes offenders that are in pretrial release or who are sentenced to probation, jail, or prison. The sample used to inform the RNR Simulation Model parameters is comprised of 35,767 defendants who were

---

[1] Bureau of Justice Statistics manuals were consulted for each data set to exclude certain subsets of data for which there were large amounts of missing data (more than 40% missing for any one variable). Means replacement was used to adjust other variables with minimal amounts of missing data.

[2] While the survey questions for these two samples were nearly identical, they were assigned different variable numbers in each data set. To combine the data from the two sources, the variables of interest in each were recoded with the same name. The SILJ data was then added to the SISFCF data using the "Add Data" function in SPSS.

Table 5.2 Demographic profiles by data source for RNR Simulation Tool and Model

Variable	SISFCF (%)	SILJ (%)	SCPS (%)	1994 Recidivism (%)
Sex				
Male	78.6	70.7	83.2	93.9
Female	21.4	29.3	16.8	6.1
Age at current offense				
16–27	25.8	47.1	46.8	29.6
28–35	26.1	21.5	24.9	34.4
36–42	22.9	17.4	15.6	20.6
43+	25.2	14.0	12.7	15.4
Race/ethnicity				
White	34.8	32.8	28.6	39.3
Black	40.0	36.1	47.7	43.5
American Indian/Alaskan Native	4.1	3.4	0.3	1.2
Asian/Hawaiian/Pacific Islander	1.4	1.0	1.4	0.4
Hispanic	18.9	16.1	19.3	14.8
Other	0.8	10.7	2.8	0.8

sentenced to probation. Variables available in this data include demographics, number of prior arrests, age at first arrest, and type of current offense. The SCPS data does not contain information regarding criminogenic needs or recidivism.

- *Recidivism of Inmates Released in 1994* (Recidivism) data contains information on demographics, current offenses, prior offenses, and recidivism for 33,796 offenders released from correctional facilities in 15 states in 1994. Follow-up information for released offenders was recorded for 3 years post release, including rearrest, reconviction, and reincarceration information. The Recidivism data does not contain any information regarding criminogenic needs.

The *Survey of Inmates* data sets provide risk and need information for prison and jail samples, while the SCPS data provides information for the community sample of pretrial releases, probationers, parolees, etc. The *Recidivism* data provided information regarding new offenses to supplement all of the offender samples. Common among the *SISFCF, SILJ, SCPS*, and *Recidivism* databases were variables related to prior and current offense information, and demographic information such as gender, age, and race/ethnicity. An algorithm was used to match cases across the 4 data sets to create a picture of the profiles of offenders as described below. The details of this matching process are discussed later in this chapter.

Table 5.2 describes the demographic profiles in the data sources. Not surprisingly, there is variability across correctional settings in these demographic factors. As expected given the composition of the justice population, the data sources used in the creation of the RNR Simulation Tool were comprised of samples that were predominately young, African-American, and male. This is reflective of national jail and prison populations. However, as described in Chap. 8, we can use the simulation model to adjust the profiles to reflect the parameters for different correctional populations.

Combining Data Sets from Multiple Sources

In this section, we describe the process used to combine the data sets to create the "ideal file" of key variables to measure the RNR factors. This is a critical process since it allows us to develop the risk and need profiles used in the simulation model. The process for combining data sets also included creating key measures and ensuring that appropriate statistical techniques were used to validate the measures and each process. The procedures used are as follows:

- Each extant BJS data set was randomly divided into a construction and validation sample (using SPSS random number generator). This was to allow us to develop the construct for each RNR measure and then validate the measure to verify the robustness of the tool. This process is critical in developing sound measures, as described in Gottfredson, (1987).
- Each variable was constructed and then validated in the "validation file." Any validation problems were used to reexamine the process to measure the variable and make adjustments. Based on these results, we integrated the measure into the data set.
- For each extant data set, we used the traditional receiver operating change (ROC) methods to determine appropriate cutoff points for the risk, need, and destabilizer scales. The ROC analysis is a calculation of the probability that the designated predictor variable will accurately assign individuals to an outcome. In cases where a ROC analysis would have been inappropriate, we used the appropriate statistical method such as using a chi-square for dichotomous variables. This method is consistent with best practice for the creation of validated risk assessment instruments (Dal Pra, 2004; Gottfredson, 1987; Glaser, 1987; LeCroy, Krysik, & Palumbo, 1998). The ROC analysis was run for each risk and need scale by using a proxy recidivism variable.[3] We also used the data that was merged with the actual recidivism variables[4] to confirm the pattern.
- The area under the curve (AUC) was also calculated and tested for significance to determine if each scale was a better predictor of recidivism than chance. Simply put, the AUC is a calculation of the probability that the predictor variable will accurately rank each positive case higher on the outcome variable than each negative case. For the purpose of the RNR Simulation Tool, the AUC reflects the probability that each scale will accurately rank each recidivist as higher for each risk and need scale than each non-recidivist offender (see Gottfredson, 1987).

---

[3] History of arrest was used as a proxy measure of recidivism in the Survey of Inmates data prior to matching with the actual Recidivism data.

[4] Numbers reported here reflect the results of the ROC analyses performed using rearrest after 3 years as the outcome variable. Results of the initial ROC analyses can be obtained from the author.

Preparing the SISFCF and SILJ Data for Joining with Recidivism Files

The *1994 Recidivism* data set was used to create the recidivism measures since recidivism was not captured in any of the other files. Before beginning the process of importing the rearrest, reconviction, and reincarceration variables from the *1994 Recidivism* data set into the *SISFCF* and *SILJ* data, the latter two data sets were combined into one larger data set. As previously mentioned, both data sets contain the same variables which made adding the files together a simple process. The SILJ cases were thus added to the SISFCF cases for a combined total of 25,935 individuals. Once the SISFCF and SILJ samples were combined, risk and need scales were created for each of the 25,935 cases. This process is discussed in depth later in this chapter.

In preparation for matching with the *Survey of Inmates* data, the 1994 Recidivism data was cleaned and restructured. Each offender's release date was created based on the day, month, and year variables provided. A 1- and 3-year post-release date was then created for each individual by adding 1 or 3 years to the release date, respectively. This process truncated each individual's follow-up period to ensure that it would be no longer than 1 or 3 years, respectively. This corrected a discrepancy with the original calculations that allowed up to 11 months additional follow-up for individuals released in January of 1994, based on the release year.

Next, all invalid arrest months, days, and years for each arrest cycle (those which were coded numerically as not applicable or as missing) were recoded to system missing so they would not be improperly included in the analysis. Arrest dates were calculated for each cycle with valid day, month, and year information. A flag was created for each arrest cycle to indicate if that cycle included a prior arrest. The prior arrests were summed to calculate the total number of prior arrests, once again subtracting 1 for the instant offense. A rearrest flag was created to indicate if any of the new arrest dates fell after the release date. The process was then repeated using adjudication dates to calculate reconviction and reincarceration rates for the recidivism data.

Identifying Match Criteria

Before joining the data sets, we identified a core set of variables that are theoretically relevant in linking offender profiles with recidivism. These variables, as described in Chap. 3, are known for their importance in predicting recidivism. For continuous variables, we tried to use the original variable but then had to aggregate in categories when we exhausted the match potential. These core variables are gender, race, age (continuous, 35 groups, 14 groups, 4 groups), current offense (personal, property, drug, others), population type (3 or more of any of the following offenses: violent, sexual, drug; all others are general offenders), number of prior

Prison, Jail, or Community Sample		Recidivism Sample
Male, Hispanic, Age 35, 16 at First Arrest, 8 Prior Arrests, 3+ Drug Crimes, Current Offense: Drug	NOT A MATCH	Male, Hispanic, Age 21, 16 at First Arrest, 8 Prior Arrests, General Offender, Current Offense: Drug
Male, Black, Age 35, 16 at First Arrest, 8 Prior Arrests, 3+ Drug Crimes, Current Offense: Drug	MATCH	Male, Black, Age 35, 16 at First Arrest, 8 Prior Arrests, 3+ Drug Crimes, Current Offense: Drug

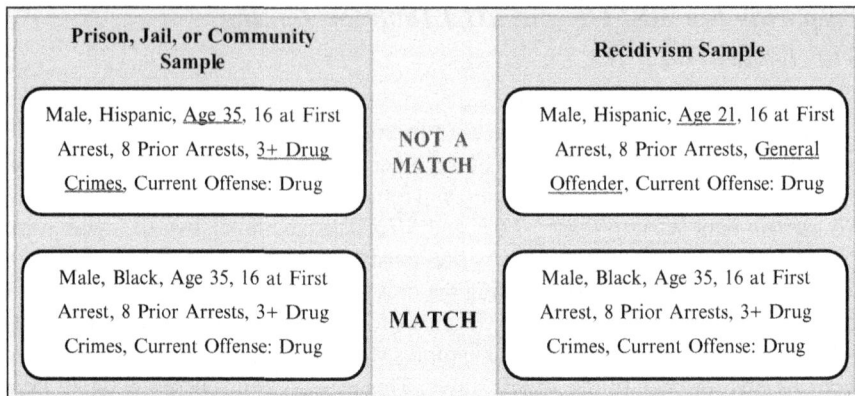

Fig. 5.1 Match process algorithm example

arrests (continuous, 50 groups, 10 groups, 5 groups, 3 groups), and age at first arrest (continuous, 35 groups, 14 groups). These specific variables were chosen because they reflect risk items which have been shown to predict recidivism (Andrews & Bonta, 2010). The demographic and criminal history variables afforded the opportunity to match individuals based on known variables in the prison, jail, and community samples to their counterpart in the Recidivism data. The match process is characterized in Fig. 5.1 which shows how we handled different situations.

Since several of the variables used in the match were scale items, such as current age, age at first arrest, and number of prior arrests, a one-to-one match was extremely specific and not always possible when matching with several scale variables at a time. For example, a 35-year-old drug offender and a 21-year-old general offender who are alike in every other way would not be matched using continuous variables (see Fig. 5.1). Rather than simply dropping each of these variables after attempting to match them in continuous form, the items in each scale variable were grouped into categories for at least two more iterations of the match process. This allowed for a gradual relaxation of the match specificity to continue to ensure that like individuals from each data set would be matched to each other (see Table 5.3 for a list of variables and groupings). This means that some variables had multiple groupings of values to improve the alignment of the data sets.

Merging the SISFCF and SILJ Data with Recidivism Data

Recidivism data was obtained from the *Recidivism of Inmates Released in 1994* data file, which provided 3-year post-release recidivism rates for inmates. To ensure that the 1994 BJS data were matched to the appropriate cases in the *Survey of Inmates in State and Federal Correctional Facilities (2004)* and *Survey of Inmates in Local Jails (2002)*, cases from each data set were matched based on a combination of

Table 5.3 Variable groupings for scale items used in match process in RNR Simulation Tool and Model

Variable	Continuous	Grouped
Age	13–84	2-year intervals, 5-year intervals, 4 groups (16–27, 28–35, 36–42, 43+)
Age at first arrest	7–50	2-year intervals, 5-year intervals
Number of prior arrests	0–50	50 percentiles, 10 percentiles, 5 percentiles, 3 percentiles

seven distinct variables (described above) which were present in all three data sets. This process was then repeated with the *SCPS data* to match it with the prison and jail data to import information on offender needs and recidivism. The SCPS data was matched to the prison and jail samples rather than the *Recidivism* data directly in order to import both needs and recidivism at one time (needs are not available in the *Recidivism* data).

Rather than importing the mean recidivism rates for individuals matched on a group of variables, unique random numbers were assigned to cases within each match group. The cases from both data sets were then matched one to one whenever possible. In instances when there were fewer combined prison and/or jail cases from the *Survey of Inmates* data sets than in the *Recidivism* cases, individuals were randomly matched one to one, with excess *Recidivism* cases left unmatched. In instances when there were more combined prison and jail cases than *Recidivism* cases, each *Recidivism* case was first randomly matched with one prison/jail case and then the *Recidivism* cases were randomly duplicated to match prison/jail cases within the same group until all cases were matched. At the end of the first iteration of this process (using all eight matching variables listed above), a total of 1,692 (6.5%) of the *Survey of Inmates* cases had been matched to the *Recidivism* cases, as shown in Table 5.4 below. Subsequent iterations of matching were then completed with increasingly relaxed match criteria. A total of 17 iterations were completed to fully match the *Survey of Inmates* data to the *Recidivism* data. The percent of cases matched in each iteration and variables used to match cases can be found in Table 5.4.

Table 5.5 illustrates the recidivism rates based on each data source. After importing the *Recidivism* data, the 1-year rearrest rate for the combined *SISFCF, SILJ*, and *Recidivism* samples (hereafter referred to as the synthetic data) was 33.9%, and the 3-year rearrest rate for the sample was 57.6%. For the *SISFCF* sample alone, the 1- and 3-year rearrest rates were 34.1% and 58.3%, respectively. These rates were similar for the *SILJ* sample, with 1- and 3-year rearrest rates of 33.6% and 56.1%, respectively. We compared these data with recidivism rates from the *Recidivism* data set in order to verify the distributions after the match process. As seen in Table 5.5, within each measure of recidivism, there is less than 5% difference in rates between each data sample. Recidivism rates for the community sample are slightly higher in all measures of recidivism, which may be in part a result of the matching process. This inflation may also be a product of different sentencing practices which do not necessarily take risk and need factors into account when making sentencing decisions.

Table 5.4 Survey of Inmates and Recidivism data match process

Iteration	Variables included in match iteration	% matched	% cumulative
1	Sex, race, age category (4 groups), current offense, population type, number of prior arrests (scale), age (scale), age at first arrest (scale)	6.5	6.5
2	Sex, race, age category (4 groups), current offense, population type, number of prior arrests (scale), age (scale), age at first arrest (2-year groups)	3.3	9.8
3	Sex, race, age category (4 groups), current offense, population type, number of prior arrests (scale), age (5-year groups), age at first arrest (scale)	6.0	15.8
4	Sex, race, age category (4 groups), current offense, population type, number of prior arrests (scale), age (5-year groups), age at first arrest (2-year groups)	5.7	21.5
5	Sex, race, age category (4 groups), current offense, population type, number of prior arrests (50 percentiles), age (5-year groups), age at first arrest (2-year groups)	12.3	33.8
6	Sex, race, age category (4 groups), current offense, population type, number of prior arrests (10 percentiles), age (5-year groups), age at first arrest (2-year groups)	17.9	51.7
7	Sex, race, age category (4 groups), current offense, population type, number of prior arrests (10 percentiles), age (5-year groups), age at first arrest (5-year groups)	10.6	62.3
8	Sex, race, age category (4 groups), current offense, population type, number of prior arrests (10 percentiles), age at first arrest (5-year groups)	11.0	73.3
9	Sex, race, age category (4 groups), current offense, population type, number of prior arrests (50 percentiles), age at first arrest (10-year groups)	1.6	74.9
10	Sex, race, age category (4 groups), current offense, population type, number of prior arrests (50 percentiles), age at first arrest (4 groups)	5.7	80.6
11	Sex, race, age category (4 groups), current offense, population type, number of prior arrests (10 percentiles), age at first arrest (4 groups)	5.0	85.6
12	Sex, race, age category (4 groups), current offense, population type, number of prior arrests (5 percentiles), age at first arrest (4 groups)	7.0	92.6
13	Sex, race, age category (4 groups), current offense, number of prior arrests (5 percentiles), age at first arrest (4 groups)	4.3	96.9
14	Sex, race, age category (4 groups), current offense, number of prior arrests (3 percentiles), age at first arrest (4 groups)	0.3	97.2
15	Sex, race, age category (4 groups), number of prior arrests (3 percentiles), age at first arrest (4 groups)	2.2	99.4
16	Sex, race, age category (4 groups), number of prior arrests (3 percentiles)	0.3	99.7
17	Sex, age category (4 groups), number of prior arrests (3 percentiles)	0.3	100

Table 5.5 Recidivism patterns by data source for simulated data

Data set	1-year new arrest (%)	3-year new arrest (%)	1-year new conviction (%)	3-year new conviction (%)	1-year new incarceration (%)	3-year new incarceration (%)
1994 Recidivism	36.5	59.1	20.1	38.8	16.6	30.4
SISFCF	34.1	58.3	19.2	38.2	15.8	29.5
SILJ	33.6	56.1	20.2	37.1	17.4	31.0
SCPS	37.5	60.8	23.1	41.0	19.0	32.5

Measuring the Key Variables in the Model

In this section, we will discuss how each key variable was constructed for the RNR model. The construction of the key variables relied upon data in the extant data sets. These variables can be grouped into five larger categories: recidivism, static risk, criminogenic needs, clinical destabilizers, and lifestyle stabilizers/destabilizers. This section contains a detailed discussion of each category and the variables within them.

Recidivism

The *Recidivism* data set includes three measures of recidivism: new arrest, new conviction, and new incarceration. We included all three variables in the RNR Simulation Tool, which allows the model to be flexible to the needs of various jurisdictions that may measure recidivism in different ways. Each of these variables was recorded in such a way that for each measure of recidivism a 1- and 3-year recidivism rate could be calculated.

Static Risk

Since future behavior is often a product of past behavior, an offender's criminal history can be very informative in predicting risk of future recidivism (Austin, 2006; Hoffman, 1983). There are a number of instruments available to measure static risk; however, they all use similar types of information to construct static risk scales. Figure 5.2 lists static risk items in three commonly available risk assessment instruments and the four BJS data sets.

The variables selected to create the criminal history are listed in Table 5.6. A series of logistic regressions was run to examine the strength of the relationship between each static risk item and the recidivism variables. Each response was then assigned a point value based on the response weight in the multivariate model (see Table 5.6).

Item	Risk Instruments			Data Sets Used in this Study		
	LSI-R[1]	ORAS[2]	Wisconsin[3]	Survey of Inmates[4]	State Court Processing[5]	Recidivism[6]
Prior Arrests (convictions)	✓	✓	✓	✓	✓	✓
Age of First Arrest (conviction)	✓		✓	✓		✓
Prior Incarceration	✓	✓			✓	✓
History of Escape (or attempt) from Correctional Facility	✓			✓	✓	
History of Institutional Misconduct	✓	✓				
Number of Prior Periods of Probation/Parole Supervision		✓	✓	✓	✓	
Probation/Parole Revocations	✓	✓	✓	✓		
History of Assault or Violence	✓		✓		✓	✓
Juvenile Conviction for Burglary, Theft, Auto Theft, Robbery, or Forgery			✓			✓

[1]Lowenkamp, C.T., Holsinger, A.M., Brusman-Lovins, L., & Latessa, E.J. (2004). Assessing the Inter-Rater Agreement of the Level of Service Inventory Revised. *Federal Probation, 68* (3): 34-38.
[2]Latessa, E., Smith, P., Lemke, R., Makarios, M., & Lowenkamp, C. (2009).*Creation and Validation of the Ohio Risk Assessment System Final Report* . Cincinnati, OH: University of Cincinnati.
[3]Eisenberg, M., Bryl, J., & Fabelo, T. (2009). *Validation of the Wisconsin Department of Corrections Risk Assessment Instrument.* New York, NY: Council of State Governments Justice Center, PEW Center on the States, Bureau of Justice Assistance, & Public Welfare Foundation.
[4]Bureau of Justice Statistics. Survey of Inmates in State and Federal Correctional Facilities, 2004; Survey of Inmates in Local Jails, 2002.
[5]Bureau of Justice Statistics. State Court Processing Statistics, 1990-2006.
[6]Bureau of Justice Statistics. Recidivism of Inmates Released in 1994.

Fig. 5.2 Risk items included in various instruments and data sets

The criminal history risk scale assigned points to each item. The points were then summed to reflect the range of static risk scores. The scores range from 1 to 27 with an average score of 15.7 (SD = 5.79). The AUC was 0.588 ($p < 0.001$), which indicates that the criminal history domain is a stronger predictor of recidivism than chance. Based on the analysis, risk scores of 5 and 19 were chosen as cutoffs for the high- and low-risk groups with 85% certainty that each individual would be appropriately classified as either likely to recidivate or not. This process resulted in 14.0% of the offenders classified as low risk, 62.3% as moderate risk, and 23.7% as high risk for the prison sample. For the jail sample, 11.8% were classified as low risk, 72.7% as moderate risk, and 15.5% as high risk. In the community sample, 30.1% were classified as low risk, 49.7% as moderate risk, and 20.2% as high risk. The

Table 5.6 Available static risk items in SISFCF and SILJ data[a]

Variable name	Value	Score	%
Age at first arrest	43+	1	0.8
	36–42	2	1.9
	28–35	3	6.8
	16–27	4	55.7
	7–16	6	34.8
Number of prior arrests	0	0	19.1
	1–3	3	39.1
	4–6	4	24.6
	7–10	5	5.2
	11+	6	12.0
Ever on probation as a juvenile	N/A	0	19.1
	No	1	58.4
	Yes	2	22.5
Ever on probation as an adult	N/A	0	19.1
	No	1	10.6
	Yes	2	70.4
Probation revocations	N/A	0	19.8
	No	1	5.2
	Yes	2	75.1
Ever on parole	N/A	0	19.1
	No	1	4.5
	Yes	2	76.4
Parole revocations	N/A	0	23.3
	No	1	0.7
	Yes	2	76.0
Ever written up or found guilty of	N/A	0	19.1
escape or attempt to escape	No	1	80.5
	Yes	2	0.5
Number of times written up or found	N/A	0	99.5
guilty of escape or attempt to escape	1	2	0.4
	2+	3	0.1
Highest possible total score		27	

[a]Some items were not available in the SCPS data

recidivism rates by static risk level for each sample are reported in Table 5.7. The distribution of risk scores and resulting risk levels is consistent with the measurement of recidivism, with higher rates for individuals who have more points on the risk scale (high risk).

Criminogenic Needs and Stabilizers/Destabilizers

Dynamic risk refers to those psychosocial needs that are related to offending as well as factors that affect how stable the offender is in the community. Research suggests

Table 5.7 Recidivism rates by risk level in the RNR Simulation Tool model

Data set	1-year new arrest (%)	3-year new arrest (%)	1-year new conviction (%)	3-year new conviction (%)	1-year new incarceration (%)	3-year new incarceration (%)
Prison						
High	46.4	71.7	28.5	50.4	23.0	38.1
Moderate	30.3	54.7	16.3	35.1	13.6	27.4
Low	26.6	47.8	13.2	27.9	10.9	21.6
Jail						
High	48.8	72.3	30.3	51.4	25.6	42.6
Moderate	31.8	54.2	19.5	35.5	17.0	29.9
Low	24.7	46.6	11.3	28.3	08.8	22.4
Community						
High	47.3	70.4	31.0	52.7	23.8	39.6
Moderate	34.5	58.0	21.7	38.2	18.7	31.2
Low	30.2	53.3	14.6	30.5	12.9	25.1

that identifying and targeting criminogenic needs should be the focus of treatment programming to produce the greatest improvements in offender outcomes and reductions in recidivism (Andrews & Bonta, 2010). One major distinction between the RNR Simulation Model and prior research is how we classify these need factors, as discussed above. While other models combine many factors in a single need score, the RNR Simulation Model makes a distinction between those need factors that research shows are directly related to offending behavior (criminogenic) and those that are not (non-criminogenic) but are clinically relevant to how well a person is likely to do in programming.

The RNR Simulation Tool classifies non-criminogenic needs into two categories: clinically relevant stabilizers/destabilizers that can effect engagement and successful completion of treatment and supervision, and lifestyle stabilizers/destabilizers that have an impact on an offender's ability to participate in daily activities such as employment as well as meet basic needs of food and shelter. Each of these variables represents a continuum of healthy functioning to unhealthy daily functioning for an individual. Offenders who have healthy daily functioning are characterized by having few (less than four) destabilizers, while offenders with four or more destabilizers are likely to exhibit unhealthy daily functioning patterns. These individuals are less likely to engage in treatment and more likely to fail to complete treatment and supervision (Broner, Lang, & Behler, 2009; DeLisi, 2000). The concept of stabilizers and destabilizers is discussed above.

Substance Use

Andrews and Bonta (2010) acknowledge substance abuse as a lesser criminogenic need in terms of affecting outcomes. The relationship between substance abuse and recidivism is convoluted and may not appropriately measure the linkage between substance abuse patterns and offending. As discussed in Chap. 4, one major measurement issue is the distinction between substance dependence and substance abuse. Even more so, many researchers refer to any lifetime use of drugs and alcohol as substance abuse (even going back to onset of use or use during teenage years). As noted by Taylor et al. (2001), this results in a large percentage of the offender population being classified as having an abuse problem even if the person is not currently abusing. The measurement problems weaken the association with recidivism because the measurement does not capture the drug-crime nexus. Research indicates that it is not only important to identify the specific substance of choice an offender is using (Bennett, Holloway, & Farrington, 2008), but it is also important to identify the severity of use (Taxman, 1998). Bennett, Holloway, and Farrington (2008) used meta-analytic techniques to examine the association between substance use and crime. They found that individuals who used hard drugs such as crack, cocaine, heroin, and amphetamines were more likely to offend than nondrug users and users of other recreational drugs such as marijuana (Bennett et al., 2008).

Within the RNR Simulation Tool, a distinction is made between substance abuse and dependence. Table 5.8 lists the variables included in the substance use scale used to determine offenders' severity of use. Many items are also used by the American Psychological Association to diagnose substance abuse or dependence (American Psychiatric Association, 2000). While both abuse and dependence are characterized by a number of lifestyle disruptions that are a result of chronic substance use, individuals who are dependent on a substance will face additional challenges when attempting to quit use, including experience of withdrawal symptoms and the need to address daily triggers for substance use (American Psychiatric Association, 2000; Monti, Rohsenow, Michalec, Martin, & Abrams, 1997). The substance use scale created for use in the RNR Simulation Tool measures individuals' severity of use by the number of disruptive factors in their lives. We tried to replicate the clinical abuse or dependence and some additional items which indicate risky behaviors associated with substance use (substance use at time of offense, use of a needle to inject drug). Drug use scale scores ranged from 0 to 12 with an average score of 9 (SD = 1.84).

Regression analyses were run to measure the strength of the relationship between the need items in each scale and the proxy recidivism measure, and an ROC analysis was used to determine cutoffs. The ROC analysis produced an AUC of 0.670 ($p < 0.001$), which indicates that although substance use is not a particularly strong predictor of recidivism, it was statistically significant. This is consistent with the above literature which states that not all forms of substance abuse are correlated with recidivism (Bennett et al., 2008). Based on the analysis, offenders who had less than three substance use factors or who had not regularly or ever used a substance were classified as no/low substance users. Offenders who reported more than three substance use risk or disruptive factors and had not reported current (within the month prior to their arrest) or regular (once a week or more for at least a month) use of a drug associated with offending behavior (crack, cocaine, heroin, other opiates, methamphetamine, or other amphetamines) were classified as abusers of a

Table 5.8 Available substance use variables in SISFCF and SILJ data[d]

Variable name	Value	Score	%
Ever use any substance	No	0	22.0
	Yes	1	78.0
Ever use opiates (heroin), amphetamines (meth), or crack/cocaine[a]	No	0	46.8
	Yes	1	53.2
Use substance once a week or more for at least a month	No	0	35.0
	Yes	1	65.0
Regular use of opiates (heroin), amphetamines (meth), or	No	0	59.4
crack/cocaine	Yes	1	40.6
Use of any substance in prior month	No	0	52.0
	Yes	1	48.0
Use of opiates (heroin), amphetamines (meth), or crack/cocaine	No	0	70.1
in prior month	Yes	1	29.9
Age at first time of use of any substance	43+	0	23.4
	36–42	1	0.8
	28–35	3	3.0
	17–27	4	25.7
	7–16	6	47.0
In the year before admission, were you ever arrested or held at	No	0	77.9
a police station because of substance use[c]	Yes	1	22.1
Under the influence of any substance at the time of arrest	No	0	73.1
	Yes	1	26.9
Have you ever driven a car or any other vehicle while under	No	0	49.1
the influence of a substance[c]	Yes	1	50.9
Under the influence of opiates (heroin), amphetamines (meth),	No	0	95.1
or crack/cocaine at the time of arrest	Yes	1	4.9
Ever committed an offense trying to get money to buy a	No	0	84.7
substance or otherwise obtain a substance for use[b]	Yes	1	15.3
Ever lost a job because of substance use[b]	No	0	86.6
	Yes	1	13.4
In year before admission, have you had any other job or school	No	0	82.8
trouble because of substance use[b]	Yes	1	17.2
Have you ever gotten into a physical fight while or right after	No	0	78.2
using a substance[c]	Yes	1	21.8
Have you ever had arguments with your spouse, family, or friends	No	0	68.3
while or right after using a substance[c]	Yes	1	31.7
Have you ever used a needle to inject any substance for nonmedical	No	0	83.5
reasons	Yes	1	16.5
Highest possible total score		22	

[a] These substances have been specifically shown to be related to criminal offending (Bennett et al., 2008; Nurco, Hanlon, & Kinlock, 1991)
[b] Item used in the American Psychiatric Association's Diagnostic and Statistical Manual (DSM-IV) to diagnose substance dependence
[c] Item used in the American Psychiatric Association's Diagnostic and Statistical Manual (DSM-IV) to diagnose substance abuse
[d] Some items were not available in the SCPS data

Table 5.9 Severity of substance use by data sample in the RNR Simulation Tool and Model

Severity	SISFCF % (rearrested)	SILJ % (rearrested)	SCPS % (rearrested)
Dependence on a substance associated with offending behavior (i.e., crack, cocaine, heroin, other opiates, methamphetamine, other amphetamines)	10.0 (60.4)	6.5 (60.9)	8.4 (66.2)
Abuse of a substance associated with offending or abuse/dependence of a substance not associated with offending	73.7 (60.0)	65.6 (62.5)	68.7 (63.8)
No abuse or dependence	16.3 (53.4)	27.9 (44.0)	22.9 (54.2)

criminogenic drug or dependent on a non-criminogenic drug (68.0%). Finally, offenders who reported recent and frequent use of a drug associated with offending behavior as well as more than three risky or disruptive behaviors were classified as substance dependent on a criminogenic drug. The distribution and recidivism rates for each substance use level are presented for each of the data samples below (Table 5.9). The Substance Abuse and Mental Health Services Administration reports that among the general population, 8.7% of individuals either abuse or are dependent on a substance (Substance Abuse and Mental Health Services Administration, 2011). These rates were significantly higher among criminal justice populations, with 38.2% of individuals on supervised release from prison or jail reported as either abusers or dependent.

Mental Health

While rates of mental illness among justice-involved individuals are significantly higher than that of the general population (BJS, 2006), research fails to show a causal link between mental illness and offending (Baillargeon, Binswanger, Penn, Williams, & Murray, 2009; Martin, Dorken, Wamboldt, & Wootten, 2012). For this reason, within the RNR Simulation Model, mental illness is not considered to be a criminogenic need but an indicator of destabilizer. Presence of a mental health problem has been shown to impact treatment engagement and completion and the ability to comply with the conditions of supervision (Wormith & Olver, 2002). An offender's mental health is considered to be a clinically relevant indicator of stability. Within the synthetic data set, the mental health variables include whether or not a person has been diagnosed with a mental disorder, has been hospitalized for a mental condition, and ever received treatment or taken medication for a mental disorder (see Table 5.10). The availability of this data in the extant data sets provides more detailed information that can be used to gauge the degree to which the individual might have difficulties adjusting in the community.

The mental health scores for the RNR Simulation Model ranged from a low of 1 to a high of 5 with an average score of 3 (SD = 1.11) and an AUC of 0.510 ($p < 0.05$). An indicator variable was created to separate those offenders with a history of

Table 5.10 Available mental health variables in SISFCF and SILJ data[a]

Variable name	Value	Score	%
Ever diagnosed with a mental disorder	No	0	73.3
	Yes	1	26.7
Ever taken medication for mental conditions a week for more than a month	No	0	75.5
	Yes	1	24.5
Ever received counseling from a trained professional	No	0	77.7
	Yes	1	22.3
Ever received other mental health treatment services	No	0	96.4
	Yes	1	3.6
Ever admitted to a mental hospital, stayed overnight	No	0	88.0
	Yes	1	12.0
Highest possible total score		5	

[a]Some items were not available in the SCPS data

mental illness as a clinically relevant destabilizer from those who do not have a history of mental illness. Just over one quarter of the prison and jail samples (26.9% and 26.3%, respectively) were classified as having a mental health destabilizer. This rate was significantly lower for the community sample (5.8%).[5]

Antisocial Associates/Peers

Research indicates criminally involved peer networks contribute to greater involvement in offending behavior (Laub & Sampson, 2001; Warr, 1993; Wright & Cullen, 2004). It is unknown whether these associations lead to continued criminal involvement, or if these associations are the product of criminal thinking patterns or subcultural systems which lead offenders to associate with other like-minded individuals (Gottfredson & Hirschi, 1990).

Associations with criminal peers is identified by Andrews and Bonta (2010) as one of the "Big Four" risk/need factors due to its demonstrated link to offending behavior. To measure the presence of criminal associates in the synthetic data, items regarding the criminal involvement of the offender's friends and family were included in a scale measuring antisocial associates/peers (Table 5.11). The extant data included information regarding their childhood family and associations. The use of these variables is supported by research on social learning theory, which states that offending behavior is learned over an individual's life from influences such as their family and peers (Akers & Sellers, 2003; Andrews & Bonta 2010; Gottfredson & Hirschi, 1990).

Antisocial peers and family scores ranged from 0 to 13 with an average score of 4 (SD = 3.08) and an AUC of 0.662 ($p < 0.001$). The ROC analysis suggested that a two-group distinction was appropriate: individuals with fewer than three indicators

---

[5]An algorithm was applied to the synthetic data to adjust for the low prevalence of mental illness within the community sample.

Table 5.11 Available antisocial associates/peers variables in SISFCF and SILJ data[a]

Variable name	Value	Score	%
Parent(s) abused drugs or alcohol when offender was a child	No	0	68.7
	Yes	1	31.3
Friends abused drugs or alcohol when offender was a child	No	0	37.6
	Yes	1	62.4
Parent(s) served time in prison or jail when offender was a child	No	0	80.5
	Yes	1	19.5
Family member served time in prison or jail when offender was a child	No	0	64.2
	Yes	1	35.8
Boyfriend/girlfriend served time in prison or jail when offender was a child	No	0	94.9
	Yes	1	5.1
Friends stole property when offender was a child	No	0	72.0
	Yes	1	28.0
Friends sold drugs when offender was a child	No	0	66.8
	Yes	1	33.2
Friends damaged property when offender was a child	No	0	72.5
	Yes	1	27.5
Friends robbed people when offender was a child	No	0	85.7
	Yes	1	14.3
Friends shoplifted when offender was a child	No	0	61.5
	Yes	1	38.5
Friends stole a vehicle when offender was a child	No	0	73.2
	Yes	1	26.8
Friends broke into a home or establishment when offender was a child	No	0	75.2
	Yes	1	24.8
Friends engaged in other illegal activities when offender was a child	No	0	97.6
	Yes	1	2.4
Highest possible total score		13	

[a]Some items were not available in the SCPS data

of antisocial associates or those with greater than three indicators of antisocial associates. Over half of the prison sample (54.3%) and almost half of the jail sample (47.8%) were classified as having antisocial associates as a lifestyle destabilizer. The percent within the community sample was slightly lower, at 42.0%.

Employment

Employment can be measured in many ways, ranging from whether the offender has a job, works full-time or is satisfied with their employment. While there is an evolving literature on employment and offending behavior, the general findings have mixed results (Apel & Sweeten, 2010). In general, employment does not significantly reduce offending behavior, but job stability appears to be an important factor to increasing stability in the community and reducing criminality (Laub &

Sampson, 2001; Sampson & Laub, 1992; Wright & Cullen, 2004). Within the synthetic data set, job stability is measured by number of hours worked per week prior to instant offense. Employment is considered a stabilizing factor for those individuals with at least 30 or more hours a week but a destabilizing factor for those individuals with part time (less than 30 h per week), intermittent, or no employment. The rates of offenders with employment as a destabilizer were highest among the jail sample, at 55.9%, followed by the community sample with 47.2%, and the lowest rate among the prison sample (42.4%).

Housing

While offenders have higher rates of homelessness than the general public (BJS, 2006), there is not a direct causal link between homelessness and criminality (DeLisi, 2000). Homelessness is considered a lifestyle factor in the RNR Simulation Model due to its implications for disrupting an offender's day to day activities as well as their ability to successfully complete treatment and supervision (Broner et al., 2009; DeLisi, 2000). The synthetic data thus included a housing variable to indicate if an offender had been homeless at any point in the year prior to their arrest. Only 1.7% of the prison sample and 2.4% of the jail sample were classified as having housing as a lifestyle destabilizer. This rate was similar for the community sample, with 1.5% of the sample classified with a housing destabilizer. A chi-square analysis between the two variables showed a weak but statistically significant correlation between the housing indicator and recidivism ($r=0.031$, $p<0.01$).

Education

Education can be measured in several ways, including whether or not an offender has completed high school, received their graduate education degree (GED), has some postsecondary education, or is currently enrolled in school. Research regarding the link between education and offending has mixed results. Lochner and Moretti (2004) found that individuals who have completed high school are less likely to engage in offending behavior than those who did not complete high school or who have a GED. Andrews and Bonta (2010) consider education to be a lesser criminogenic need since it has a weak correlation with recidivism. Researchers note that the causal link between education and offending is not direct (Hansen, 2003). In the RNR Simulation Tool synthetic data, offenders' education is considered a lifestyle indicator and is measured as a single variable indicating the highest level of education completed. This variable was treated as a scale item and entered into an ROC analysis. The AUC for the education scale was 0.518 ($p<0.001$), which indicates a weak but statistically significant relationship with recidivism. The education variable was created from the scale indicating that individuals with a GED or

Table 5.12 Available employment, housing, and education variables in SISFCF and SILJ data[a]

Variable name	Value	Score	%
Employment	Yes	0	53.6
Working full time prior to arrest	No	1	46.4
Housing	No	0	98.1
Homeless any time during year prior to arrest	Yes	1	1.9
Education	Yes	0	65.0
Received high school diploma	No	1	35.0

[a]Some items were not available in the SCPS data

Table 5.13 Family support variables in SISFCF and SILJ data[a]

Variable name	Value	Score	%
Received mail from children while incarcerated	No	0	56.7
	Yes	1	43.3
Received phone calls from children while incarcerated	No	0	61.7
	Yes	1	38.3
Have visits from children while incarcerated	No	0	72.9
	Yes	1	27.1
Received financial support from family member(s) prior to incarceration	No	0	87.5
	Yes	1	12.5
Highest possible total score		4	

[a]Some items were not available in the SCPS data

less than a high school diploma were categorized as having a destabilizer; offenders with at least a high school diploma are considered a stabilizer. Individuals with education destabilizer comprise 32.4% of the prison sample, 41.0% of the jail sample, and 41.0% of the community sample (Table 5.12).

Marital/Family Support

The relationship between marital and other social supports and offending behavior is rather unclear, even though Andrews and Bonta refer to it as a lesser criminogenic need. While some studies report that prosocial relationships help to reduce offending (Warr, 1993; Wright & Cullen, 2004), other research indicates that the effects of these relationships diminish when controlling for other factors (Sampson & Laub, 1992). The family support scale in the synthetic data includes three items measuring offender contact with their children while incarcerated through mail, phone calls, visits to the prison/jail, and financial support from family prior to incarceration (Table 5.13). The ROC analysis for this three item scale produced an AUC of 0.510 ($p < 0.05$), indicating that this variable has only a weak association with recidivism. Within the prison sample, 50.7% of the offenders are classified as having a lack of

Table 5.14 Available unstable lifestyle variables in SISFCF and SILJ data[a]

Variable name	Value	Score	%
Has been diagnosed with or treatment for a mental illness	No	0	73.3
	Yes	1	26.7
Has three or more antisocial indicators of associates	No	0	47.6
	Yes	1	52.4
Has less than full-time employment (<30 h per week, or	No	0	53.6
unemployed)	Yes	1	46.4
Has been homeless in the year prior to arrest	No	0	98.1
	Yes	1	1.9
Has less than a high school diploma (GED or no diploma)	No	0	65.0
	Yes	1	35.0
Had no contact with children while incarcerated and/or	No	0	79.0
received no financial support from relatives prior to incarceration	Yes	1	21.0
Highest possible total score		6	

[a]Some items were not available in the SCPS data

social supports. This rate was slightly higher in the community sample at 57.5% and in the jail sample (67.2%).

Unstable Lifestyle

The risk and needs model focuses on the "Big Four" which refers to antisocial behavior, antisocial values and attitudes, and low self-control. Presently, however, there is no national data on these criminogenic factors. In the absence of such data, a proxy variable was created to identify individuals with what we call a "criminal lifestyle." While factors such as mental illness, substance use, homelessness, and antisocial peers are not themselves criminogenic (each is discussed in detail above), the prevalence rates of these factors among offenders allow us to consider the presence of these variables as contributing to more serious behavioral problems. We use this variable to identify offenders that have a compilation of factors. We developed this as a proxy measure to identify offenders who are likely to have one or more criminogenic needs that we were not able to directly measure. None of the extant databases had variables to measure criminal attitude or variables. We refer this as a presence of multiple destabilizers. Table 5.14 lists the destabilizers used to create the lifestyle scale.

A point was assigned to each variable to create a scale that ranged from 0 to 6 with a mean of 1.91 (SD = 1.10). The ROC analysis indicated that the scale was a statistically significant predictor of recidivism with an AUC of 0.569 ($p < 0.000$). Based on this analysis, it was determined that an appropriate cutoff would distinguish individuals with three or more lifestyle indicators as having a criminogenic need. The prevalence rate of individuals with a lifestyle need was highest within the jail sample (33.2%) and lowest within the prison sample (23.2%). The rate within the community sample was 28.9%.

Validating the Model

We used two different strategies to validate each of these measures for the RNR Simulation Tool. As indicated earlier, we created a construction and validation sample of the extant data sets. That is, as discussed above, the original data sets were each cut in half so that we could develop the measures on one data set and then test the robustness on the same data sets. This was done within each step of constructing the measure. In every case, we were able to reconstruct the variable and received roughly the same distributional patterns. For example, the static criminal justice scale for the prison validation sample was 14.8% in low, 58.5% in moderate, and 26.7% in high. The same distribution rates for the validation sample were 14.7% in low, 58.0% in moderate, and 27.3% in high.

We also obtained data from other data sets from specific jurisdictions as part of the validation of this simulation model. We have three jurisdictions that provided data to the research team to assess individual parameters and the model overall. This process was used to (1) explore how the national data sets might compare with local data sets, (2) examine the construct validity of the key measures, and (3) assess how the model performs both within a jurisdiction and at the national level. The process begins with the key risk, need, and destabilizer measures, as seen in Table 5.15.

Table 5.15 Prevalence rates by risk, needs, and stabilizers/destabilizers in validation samples

Variable	Synthetic data %	Jurisdiction A %	Jurisdiction B %	Jurisdiction C %
Static risk				
High	23.7	23.7	10.5	20.7
Moderate	62.3	48.9	49.7	64.3
Low	14.0	27.4	39.8	14.9
Criminogenic needs				
Substance dependence[a]	9.0	25.3	14.8	37.7
Criminal lifestyle/thinking	26.2	22.7	25.5	12.3
Clinically relevant destabilizers				
Substance abuse[b]	71.3	24.9	–	65.4
Mental health	26.7	11.1	–	–
Indicators of lifestyle destabilizers				
Housing	1.9	18.3	10.1	23.9
Education	35.0	54.0	36.2	38.7
Employment/financial	46.4	15.0	23.4	46.8
Marital/social support	21.0	24.4	19.5	81.5
Antisocial associates	52.4	39.4	32.7	42.4

[a]Substance dependence is defined as dependence on a substance for which there is a clear drug-crime nexus such as cocaine, crack, heroin, methamphetamine, other amphetamines, and other opiates.
[b]Substance abuse refers to abuse of a substance for which there is a clear drug-crime nexus, such as cocaine, crack, heroin, methamphetamine, other amphetamines, and other opiates, or dependence on a substance for which there is no clear drug-crime nexus such as marijuana or alcohol.

Table 5.16 Three-year rearrest by demographic traits and data sample

Variables included in estimate	SISFCF (%)	SILJ (%)	SCPS (%)
Age 16–27			
Total	76.5	60.8	66.0
Male	78.4	61.3	66.1
Female	68.2	59.5	65.3
Age 16–27, male			
White	74.0	70.7	60.6
Black	81.7	79.7	69.2
Hispanic	76.9	69.3	64.4

Besides core risk, need, and stabilizer variables, the model uses key demographic variables to create profiles that are appropriate for each age, gender, and ethnicity to allow individual jurisdictions to obtain accurate estimates based on the specific demographic composition of their population. Table 5.16 provides an example of the variability associated with a single recidivism rate (3-year rearrest) by looking at different layers of demographic categories. As more variables are included, the estimated recidivism rate becomes even more specified. This allows the model to differentiate between categories of offenders that may be more or less prominent within a given jurisdiction. In this way, the RNR Simulation Model is able to provide highly specified recidivism estimates for jurisdiction-specific outputs.

Conclusion

In this chapter we defined how the measures of risk and need were developed for the RNR Simulation Tool. The goal of a synthetic data set is to approximate data that help us examine patterns of recidivism. The database used data from four existing sources to generate a nationally representative model of risk and needs. This model is designed to assist jurisdictions in estimating the risk and need levels of individuals within their population and of their population as a whole. It allows for each jurisdiction to enter population-specific information in order to provide the most accurate estimations possible; however, it also draws upon national estimates whenever jurisdiction-specific information is not available. The measures used in creating the RNR Simulation Tool and Model parameters are based on the solid theoretical framework of the RNR principles and a growing body of research supporting the importance of risk, need, and responsivity in reducing offending behaviors. Furthermore, data from a number of jurisdictions has been used to validate the model assumptions and data parameters. This chapter has described the assumptions and data elements of the RNR Simulation Tool and Model. Chapter 5 discusses how these data elements are used to inform responsivity decisions and recommendations, and Chap. 8 demonstrates how the data is used to provide national- and jurisdiction-specific estimations of risk, need, and recidivism. The use of national data sets allows us to create a mix of profiles, and the process of validation with data from specific jurisdictions adds to the profile mix by providing more information about risk-need profiles of offenders.

References

Akers, R. L., & Sellers, C. S. (2003). *Criminological theories: Introduction, evaluation, and application* (4th ed.). Los Angeles: Roxbury.

American Psychiatric Association. (2000). *Diagnostic and statistical manual of mental disorders DSM-IV-TR* (4th ed.). Washington, DC: APA.

Andrews, D. A., & Bonta, J. (2010). *The psychology of criminal conduct*. Burlington: Elsevier Science.

Andrews, D. A., Bonta, J., & Wormith, J. S. (2006). The recent past and near future of risk and/or need assessment. *Crime & Delinquency, 52*(1), 7–27.

Apel, R., & Sweeten, G. (2010). The impact of incarceration on employment during the transition to adulthood. *Social Problems, 57*(3), 448–479.

Austin, J. (2006). How much risk can we take—The misuse of risk assessment in corrections. *Federal Probation, 70*, 58.

Baillargeon, J., Binswanger, I. A., Penn, J. V., Williams, B. A., & Murray, O. J. (2009). Psychiatric disorders and repeat incarcerations: The revolving prison door. *The American Journal of Psychiatry, 166*(1), 103–109.

Bennett, T., Holloway, K., & Farrington, D. (2008). The statistical association between drug misuse and crime: A meta-analysis. *Aggression and Violent Behavior, 13*(2), 107–118.

Broner, N., Lang, M., & Behler, S. A. (2009). The effect of homelessness, housing type, functioning, and community reintegration supports on mental health court completion and recidivism. *Journal of Dual Diagnosis, 5*(3–4), 323–356.

Dal Pra, Z. (2004). In search of a risk instrument. In M. Thigpen & G. Keiser (Eds.), *Topics in community corrections: Assessment issues for managers* (pp. 9–12). Washington DC: Community Corrections Division, National Institute of Corrections, U.S. Department of Justice.

DeLisi, M. (2000). Who is more dangerous? Comparing the criminality of adult homeless and domiciled jail Inmates: A research note. *International Journal of Offender Therapy and Comparative Criminology, 44*(1), 59–69.

Glaser, D. (1987). Classification for risk. *Crime and Justice, 9*, 249–291.

Gottfredson, D. (1987). Prediction and classification in criminal justice decision making. *Crime and Justice: A review of research, 9*(1).

Gottfredson, M., & Hirschi, T. (1990). *A general theory of crime* (1st ed.). Stanford: Stanford University Press.

Hansen, K. (2003). Education and the crime-age profile. *British Journal of Criminology, 43*(1), 141–168.

Hoffman, P. B. (1983). Screening for risk: A revised salient factor score (SFS 81). *Journal of Criminal Justice, 11*(6), 539–547.

Landenberger, N. A., & Lipsey, M. W. (2005). The positive effects of cognitive-behavioral programs for offenders: A meta-analysis of factors associated with effective treatment. *Journal of Experimental Criminology, 1*(4), 451–476.

Laub, J. H., & Sampson, R. J. (2001). Understanding desistance from crime. *Crime and Justice: A Review of Research, 28*, 1.

LeCroy, C. W., Krysik, J., & Palumbo, D. (1998). *Empirical validation of the Arizona risk/needs instrument and assessment process*. Phoenix, AZ: Arizona Supreme Court, Administrative Office of the Courts, Juvenile Justice Services Division.

Lipsey, M. W., & Cullen, F. T. (2007). The effectiveness of correctional rehabilitation: a review of systematic reviews—Annual review of law and social science. *Annual Reviews of Law and Social Science, 3*(1), 297–320.

Lochner, L., & Moretti, E. (2004). The effect of education on crime: Evidence from prison inmates, arrests, and self-reports. *The American Economic Review, 94*(1), 155–189. Retrieved from http://www.jstor.org/stable/3592774.

Lowenkamp, C. T., & Latessa, E. J. (2005). Increasing the effectiveness of correctional programming through the risk principle: Identifying offenders for residential placement*. *Criminology & Public Policy, 4*(2), 263–290.

Lowenkamp, C. T., Latessa, E. J., & Holsinger, A. M. (2006). The risk principle in action: What have we learned from 13,676 offenders and 97 correctional programs? *Crime & Delinquency, 52*(1), 77–93.

Lowenkamp, C. T., Latessa, E. J., & Smith, P. (2006). Does correctional program quality really matter? The impact of adhering to the principles of effective intervention. *Criminology & Public Policy, 5*(3), 575–594.

Martin, M. S., Dorken, S. K., Wamboldt, A. D., & Wootten, S. E. (2012). Stopping the revolving door: A meta-analysis on the effectiveness of interventions for criminally involved individuals with major mental disorders. *Law and Human Behavior, 36*(1), 1–12.

Monti, P. M., Rohsenow, D. J., Michalec, E., Martin, R. A., & Abrams, D. B. (1997). Brief coping skills treatment for cocaine abuse: substance use outcomes at three months. *Addiction, 92*(12), 1717–1728.

Nurco, D. N., Hanlon, T. E., & Kinlock, T. W. (1991). Recent research on the relationship between illicit drug use and crime. *Behavioral Sciences & the Law, 9*(3), 221–242.

Sampson, R. J., & Laub, J. H. (1992). Crime and deviance in the life course. *Annual Review of Sociology, 18*(1), 63–84.

Substance Abuse and Mental Health Services Administration. (2011). *Results from the 2010 National survey on drug use and health: Summary of national findings*. Rockville, MD: Substance Abuse and Mental Health Services Administration, Center for Behavioral Health Statistics and Quality. NSDUH Series H-41 No. (SMA) 11-4658.

Taxman, F. S. (1998). *Reducing recidivism through a seamless system of care: Components of effective treatment, supervision, and transition services in the community*. Washington, DC: Office of National Drug Control Policy.

Taylor, B., Fitzgerald, N., Hunt, D., Reardon, J., & Brownstein, H. (2001). *Preliminary 2000 Findings on Drug Use and Drug Markets—Adult Male Arrestees*. Washington, D.C.: U.S. Department of Justice, Office of Justice Programs, National Institute of Justice.

Warr, M. (1993). Age, peers, and delinquency*. *Criminology, 31*(1), 17–40.

Wormith, J. S., & Olver, M. E. (2002). Offender treatment attrition and its relationship with risk, responsivity, and recidivism. *Criminal Justice and Behavior, 29*(4), 447–471.

Wright, J. P., & Cullen, F. T. (2004). Employment, peers, and life-course transitions. *Justice Quarterly, 21*(1), 183–205.

Chapter 6
The Responsivity Principle: Determining the Appropriate Program and Dosage to Match Risk and Needs

Erin L. Crites and Faye S. Taxman

Introduction

In this chapter, we describe the rationale for the matching of risk–need configurations to appropriate programming. The concept of responsivity, in the RNR model, is not well defined. The Andrews and Bonta RNR model suggests that individuals who are at higher risk should be targeted for more intensive interventions (risk principle) and interventions should address clearly identified needs associated with criminal behavior (needs principle) and should be consistent with individuals' abilities, gender, culture, and motivation (responsivity principle). The responsivity principle is the most vague component of the RNR model where the emphasis is on matching offenders to appropriate services to reduce the likelihood of recidivism.

The responsivity component, more so than the risk and need principles, requires a convergence of clinical practice and empirical evidence. While the original crafters of the model focus on programming, without attention to the facets of the justice system, it is important to recognize that these programs operate in the context of the criminal justice system. The implications are that (1) the programs can use the liberty restrictions imposed on individuals to create more intensive programs and structures and (2) the level of programming should address non-criminogenic components of a person's life, which contribute to negative treatment and justice outcomes, to stabilize the person and to enhance outcomes. Given that there are no agreed upon clinical standards regarding the appropriate placement of offenders into programs based on the RNR principles, it is necessary that such standards be discussed and tested.

One method for testing standards for program placement is simulation modeling. Simulation modeling can offer a number of insights to assist criminal justice agents in

E.L. Crites, M.A. (✉) • F.S. Taxman, Ph.D.
Department of Criminology, Law and Society, George Mason University,
10900 University Boulevard, Fairfax, VA 20110, USA
e-mail: ecrits@gmu.edu; ftaxman@gmu.edu

F.S. Taxman and A. Pattavina (eds.), *Simulation Strategies to Reduce Recidivism:*   143
Risk Need Responsivity (RNR) Modeling for the Criminal Justice System,
DOI 10.1007/978-1-4614-6188-3_6, © Springer Science+Business Media New York 2013

determining what types of programming individuals' need and the capacity of the jurisdiction to provide this programming. In order to create a simulation model for responsivity that can test the effects of using clinical standards for placing justice-involved persons into appropriate programming, a number of different types of information and data must be identified. First, developing clinical standards requires identifying individual-level characteristics that are related to risk and needs that affect the likelihood of engaging in criminal behavior. These individual-level factors can be assessed using a number of validated risk and needs assessments. Many of these assessments' items can be matched to the profile features used in this simulation model (see Chaps. 4 and 5). Second, we can identify program features that specifically target these risk and needs, leading to the best possible chance for reducing individual factors that increase the likelihood of recidivism. These features help to identify the types of needs the program targets and whom the program might best serve, based on its targets, content, dosage, and implementation. With the limited resources available to criminal justice agencies, matching individuals to the programs that best meet their risk and needs, having the highest likelihood of reducing recidivism, is an important goal. Simulation modeling can be used to evaluate a set of clinical standards, or decision rules, allowing for the best use of limited resources. Additionally, simulation modeling can also help to identify gaps in program availability.

This chapter will detail the process used to develop two components of a simulation model for responsivity—individual program-group assignment and classification criteria for programs. First, some of the foundational literature guiding these two features of the model will be discussed. Next, the process used to develop placement criteria that incorporates the risk level and needs of individuals is described. Then, the process for developing the key criteria of programs associated with each type of care is discussed. To validate this model, we used the synthetic data (described in Chap. 5) and data from a number of state and local corrections agencies to test the assumptions made for classifying individuals and programs into groups. The chapter will conclude with a discussion of the implications of this simulation model of responsivity.

Background

Responsivity is defined as matching the correct type (behavioral target and intensity) of programming to an individual based on his or her risk and needs profile. The needs include criminogenic factors, in addition to stabilizers and destabilizers that affect the overall functioning of the individual. This differs from the original definition used by Andrews and Bonta (2010) who define responsivity as "delivering treatment in a style and mode that is consistent with the ability and learning style of the offender" (p. 49). Since learning style and ability are difficult to measure and the availability of programs for incarcerated individuals and people under community corrections is limited (Taxman, Young, Wiersema, Mitchell, & Rhodes, 2007), we defined the concept to fit within the risk and need framework.

Using this alternative definition of responsivity, program placement will differ slightly from the recommendations of Andrews and Bonta (2010) in that individuals will be matched to specific programs (within a group) based on what the programs target, not on whether the method of treatment matches the individual's learning style. The definition used is similar to the concept of treatment matching in the substance abuse and mental health treatment fields (see Mee-Lee, Shulman, Fishman, Gastfriend, & Griffith for the American Society of Addiction Medicine (ASAM) Patient Placement Criteria (PPC), 2001; American Association of Community Psychiatrists (AACP) Level of Care Utilization for Psychiatric and Addiction (LOCUS), 2009; Gastfriend & McLellan, 1997; Thornton, Gottheil, Weinstein, & Kerachsky, 1998). In substance abuse and mental health fields, the emphasis is on the content, dosage, implementation fidelity, and restrictiveness of treatment in response to the risk (for relapse or continued symptoms) and needs (supportive recovery environment or triggers for substance use). Identifying the appropriate amount and targets of services an individual could receive to reduce the likelihood of recidivism is an important concern for criminal justice professionals. To match individuals to appropriate services, it is important to identify the profiles of individuals most likely to need a certain type and intensity of services and to identify which services can address the risk and needs of these individual profiles.

The Notion of Individual Program Placement

Treatment matching is conducted on the basis of agreed upon clinical criteria for assigning people to programs and services at a level of intensity that best meets the specific needs of the patient. ASAM PPC recommends that treatment should occur in the least restrictive environment available where an individual's highest level of need can be addressed (ASAM, 2001). The highest severity problem should determine the patient's initial placement (in this case hospitalization vs. outpatient), and then the individual can be transferred to a lesser level of care as he or she improves.

A number of mental health and substance abuse treatment matching protocols can be guides for developing placement criteria for justice-involved persons. They provide a foundation for examining how to consider problem severity and type of interventions; the drawback of these approaches is that they tend to be focused on one problem behavior (such as mental health or substance abuse), while justice-involved individuals tend to have multidimensional deficits in intertwining areas (such as cognitive, social, and interpersonal). The LOCUS, developed by the AACP, has six dimensions on which individuals are evaluated—risk of harm, functional status, comorbidity, recovery environment (level of stress and level of support), treatment and recovery history, and engagement and recovery status. These six dimensions feed into six treatment levels, ranging from recovery maintenance (level 1) to medically managed residential services (level 6). For example, an individual placed into a recovery maintenance program would be classified as low risk of harm, only minor impairment in interpersonal interactions or ability to care for oneself, minor medical or mental health concerns, mildly stressful but supportive recovery environment, history of responding positively to treatment, and positive attitude towards recovery.

ASAM, for addiction disorders, includes six dimensions—acute intoxication/withdrawal; biomedical conditions and complications; emotional, behavioral, or cognitive conditions and complications; readiness to change; relapse, continued use, or continued problem potential; and recovery environment—which feed into four treatment levels ranging from outpatient services (level 1) to medically managed intensive inpatient services (level 4). An individual would be assigned to intensive outpatient treatment (level 2) if he or she was diagnosed with a substance use disorder based on DSM-IV-TR criteria, shows no symptoms of withdrawal, has no biomedical conditions that would interfere with treatment, has no cognitive or emotional disorders, is willing to participate in treatment but not yet ready to maintain treatment progress outside of a structured environment, and shows signs of a reduced level of functioning or increased drug-seeking behavior. Criminal justice settings, recommendations, and placement into treatment programs may not always fit these clinical criteria. ASAM (2001) recognized that court-ordered participation in a treatment program may result in a referral ill-matched to the needs of the individual and encourages practitioners to seek to evaluate, and when necessary amend, the order into treatment to better target an individual's needs.

Based on LOCUS and ASAM, the key elements for assessing individuals to identify the appropriate level of care include risk for recidivism based on history (similar to risk of harm), criminogenic needs (similar to functional status), clinical destabilizers (comorbidity), and lifestyle destabilizers and stabilizers (similar to recovery environment). These factors can all be identified using available assessments and can be targeted with programming to reduce the influence of these factors on the individuals' likelihood of committing future crimes.

Program Classification

Classifying programs into group is another component of responsivity. The same technique used in the LOCUS and ASAM (discussed above) for identifying the appropriate level of treatment for an individual can be used to classify programs. LOCUS considers care environment, clinical services, supportive services, and crisis stabilization and prevention services as key components of treatment programs for substance use disorders. ASAM considers setting, support systems, staff, type of interventions included, level of assessment used to define eligibility, and degree of documentation patient's progress. These tools identify the defining features of treatment at each level of care to meet the needs of individuals placed into that level. While the ASAM and the LOCUS focus predominately on drug and alcohol use disorders, criminal justice programs must address a number of issues beyond these, and therefore, the defining features of different groups of programming within the criminal justice system are slightly different from those in the ASAM and LOCUS.

Using a similar strategy, we identified key programming features for criminal justice-involved individuals. These features are grouped into four major categories—target, content, dosage, and implementation quality—and will be discussed in detail below. Within each of these categories, there are "essential features" necessary for determining which of these groups a program fits into and "additional features" which help to further refine the program classification. We have developed an RNR Program Tool, based on these essential and additional features, as a resource for assisting programs in determining which group of individuals a program or service is best suited for treating. This is a web-based tool that asks a series of questions starting with the target population, primary target behavior the program is designed to address, and program content, dosage, and implementation. Based on responses to these items, programs are placed into six groups. The tool is organized along the four major domains discussed below.

Target: For the justice-involved person dealing with multidimensional problem areas ranging from acute to no behavioral health issues and other social and interpersonal dysfunctions, there is a need for programs that can deal with this diversity. Considering that needs should be defined around the severity of disorder and other social issues identified using individual assessments, program placement should focus on the most debilitating issue faced by the individual. Therefore in this conceptualization, six primary targets of programming aimed at addressing the individual's most pressing problem are identified:

1. Dependence on "hard" drugs—heroin, cocaine, amphetamines, and methamphetamine where the linkage between criminal behavior and drug use is clearer (Holloway, Bennett, & Farrington, 2006). Programming for dependence on these highly addictive drugs should occur before other issues, such as criminal thinking or social skills, are addressed. Individuals who use these substances and engage in criminal acts on a consistent basis can benefit from treatment targeting their drug use yielding both reduced drug use and criminal behavior (Holloway et al., 2006; Prendergast, Huang, & Hser, 2008).
2. Criminal thinking/Cognitive restructuring—Big 4 criminogenic needs—history of antisocial behavior, antisocial personality pattern, antisocial associates, and antisocial cognitions (Andrews & Bonta, 2010). Criminal thinking patterns drive how individuals interact with others and are strongly correlated with continued criminal behavior. Adjusting these patterns by increasing self-control, reducing antisocial thinking, and increasing prosocial connections provides a foundation for improved functioning and reducing future criminal behavior (Andrews & Bonta, 2010).
3. Self-improvement and self-management—substance abuse, family issues, and mental health. Research with juveniles suggests that programs that increase social competence, problem solving, and self-control were associated with fewer problem behaviors in young people (Ang & Hughs, 2001). Increasing social problem solving skills can help individuals resist social pressures to become

involved in undesirable behaviors including drug use, delinquency, and crime (Botvin, Griffin, & Nichols, 2006; Botvin & Wills, 1984).

4. Social and interpersonal skills—family issues, relationships, etc. Family and marital circumstances are one of Andrews and Bonta's (2010) "Moderate Four" dynamic needs. Improving relationships by reducing interpersonal conflict and developing more positive relationships through structured counseling where clinicians model appropriate behavior can be effective at improving relationships and reducing criminal offending (Andrews & Bonta, 2010).

5. Physical/life needs—employment, education, and housing. School and work are also part of the Moderate Four dynamic needs described by Andrews and Bonta. When an individual has been deeply involved in a criminal lifestyle, stressors such as inability to find employment, low education, and unstable housing can make going back to a criminal lifestyle more appealing or make it more difficult to maintain a crime-free lifestyle.

6. Punishment only—this is reserved for low-risk, low-need individual for whom none of the above targets are indicated.

These program groups are designed to reflect the differing dynamic needs of individuals within the criminal justice system, regardless of setting (e.g., community, jail, prison). Many individuals with drug dependence or substantial criminal thinking patterns also experience difficulties with social, interpersonal, and life skills. Therefore, these program groups represent a continuum of care, where each of these intermediate targets can be addressed by programming with the ultimate goal of reducing recidivism (Andrews & Bonta, 2010; Andrews, Dowden, & Gendreau, 1999; Lipsey & Cullen, 2007). Of utmost importance is ensuring that the target of programming reflects the most serious need of an individual.

Dosage: The dosage of a treatment affects the likelihood of longer-term positive outcomes. In this model, program dosage is designed to differentiate between treatment activities that occur infrequently from those sustained over a period of time and includes the frequency of treatment or program sessions, length and duration of sessions, and the availability and duration of aftercare as recommended by Huber, Hall, and Vaughn (2001). Each of these four components should be measured to accurately account for dosage in behavioral (or nonmedical/pharmacological) interventions. Programs that meet more frequently, have a greater amount of treatment per meeting, and have a longer duration are considered high dosage.

References to specific treatment dosage in correctional interventions are limited. Dosage measures are still applicable to correctional treatment programs, and information regarding the appropriate dosage of behavioral interventions can be inferred from substance abuse and mental health treatment studies (e.g., Bourgon & Armstrong, 2005; Lipsey & Landenberger, 2005; Simpson, 1979; Taxman, Byrne & Thanner, 2002; U.S. Department of Health and Human Services, 2009). For example, Lipsey and Landenberger (2005) in a meta-analysis of cognitive–behavioral therapy (CBT) programs for offenders found that the number of sessions of CBT per week, the number of hours per week, and the overall number of hours of treatment received were positively related to the effect size for the CBT intervention on

---

Box 6.1   Characteristics of Dosage
_____

Dosage: How much treatment is an individual receiving?
1. Amount—total number of clinical hours
2. Duration—number of weeks
3. Frequency—number of times per week
4. Quantity—number of hours per week

---

recidivism (defined as rearrest or reconviction). Similarly in a study of "comprehensive structured cognitive behavioral programs," Bourgon and Armstrong (2005) found a decrease in recidivism ranging from 1.2 to 1.7 % for each week (equating to about 20 h) of treatment received after controlling for risk and need. In the Washington–Baltimore HIDTA (discussed in Taxman et al., 2002), the highest level of placement provided 20–30 h per week of therapy in predominately residential settings for 30–90 days. The next level included 20 h per week, and the final outpatient or aftercare level involved treatment on a monthly basis for approximately 6 months each. They found reductions in recidivism for higher-risk offenders (Thanner & Taxman, 2003). In a review of studies of psychotherapy (typically CBT) for patients with psychiatric disorders, Hansen, Lambert, and Forman (2002) found that between 13 and 18 sessions (approximately one session per week for 3–5 months) were usually needed in order to see a 50 % reduction in symptoms. The National Institute of Corrections recommends that higher-risk offenders should spend 40–70 % of their time for 3–9 months in scheduled programs (National Institute of Corrections [NIC], 2005).

The dose of treatment required to gain clinical or statistically significant change may also be affected by the type and degree of needs. In the psychotherapy literature, Kopta, Howard, Lowry, and Beutler (1994) found that individuals with more acute symptoms responded to shorter lengths of treatment (five sessions) compared to those with more "characterological" symptoms requiring 104 sessions to see a 50 % reduction in clinical symptoms. Bourgon and Armstrong (2005) found that offenders scoring low on risk assessments participating in 5 weeks' worth of programming had lower recidivism rates (12 %) than offenders who scored high on the risk assessment in the same 5-week program (62 %). An emerging literature in the criminal justice system suggests that higher-risk individuals should receive at least 300 h of cognitive–behavioral interventions, moderate-risk individuals about 200 h, and lower risk less than 100 h (Bourgon & Armstrong, 2005). Box 6.1 provides a summary of each component of dosage.

Content: Treatment and services are about the content of the program. The content of the program is based on the treatment orientation, the primary treatment focus, relevant additional services or components, and program tools such as rewards and punishments to reinforce compliance. Programs that target criminogenic factors, such as antisocial thinking or peers and substance dependence, have been shown to have larger impacts of reducing recidivism and are therefore weighted more heavily in the content domain than programs that address non-criminogenic factors, such as housing, education, or employment.

As discussed in Chap. 7, the synthesis research (meta-analyses and systematic reviews) identifies programs and treatment services that are effective at either reducing recidivism or addressing needs that influence individuals' abilities to abstain from engaging in criminal activity. For example, cognitive–behavioral interventions that target criminal thinking or substance dependence are especially effective in reducing recidivism (Aos, Miller, & Drake, 2006a; Bouffard & MacKenzie, 2005; Lipsey, Landenberger, & Wilson, 2007; Mitchell & MacKenzie, 2007; Pearson, Lipton, Cleland, & Yee, 2002). Therapeutic communities are also an effective method for targeting drug dependence (Aos et al., 2006a; Mitchell & MacKenzie, 2007). Other content (i.e., intensive supervised probation (ISP) without treatment, boot camps, GED programs) has less demonstrated effectiveness at reducing recidivism. Meta-analyses on ISP have consistently found limited reductions in recidivism (Drake, Aos, & Miller, 2009). When treatment is provided in addition to supervision, the effects on recidivism reductions are much better—nearly 18 % (Drake et al., 2009). Supervision based on the RNR model was found to produce a 16 % reduction in recidivism (Drake, 2011). Mitchell and MacKenzie (2007) found boot camps to have a limited effect on recidivism. While more is known about treatment focused on substance use and dependence, there is little research available demonstrating the effectiveness of treatment modalities on the Big Four criminogenic needs (antisocial attitudes, personality pattern, behavior, and peers). Table 6.1 provides a summary of the result of this synthesis research for interventions targeting justice-involved persons.

Social controls in programs are also useful to enhance the impact of the content and dosage of programs. Social controls are tools used by criminal justice agents to assist in stabilizing, monitoring, and supporting offenders in their efforts to change negative behaviors. Restrictions on liberty, therefore, can serve both a punitive and therapeutic purpose. Constraining the movement of individuals through the use of electronic monitoring, curfews, or day programs can provide the structure and reduce access to substances necessary for some clients to succeed on supervision (Drake et al., 2009; Padgett, Bales, & Blomberg, 2006; Pattavina, Tusinski-Miofsky, & Byrne, 2009). Barber and Wright (2010) in a study of batterer intervention programs found that those offenders under increased monitoring (check-ins, court orders, drug testing, and court appearances) were more likely to complete the program. While restrictions on individual freedoms can be punitive in response to non-compliance, they can also be used to facilitate treatment. As important as the content of correctional programs and interventions is, how well the programs and services are implemented impacts the effects on recidivism and other behaviors one can expect to see.

Implementation Fidelity: Not all programs are implemented as designed. Even with the best of intentions, organizational factors may negatively affect program delivery. Program implementation fidelity is linked to effectiveness (Andrews &

Table 6.1 Program content and recidivism reduction[f]

Intervention	% reduction	Source(s)
Drug dependence		
Therapeutic community	16–27	Lipton, Pearson, Cleland, and Yee (2008); Mitchell and MacKenzie (2007)
Therapeutic community (hard drugs)	45	Holloway et al. (2006)
TC (inpatient <90 days)	7	Mitchell and MacKenzie (2007)
TC (inpatient 90+ days)	18	Mitchell and MacKenzie (2007)
TC (no aftercare)	13	Mitchell and MacKenzie (2007)
TC (with aftercare)	20	Mitchell and MacKenzie (2007)
Narcotic maintenance (hard drugs)	27[c]	Holloway et al. (2006)
Narcotic maintenance	−9[e]	Mitchell and MacKenzie (2007)
Criminal thinking		
Cognitive–behavioral therapy	25	Lipsey, Landenberger, and Wilson (2007)
Moral Reconation Therapy	16[c]–35	Little (2005), Bouffard and MacKenzie (2005)
Reasoning and rehabilitation	14	Tong and Farrington (2006), Bouffard and MacKenzie (2005)
CBT for anger management	51	Beck and Fernandez (1998)
ISP with treatment orientation	17.9	Drake et al. (2009)
Intensive supervision program	33[c]	Perry, Coulton, Glanville, Godfrey, Lunn, McDougall, and Neale (1996)
EM for moderate to high risk	2[c]	Renzema and Mayo-Wilson (2005)
Social/interpersonal skills		
General drug treatment	12[c]–22[c]	Holloway et al. (2006); Prendergast, Poduc, Chang, and Urada (2002)
Counseling (general)	20	Mitchell and MacKenzie (2007)
Counseling (<90 days)	22	Mitchell and MacKenzie (2007)
Counseling (90+ days)	18	Mitchell and MacKenzie (2007)
Counseling (no aftercare)	18	Mitchell and MacKenzie (2007)
Counseling (with aftercare)	29	Mitchell and MacKenzie (2007)
Restorative justice	14[a]	Latimer, Dowden, and Muise (2005)
Post-release supervision	26[d]	Dowden, Antonowicz, and Andrews (2003)
Post-release supervision (hard drugs)	33[c]	Holloway et al. (2006)
Mental health treatment	17[c]	Martin, Dorken, Wamboldt, and Wootten (2012)
Incarceration (vs. community)	−14[c]	Smith, Goggin, and Gendreau (2002)
Intermediate sanctions	2	Smith, Goggin, and Gendreau (2002)
Boot camp	1	Wilson, MacKenzie, and Mitchell (2008)
Boot camp	5	Mitchell and MacKenzie (2007)
Life skills		
Counseling (12 steps)	21	Mitchell and MacKenzie (2007)
General vocation/education	21	Wilson, Gallagher, and MacKenzie (2000)
Ex-offender employment	3[c]	Visher, Winterfield, and Coggeshall (2005)
Academic/educational	18	Wilson, Gallagher, and MacKenzie (2000)

(continued)

Table 6.1 (continued)

Intervention	% reduction	Source(s)
Postsecondary correctional education	27	Wilson, Gallagher, and MacKenzie (2000)
Vocational	22	Wilson, Gallagher and MacKenzie (2000)
Life skills training	27	Beckmeyer (2006)
Correctional industries	19	Wilson, Gallagher, and MacKenzie (2000)

[a]Standardized mean difference was converted to odds ratio. Phi coefficient was converted to an odds ratio with an assumed 0.50 control recidivism. Success/failure rates for treatment and control groups were used to calculate odds ratio
[b]Insufficient information to calculate confidence interval
[c]Calculation assumed 0.50 control recidivism
[d]Treatment and control group recidivism rates were converted to percent reduction
[e] Negative percent reductions represent an increase in recidivism for the treatment group
[f]Ainsworth, S. A. & Caudy, M. (2012). *Correctional Interventions*. EMTAP Review Series. Fairfax, VA: Center for Advancing Correctional Excellence (ACE!); Department of Criminology, Law & Society; George Mason University

Dowden, 2005; Lipsey & Landenberger, 2005). Lipsey and Landenberger (2005) in their meta-analysis of factors related to effective cognitive–behavioral programs found greater implementation fidelity associated with larger treatment effects ($\beta = 0.40$; $p < 0.05$). Likewise, in a meta-analysis of 273 studies, Andrews and Dowden (2005) found that when clinically appropriate treatment was being delivered, effect sizes were significantly greater in programs that had components of program integrity—adherence to a specific treatment model, staff possessed general interpersonal skills, staff were trained in the delivery of a specific program, workers received clinical supervision from an individual trained in program delivery, and an evaluator was involved in program design or delivery—than those that did not include elements of treatment integrity ($R^2 = 0.229$; $p < 0.001$). Lipsey (2009) in a meta-analysis of juvenile justice interventions found the quality of implementation to be significantly related to the effect size of counseling ($\beta = 0.13$; $p < 0.10$), skill building ($\beta = 0.25$; $p < 0.05$), and combined interventions ($\beta = 0.18$; $p < 0.05$). In the field of mental health services programs, McGrew, Bond, Dietzen, and Salyers (1994) found that programs that scored higher on program integrity (i.e., fidelity to a specified program model) measures yielded more positive benefits (measured as fewer hospital days) from program participation ($r = 0.60$; $p < 0.01$).

The training and certification of staff is an important tool for ensuring program implementation fidelity. Training staff in a specific treatment protocol can improve patient outcomes (Simons et al., 2010; Stanard, 1999). Stanard (1999) trained case managers working with severely mentally ill clients in a strength-based case management technique. Case managers who received this training had clients who reported higher quality of life and improved vocational/educational outcomes compared to those case managers who did not receive the training. Simons and colleagues (2010) provided training to clinicians in cognitive–behavioral interventions for clients with depression. Patient outcomes for a period prior to the training were

Box 6.2 Program Implementation Features

Measures of implementation captured in the RNR Program Categorization Tool include:

1. Staff training and certification
2. Adherence to a treatment agenda
3. Program eligibility criteria
4. Quality assurance measures
5. Communication between providers
6. Technical assistance
7. Use of coaching

compared with outcomes of different patients after the clinicians had received the training. Clients who received treatment following the CBT training had greater improvement in depression and anxiety symptoms compared to those clients treated prior to the training ($F(2,113) = 53.40$; $p < 0.001$).

Program implementation fidelity also includes the assignment of the appropriate individuals to appropriate treatment. The use of actuarial risk and needs assessments, as compared to professional judgment, when making predictions of behavior (Andrews, Bonta, & Wormith, 2006) leads to more accurate assessments, and agencies that use these assessments for placement into programs have seen a greater impact on recidivism (Lowenkamp, 2004).

Use of a manualized treatment protocol also improves program integrity. Mann (2009) suggests that manualized treatment encourages consistency and structure and can still maintain therapeutic relationships. He argues that within the field of psychotherapy, manualized treatments are associated with better outcomes than their fully individualized counterparts. In addition to the presence of a manual for treatment, it is essential that supervisors ensure that providers are adhering the specifications in the manual. He suggests that manualized treatment improves outcomes, increases integrity, facilitates evaluations, maintains clinician focus on goals, and promotes empirically based interventions. See Box 6.2 for a summary of the characteristics of program implementation.

A final feature of implementation is staff communication. It is important for treatment and correctional staff to communicate with one another regarding the progress of individuals assigned to programming. Communication is essential for coordinating care between systems, such as community providers and those located within facilities. However, Taxman and Bouffard (2000) found that many correctional facilities and treatment providers did not have communication structures in place to move beyond organizational boundaries. Fletcher et al. (2009) identified a number of features of low and high coordination and collaboration. Features of coordination and collaboration include sharing information, having similar eligibility requirements, joint staff meetings, joint case reporting, written protocols for information sharing, and sharing manuals and budgeting.

The Programming Groups

As presented above, a number contribute to the classification of programs into groups for the purpose of developing a simulation model for responsivity. To summarize these domains, program target refers to the most pressing need addressed by the program; dosage focuses on how much of an intervention is delivered and is measured by duration, session frequency, and amount of programming per week; content measures the primary treatment orientation, secondary program components, restrictions on individual movement, drug testing, and the use of rewards or sanctions; and implementation fidelity is an assessment of program implementation quality measured by staff certification, program evaluation, use of a research manual, and eligibility criteria. The six program groups are as follows:

Group A: Treatment focuses on addressing dependence on hard drugs, but also includes cognitive restructuring techniques for criminal thinking to strengthen cognitive processing and decision making, as well as, interpersonal and social skills interventions to target group B and C issues. These programs target predominately high and moderate risk offenders (although some low risk with clear dependence on hard drugs may also be included), have a dosage of approximately 300 clinical hours, and are implemented by staff with advanced degrees using and evidence-based treatment manual.

Group B: Programs focus on criminal thinking using cognitive restructuring techniques but also include interpersonal and social skills interventions. These programs target predominately high-risk offenders, have a dosage of approximately 300 clinical hours, and are implemented by staff with college degrees in related fields using an evidence-based treatment manual.

Group C: Programs focus on self-improvement and self-management, especially problem solving and self-control related to mental health disorders and substance abuse. It also includes some cognitive restructuring work to address developing criminal thinking patterns. These programs target predominately moderate-risk offenders, have a dosage of approximately 200 clinical hours, and are implemented by staff who are certified in the programs' evidence-based curriculum.

Group D: Programs focus on social skills and interpersonal skills targeting multiple destabilizing issues. These programs target moderate- and low-risk offenders, should aim for dosage under 200 h, and are implemented by staff possessing generic certifications (e.g., PO, CO) using an internally generated treatment manual.

Group E: These programs focus on life skills such as financial stability, occupational training, or education, target predominately low risk individuals, have a dosage of about 100 hours, and are implemented by staff with relevant experience using an internally generated treatment manual.

Group F: Few to no restrictions on behavior, punishment only, with programming/services as needed.

Attaining Responsivity

To attain responsivity, there must be a match between individuals and available programs. Once an individual has been assigned to a program group based on his or her risk and needs and after a program has been categorized, placement of participants into programming should occur based on these groups. The assignment of individuals based on risk and needs to programs accounting for the program's intensity and expected effectiveness has implications for cost and recidivism, as individuals who are placed in the best program according to their needs see better results (Andrews & Bonta, 2010; Andrews, Bonta, & Hoge, 1990; Thanner & Taxman, 2003).

Components of a Responsivity in a Simulation Model

To identify the gap in responsivity, individuals and programs must first be classified into groups. The following sections will detail the processes for identifying individual program-group assignment and for classifying programs into groups based on key program features. The result of this process is the assignment of an individual to one of six program groups and the classification of programs into one of six groups. Individuals can then be matched to a program within the appropriate group to meet the risk and needs of the individual offender. We first tested this process for individual classification in a synthetic dataset created using BJS Survey of Inmates data and the State Court Processing Statistics data (see Chap. 5).

Process for Classifying Individuals

Responsivity, based on a treatment matching approach, requires that individuals' risk and needs be considered when making programming and supervision decisions (see Fig. 6.1). In this model, individuals are placed into groups based on risk and need "profiles." These profiles were first defined and identified in the synthetic database discussed in Chap. 4. Profiles were created that account for risk level, primary need (dependence on hard drugs or criminal thinking), clinical destabilizers (substance abuse and mental health), lifestyle destabilizers (lack of social support, low education, low employment, unstable housing, financial instability, lack of family support, and associating with criminal/antisocial peers), and stabilizers (the inverse of the lifestyle destabilizers). Lack of social support is determined by whether an individual has people (including friends, family, clergy) on whom he or she can depend when life circumstances are difficult. Low education is defined as having less than a high school diploma. Unstable or part-time employment (less than 30 h per week) is considered a destabilizer. An individual with temporary housing (i.e., living in a shelter, being homeless, or moving from family to friend's houses

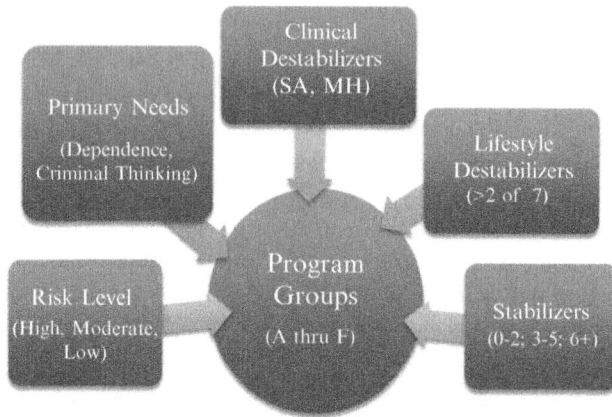

Fig. 6.1 Components of individual program-group placement. *Note*: For details on how these components are coded into groups in the data, please contact the author

regularly) is considered to have unstable housing. Financial instability is defined as the individual being unable to meet his or her financial obligations, including those required by the court. Individuals lacking family who support their prosocial lifestyle have a destabilizer in the area of family support. Finally, associating with individuals who engage in criminal behavior or possess attitudes supportive of crime or delinquency is also a destabilizer.

Decision rules for each combination of these components were created and are used to place individuals into programming groups. Individuals with dependence on hard drugs are placed into Group A programming regardless of risk level, destabilizers, or stabilizers. This is because the programming in this group for the individual is suitable for drug dependence behavior regardless of other factors. Low-risk individuals with no primary needs and no destabilizers and who have stabilizers are placed into Group F. Other combinations of risk levels, needs, destabilizers, and stabilizers are placed into groups using a detailed decision matrix.[1]

A Tool to Classify Programs

To assist in classifying programs, we developed a web-based tool that can be completed by program administrators. The first step in classifying programs into groups was to identify key features of programs that would define each of the program groups. These features were based on the four domains discussed above (target, content, dosage, and implementation fidelity). This tool has three main purposes: (1) asses programs on their use of evidence-based practice, (2) determine how well

---

[1] Copies of the decision matrix can be obtained by contacting the authors.

programs currently target risk and criminogenic needs, and (3) classify programs so they can be matched to individuals. Each domain of the tool is measured through a series of questions based on the features of the domain discussed above in the section on the RNR Program Tool.[2]

The user of the tool is asked to identify what the main target of the program is—drug dependence, criminal thinking, self-improvement and management skills, social and interpersonal skills, life skills, or punishment only. Based on the selected target area, programs are placed into an initial group classification. In addition to the target, five other items are included in the "essential features" of programming used in program classification—content, number of clinical hours, risk level targeted, use of a treatment manual, and staff certifications. These features define whether a program is operating at the level appropriate for the target (see Table 6.2). Additional features such as the duration and frequency of programming, other content, restrictiveness, evaluation, and method for screening participants for eligibility are used to rank the program within each group.

The program dosage component (discussed above) is designed to differentiate between treatment that occurs with low frequency and treatment that is sustained over a period of time and includes the frequency of treatment or program sessions, duration, aftercare, and amount of treatment hours per week. Program content reflects the program's treatment orientation supported by empirical studies (e.g., CBT), the presence of a primary focus for the program (e.g., substance dependence, criminal thinking), additional services or components, and the use of rewards and punishments as incentives for participation. Identifying the program staff's training and certification, program eligibility criteria, program adherence to a treatment curriculum, attention to quality assurance, and additional liberty restrictions define the program's implementation characteristics (see Table 6.3). The output of the tool is the classification of programs into one of six groups. Content, dosage, and implementation fidelity drive the program-group classification, and each program group has some typical features.

These typical features of the four domains of the tool are divided into six scoring categories: risk principle, need principle, responsivity principle, dosage, program integrity, and restrictiveness. These six categories are weighted based on their importance identified in the research literature about effective correctional interventions and add up to a total score. Programs are initially classified into a group based on their target (substance dependence, criminal thinking, self-improvement and management, interpersonal and social skills, life skills, and punishment only). Then items are scored relative to their assigned group on each of the six categories. All final scores in each category reflect a proportion of the number of points received out of the total number of points possible. The final score on the tool also represents a proportion of the total number of points earned out of the total possible.

For example, a medication-assisted treatment (MAT) for opiate dependence is assigned to program Group A because it targets dependence on opiates/opioids and

---

[2]The RNR Program Tool will be housed at http://www.gmuace.org/ beginning late fall of 2012.

Table 6.2 Essential features of program group[a]

	Group A	Group B	Group C	Group D	Group E
Targets	Drug dependence	Criminal thinking	Social skills	Interpersonal skills	Life skills
Content	TC	CBT-based criminal thinking interventions	CBT-based social skills interventions	Group, individual, family counseling	GED classes
	CBT-based interventions	Behavioral interventions (contingency contracting, token economies)	Behavioral interventions	Mentoring	Vocational skills training
	Medication-assisted treatments	ISP with treatment	ISP without treatment	Self-help groups (e.g., AA/NA)	ISP without treatment
	Manualized treatment	Sex, DV, and restorative justice curriculums	Individual counseling	DV education/awareness	Employment services
	Residential treatments		Group counseling	Social connections (incl. faith-based) programs)	Educational classes
					Access to support and entitlement services
Dosage (desired clinical hours)	300 h	300 h	200 h	200 h	100 h
Eligibility screening	High risk	High to moderate risk	Moderate risk	Moderate to low risk	Low risk
Manual	Evidence-based	Evidence-based	Evidence-based or internally generated	Internally generated	Internally generated
Staff credentials	Advanced degree	College degree	Certification in curriculum	Generic certification	Experience

[a]Group F has been removed from this table to improve visibility. Group F is punishment only and targets low-risk individuals

Table 6.3 Additional program-group features

		Group A	Group B	Group C	Group D	Group E
Content	Targets	Drug dependence	Criminal thinking	Social skills	Interpersonal skills	Life skills
Dosage	Tx length	Minimum of 120 days for cognitive–behavioral, longer duration for medications	180–365 days	120–180 days	60–120 days	30–60 days
	Frequency	Daily	Daily	Multiple times per week	Weekly	Weekly
	Hours per week	20+	15–19	10–14	5–9	1–4
Implementation quality	Eligibility screening	Target specific assessments for drug dependence	Target specific assessment for criminal thinking	Target specific assessment	Risk–needs assessment	Risk–needs assessment
	Completion criteria	Must successfully complete all treatment elements/phases	Must participate all treatment elements/phases	Time in treatment without relapse/new offense	Time in treatment	Time in treatment
	Quality assurance	External audit	Internal quality improvement review	Supervisor case review	Staff peer review	Case review
	Communication	Sharing of records between supervision and treatment staff	Weekly contact through progress reports or phone/email updates	Biweekly contact through progress reports or phone/email updates	Monthly progress reports from program staff to supervision staff	Contact between supervision and treatment staff as needed
	Drug testing	Random–multiple times per week	Random–multiple times per month	Random–once per month	As needed	None
Additional restrictions	Liberty restrictions	Residential requirements or areas restrictions	GPS/electronic monitoring	Weekly contact	Monthly contact	Contact as needed or required by the court

Group F has been removed from this table to improve visibility. Group F is punishment only and targets low-risk individuals uses medication-assisted treatment. Overall the program scores 60 % suggesting that it is operating at a satisfactory level but has some areas in need of improvement. The program scored 0 points in the area the risk principle because the program does not consider risk for recidivism. The program does fairly well at addressing the need principle by targeting a criminogenic need (dependence on opiates). However, the program does not use a validated assessment to identify the need. On the responsivity principle, the program scores 87 % by using an evidence-based treatment approach (MAT) and includes both rewards and sanctions in their treatment protocol. In the area of program integrity, the MAT program scores 68 %. The program is doing well by having all clinical staff (many of whom have advanced degrees (50 %) or college degrees (100 %)), using external evaluations, and coaching staff to improve performance. The program could improve by using an evidence-based manual instead of one that is internally generated or alternatively having their manual validated. Additionally, program staff members only communicate with correction's staff as needed. Dosage is less than 100 h suggesting that the program may not be meeting the needs of high-risk individuals. In addition to MAT, the program also includes a number of additional restrictions such as having daily contact with participants and, conducting drug testing all of which boost the MAT program's score.

Identifying which groups of programming are available, and for which targets, will provide guidelines for practitioners when looking for programs/services for offenders. Once programs have been classified into groups, both pieces necessary for the responsivity simulation model have been created. The individual program-group assignment and the program classification group can be combined to identify the most responsive approach to treating an individual. Additionally, information on the distribution of individuals within program groups and the number of programs available within each group can help to define the gaps between individuals' needs and program availability.

Identifying Responsivity and Defining Gaps

Responsivity involves pairing the individual program-group assignment with the group defined using the program classification tool. This process will create a best match between individuals and programming. When this match cannot be made, a gap exists. After defining these initial classification groups for individuals and programs, the assumptions underlying these models were tested using individual and program data from a number of state and local correctional agencies.

The outcomes of these two models provide interesting insights into what groups of programming individuals in the criminal justice system need and what groups of programming are most commonly available in state and local systems. First, profiles from the synthetic data were run through the individual model resulting in a somewhat normal distribution with most individuals falling into program Group B. Program Group B in the synthetic data is comprised of predominately moderate-risk

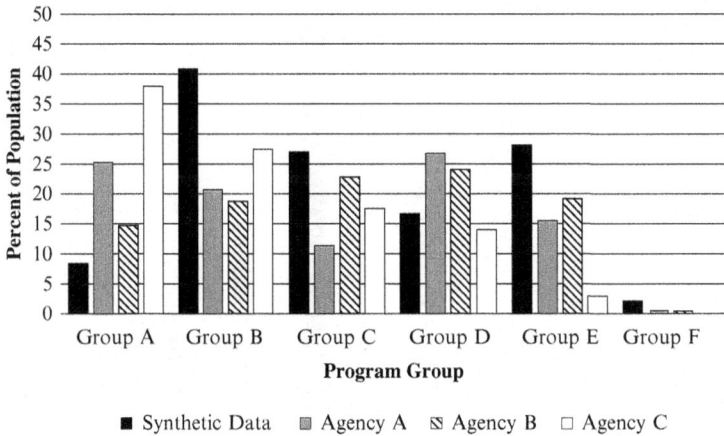

Fig. 6.2 Proportion of individuals within program levels

individuals (56.4 %), individuals with a drug abuse or alcohol dependence disorder (90.9 %), and those who have multiple lifestyle destabilizers (86.9 %).

Figure 6.2 provides a visual representation of the distribution of profiles into program groups for both the synthetic database and three specific agencies. Agency A is composed of largely incarcerated individuals, while agency B is a community corrections sample, and agency C is a community-based sample of justice-involved individuals with substance use disorders. Each of these agencies uses different risk and needs assessments, but in all cases the data provided sufficient information to define criminal history risk, primary needs, destabilizers, and stabilizers leading to the program-group assignments.

As Fig. 6.2 demonstrates, the different sites have slightly different distributions of individuals within each of the program groups. Because the synthetic data is drawn from an incarcerated sample, with high rates of destabilizers and drug abuse/alcohol dependence, it is not surprising to see this large concentration at Group B, where criminal thinking is the primary target of interventions. The variations between the sites are logical given that agencies A, B, and C represent different types of agencies (prison, community corrections, and substance abuse case management, respectively) and different geographical locations. Of most importance to note is that most of the individuals do not require intensive levels of services associated with Groups A and B. Instead, at least half of all individuals in three of these four datasets require moderate to low levels of program intensity (Groups C–F). Agency C focuses on individuals with substance use disorders; therefore, it is not surprising that nearly 40 % of individuals are assigned to Group A.

Based on the program-group assignments, expected recidivism rates for a treatment as usual group can be estimated. These rates provide a starting point for developing expectations for reductions in recidivism if and when individuals are placed into the appropriate group of programming and services. As would be expected,

Fig. 6.3 One- and three-year
rearrest rates by program
group for synthetic data

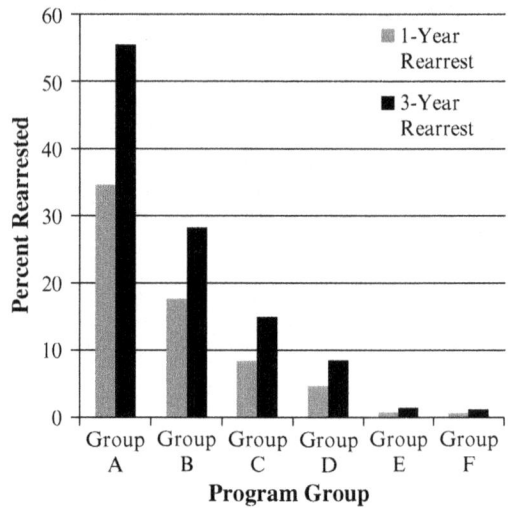

individuals in Group A have the highest expected recidivism rates, while individuals
in Group F have the lowest expected recidivism rates. Figure 6.3 presents the
recidivism rates associated with each program group in the synthetic data. Placing
individuals into the corresponding program group, especially those targeting an
individual's primary needs, can reduce these expected recidivism rates. The goal of
the model is to provide evidence for the expected recidivism reductions associated
with responsivity.

As discussed above, identifying which programs fit into these groups is not a
simple task. For example, a typical Group A program targets high-risk individuals
with drug dependence on hard drugs through residential treatment using a therapeutic community model provided by clinical professionals. Group A programming is
high dosage with treatment services occurring everyday for a total of 20 or more
hours per week for at least 90 days, but preferably longer. Individuals are placed in
Group A programs based on their scores on a validated substance use assessment
criteria such as the ASAM or DSM-IV. In contrast, a Group C program would typically target moderate-risk individuals with dependence on alcohol or who abuse
drugs. One example of a Group C program is a drug court that meets multiple times
per week for at least 6 months and focuses on drug abuse or alcohol dependence.
While Group C drug treatment programs may extend over more days than some
Group A programs, their overall dosage is lower because they occur less frequently
and for fewer hours (approximately five) per week. Few drug treatment programs
would fall into the less intensive programming groups. However, Group E drug
programs do exist. These would include self-help groups (e.g., Narcotics
Anonymous) that are attended on at most a weekly basis. These are low-dosage
programs with little direct clinical intervention.

It is often difficult for jurisdictions to estimate what groups of programming
should be in existence compared to what is currently available. Figure 6.4 provides

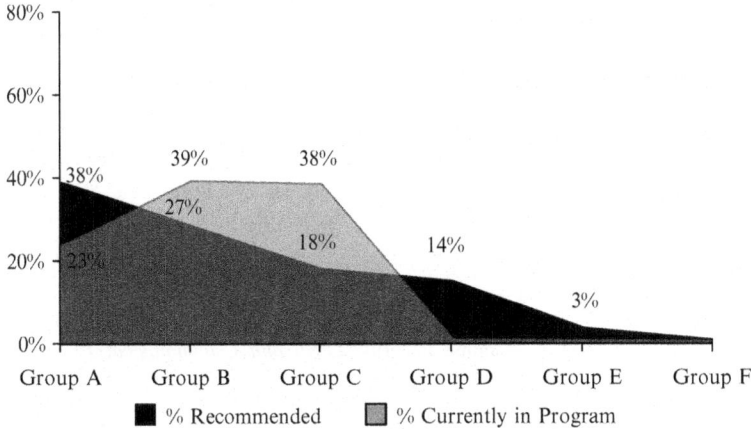

Fig. 6.4 Example of gap analysis using both individual and program groups

an example of the gaps in programming in one jurisdiction's community-based substance using sample. Of individuals who were assigned to programming in this jurisdiction, most of the programming and services were offered with Groups B and C. However, in this jurisdiction only 27 % and 18 % (respectively) of individuals required services typical for these groups. No services were offered for Groups D and E, even though these groups of services would be appropriate for 17 % of the sample. This jurisdiction is over programming individuals using services for Groups B and C while having a gap in services for individuals requiring the most intensive services in Group A. Shifting resources from those program groups with overages (groups B and C) to those where the demand exceeds availability (groups A, D, and E) can help jurisdictions close this gap while improving individual outcomes.

Conclusions

Responsivity is not an easy concept. It requires understanding individual risk, needs, and stabilizing factors compared to the available programs; programs can be targeted based on program targets, content, dosage, and implementation fidelity. One method for simplifying this process is through the development of simulation models. The models discussed above consider the important components affecting individuals' risk for reoffending—both static criminal history risk, dynamic needs, and destabilizers. Similarly, programs available to offenders can be classified into groups based on their essential features. The two pieces (individual group and program group classification) can be matched to guide responsivity. At the jurisdiction level, the models can help to identify gaps between the demand for programming from a given group assignment and the supply. This gap analysis can assist jurisdictions in determining how best to allocate resources to reduce recidivism by responding to the risk and needs of their population.

References

American Association of Community Psychiatrists [AACP]. (2009). *LOCUS: Level of care utilization system for psychiatric and addiction services—Adult version 2010.* Available from http://www.comm.psych.pitt.edu/find.html

American Society of Addiction Medicine (ASAM). (2001). *ASAM PPC-2R: ASAM patient placement criteria for the treatment of substance-related disorders.* Chevy Chase, MD: ASAM.

Andrews, D. A., & Bonta, J. (2010). *The psychology of criminal conduct* (5th ed.). New Providence, NJ: Lexis Nexis.

Andrews, D. A., Bonta, J., & Wormith, J. S. (2006). The recent past and near future of risk and/or need assessment. *Crime and Delinquency, 52*(1), 7–27.

Andrews, D., & Dowden, C. (2005). Managing correctional treatment for reduced recidivism: A meta-analytic review of programme integrity. *Legal and Criminological Psychology, 10,* 173–187.

Andrews, D., Dowden, C., & Gendreau, P. (1999). *Clinically relevant and psychologically informed approaches to reduced re-offending: A meta-analytic study of human service, risk, need, responsivity and other concerns in justice contexts.* Ottawa, ON: Carlton University.

Andrews, D. A., Zinger, I., Hoge, R. D., Bonta, J., Gendreau, P., & Cullen, F. T. (1990). Does correctional treatment work? A clinically relevant and psychologically informed meta-analysis. *Criminology, 28*(3), 396–404.

Ang, R. P., & Hughs, J. N. (2001). Differential benefits of skills training with antisocial youth based on group composition: A meta-analytic investigation. *School Psychology Review, 31*(2), 164–185.

Barber, S. J., & Wright, E. M. (2010). The effects of referral source supervision. *Criminal Justice and Behavior, 37*(8), 847–859.

Beck, R., & Fernandez, E. (1998). Cognitive-behavioral therapy in the treatment of anger: A meta-analysis. *Cognitive Therapy and Research, 22*(1), 63–74.

Beckmeyer, J. J. (2006). *Non-therapeutic correctional interventions: a meta-analysis of correctional academic, vocational, and life skills programs.* Department of Sociology/Criminal Justice & Criminology, University of Missouri--Kansas City.

Bourgon, G., & Armstrong, B. (2005). Transferring the principles of effective treatment into a "Real World" prison setting. *Criminal Justice and Behavior, 32*(1), 3–25. doi:10.1177/0093854804270618.

Botvin, G. J., Griffin, K. W., & Nichols, T. D. (2006). Preventing youth violence and delinquency through universal school-based prevention approach. *Prevention Science, 7,* 403–408.

Botvin, G. J., & Wills, T. A. (1984). Personal and social skills training: Cognitive-behavioral approaches to substance abuse prevention. In C. S. Bell & R. Battjes (Eds.), *Prevention research: Deterring drug abuse among children and adolescents* (pp. 8–49). Rockville, MD: Department of Health and Human Services (NIDA Research Monograph).

Bouffard, W., & MacKenzie, D. (2005). A quantitative review of structured, group-oriented, cognitive-behavioral programs for offenders. *Criminal Justice and Behavior, 32*(2), 172–204.

Dowden, C., Antonowicz, D., & Andrews, D. A. (2003). The effectiveness of relapse prevention with offenders: a meta-analysis. *International journal of offender therapy and comparative criminology, 47*(5), 516–528.

Drake, E. K. (2011). *"What works" in community supervision: Interim report.* Olympia, WA: Washington State Institute for Public Policy (Document No. 11-12-1201).

Drake, E. K., Aos, S., & Miller, M. (2009). Evidence-based public policy options to reduce crime and criminal justice costs: Implications in Washington state. *Victims and Offenders, 4,* 170–196.

Fletcher, B. W., Lehman, W. E. K., Wexler, H. K., Melnick, G., Taxman, F. S., & Young, D. W. (2009). Measuring collaboration and integration activities in criminal justice substance abuse treatment agencies. *Drug and Alcohol Dependence, 103S,* S54–S64.

Gastfriend, D. R., & McLellan, A. T. (1997). Treatment matching: Theoretical basis and practical implications. *Alcohol and Other Substance Abuse, 81*(4), 945–966.

Hansen, N. B., Lambert, M. J. & Forman, E. M. (2002). The psychotherapy dose-response effect and its implications for treatment delivery services. *Clinical Psychology: Research and Practice, 9*, 329–343.

Holloway, K. R., Bennett, T. H., & Farrington, D. P. (2006). The effectiveness of drug treatment programs in reducing criminal behavior: A meta-analysis. *Psicothema, 16*(3), 620–629.

Huber D. L., Hall J. A., Vaughn T. (2001). The dose of case management interventions. *Lippincott's Case Management, 6*(3), 119–126.

Kopta, S. M., Howard, K. I., Lowry, J. L., & Beutler, L. E. (1994). Patterns of symptomatic recovery in psychotherapy. *Journal of Consulting and Clinical Psychology, 62*, 1009–1016.

Latimer, J., Dowden, C., & Muise, D. (2005). The Effectiveness of Restorative Justice Practices: A Meta-Analysis. *The Prison Journal, 85*(2), 127–144.

Little, G. L. (2005). Meta analysis of moral reconation therapy recidivism results from probation and parole implementations. *Cognitive Behavioral Treatment Review, 14*(1/2), 14–16.

Lipsey, M. W. (2008). *The Arizona Standardized Program Evaluation Protocol (SPEP) for assessing the effectiveness of programs for juvenile probationers: SPEP ratings and relative recidivism reduction for the initial SPEP sample.* A Report to the Juvenile Justice Services Division, Administrative Office of the Courts, State of Arizona. Center for Evaluation Research and Methodology, Vanderbilt Institute for Public Policy Studies.

Lipsey, M. W. (2009). The Primary Factors that Characterize Effective Interventions with Juvenile Offenders: A Meta-Analytic Overview. *Victims & Offenders, 4*(2), 124–147. doi:10.1080/15564880802612573.

Lipsey, M. W., & Cullen, F. T. (2007). The effectiveness of correctional rehabilitation: A review of systematic reviews. *Annual Review of Law and Social Science, 3*(1), 297–320.

Lipsey, M. W., & Landenberger, N. A. (2005). In B. C. Welsh & D. P. Farrington (Eds.), *Preventing crime: What works for children, offenders, victims, and places.* Berlin: Springer.

Lipton, D. S., Pearson, F. S., Cleland, C. M., & Yee, D. (2008). The Effects of Therapeutic Communities and Milieu Therapy on Recidivism: Meta-analytic Findings from the Correctional Drug Abuse Treatment Effectiveness (CDATE) Study. In J. McGuire (Ed.), *Offender Rehabilitation and Treatment* (pp. 39–77). John Wiley & Sons Ltd.

Lipsey, M. W., Landenberger, N. A., & Wilson, S. J. (2007). *Effects of Cognitive-Behavioral Programs for Criminal Offenders* (Campbell Systematic Review). Oslo, Norway: The Campbell Collaboration.

Lowenkamp, C. T. (2004). *Correctional program integrity and treatment effectiveness: A multi-site, program-level analysis.* Unpublished doctoral dissertation, University of Cincinnati, Cincinnati, OH.

Mann, R. E. (2009). Sex offender treatment: The case for manualization. *Journal of Sexual Aggression, 15*(2), 121–131.

Martin, M. S., Dorken, S. K., Wamboldt, A. D., & Wootten, S. E. (2012). Stopping the revolving door: A meta-analysis on the effectiveness of interventions for criminally involved individuals with major mental disorders. *Law and Human Behavior, 36*(1), 1–12. doi:10.1037/h0093963.

McGrew, J. H., Bond, G. R., Dietzen, L., & Salyers, M. (1994). Measuring the fidelity of implementation of a mental health program model. *Journal of Consulting and Clinical Psychology, 62*(4), 670–678.

Mee-Lee, D., Shulman, G. D., Fishman, M., Gastfriend, D. R., & Griffith, J. H. (Eds.). (2001). *ASAM patient placement criteria for the treatment of substance-related disorders (ASAM PPC-2R)* (Revised 2nd ed.). Chevy Chase, MD: American Society of Addiction Medicine.

Mitchell, W., & MacKenzie, D. (2007). Does incarceration-based drug treatment reduce recidivism? A meta-analytic synthesis of the research. *Journal of Experimental Criminology, 3*, 353–375.

National Institute of Corrections [NIC]. (2005). *Implementing evidence-based practice in community corrections: The principle of effective interventions.* Washington, DC: NIC.

Padgett, K. G., Bales, W. D., & Blomberg, T. G. (2006). Under surveillance: An empirical test of the effectiveness and consequences of electronic monitoring. *Criminology & Public Policy., 5*(1), 61–92.

Pattavina, A., Tusinski-Mitofsky, K., & Byrne, J.M. (2009). Persuasive technology: Moving beyond restriction to rehabilitation. *Journal of Offender Monitoring, 23*(1), 4.

Pearson, F. S., Lipton, D. S., Cleland, C. M., & Yee, D. S. (2002). The effects of behavioral/cognitive-behavioral programs on recidivism. *Crime & Delinquency, 48*(3), 476–496.

Perry, A., Coulton, S., Glanville, J., Godfrey, C., Lunn, J., McDougall, C., & Neale, Z. (1996). Interventions for drug-using offenders in the courts, secure establishments and the community. *In Cochrane Database of Systematic Reviews*. John Wiley & Sons, Ltd.

Prendergast, M., Huang, D., & Hser, Y. (2008). Patterns of crime and drug use trajectories in relation to treatment initiation and 5-year outcomes: An application of growth mixture modeling across three data sets. *Evaluation Review, 32*(1), 59–83.

Prendergast, M. L., Podus, D., Chang, E., & Urada, D. (2002). The effectiveness of drug abuse treatment: a meta-analysis of comparison group studies. *Drug and Alcohol Dependence, 67*(1), 53–72.

Renzema, M., & Mayo-Wilson, E. (2005). Can Electronic Monitoring Reduce Crime for Moderate to High-Risk Offenders? *Journal of Experimental Criminology, 1*, 215–237.

Simons, A. D., Padesky, C. A., Montemarano, J., Lewis, C. C., Murakami, J., Lamb, K., et al. (2010). Training and dissemination of cognitive behavioral therapy for depression in adults: A preliminary examination of therapist competence and client outcomes. *Journal of Consulting and Clinical Psychology, 78*(5), 751–756.

Simpson, D. D. (1979). The relation of time spent in drug abuse treatment to posttreatment outcome. *American Journal of Psychiatry, 136*(11), 1449–1453.

Smith, P., Goggin, C., & Gendreau, P. (2002). *The effects of prison sentences and intermediate sanctions on recidivism: General Effects and Individual Differences* (No. JS42-103/2002). Canada: University of New Brunswick, Saint John.

Stanard, R. P. (1999). The effect of training in a strengths model of case management on client outcomes in a community mental health center. *Community Mental Health Journal, 35I*(2), 169–179.

Taxman, F. S., & Bouffard, J. (2000). The importance of systems in improving offender outcomes: New frontiers in treatment integrity. *Justice Policy Journal, 2*(2), 37–58.

Taxman, F. S., Byrne, J., & Thanner, M. (2002). *Evaluating the implementation and impact of a seamless system of care for substance abusing offenders-the HIDTA model*. Washington, DC: National Institute of Justice.

Taxman, F. S., Young, D., Wiersema, B., Mitchell, S., & Rhodes, A. G. (2007). The national criminal justice treatment practices survey: Multilevel survey methods & procedures. *Journal of Substance Abuse Treatment, 32*, 225–238.

Thanner, M. H. & Taxman, F. S. (2003). Responsivity: the value of providing intensive services to high-risk offender. *Journal of Substance Abuse Treatment, 24*, 131–147.

Thornton, C. C., Gottheil, E., Weinstein, S. P., & Kerachsky, R. S. (1998). Patient-treatment matching in substance abuse. *Journal of Substance Abuse Treatment, 15*(6), 505–511.

Tong, J., & Farrington, D. P. (2006). How effective is the "Reasoning and Rehabilitation" programme in reducing reoffending? A meta-analysis of evaluations in four countries. *Psychology, Crime & Law, 12*(1), 3–24.

U.S. Department of Health and Human Services, National Institute on Drug Abuse. (2009). *Principles of drug addiction treatment: A research-based guide*. Washington, DC: National Institute on Drug Abuse, National Institutes of Health (NIH Publication No. 09-4180).

Visher, C., Winterfield, L., & Coggeshall, M. (2005). Ex-offender employment programs and recidivism: A meta-analysis. *Journal of Experimental Criminology, 1*(3), 295–316.

Wilson, D. B., MacKenzie, D. L., & Mitchell, F. N. (2008). *Effects of Correctional Boot Camps on Offending* (Campbell Systematic Review). Oslo, Norway: The Campbell Collaboration.

Wilson, D. B., Gallagher, C. A., & MacKenzie, D. L. (2000). A Meta-Analysis of Corrections-Based Education, Vocation, and Work Programs for Adult Offenders. *Journal of Research in Crime and Delinquency, 37*(4), 347–368.

Chapter 7
Reducing Recidivism Through Correctional Programming: Using Meta-Analysis to Inform the RNR Simulation Tool

Michael S. Caudy, Liansheng Tang, Stephanie A. Ainsworth, Jennifer Lerch, and Faye S. Taxman

Introduction

Over the past 3 decades, scientists and evaluators have conducted empirical studies to assess the effectiveness of various correctional interventions. Generally the emphasis has been on assessing the impact of a program or intervention on recidivism or some measure of returning to offending. Meta-analytic techniques have emerged as a primary tool for synthesizing this growing body of research due to their applicability for compiling findings across different study designs, settings, and levels of methodological rigor. The result of this expanded use of meta-analytic techniques is a summary of "what works" in correctional interventions that relies on meta-analyses to define evidence-based correctional practices (see Lipsey & Cullen, 2007 for one review of meta-analyses examining correctional interventions).

The relatively consistent finding from the expanding body of empirical research on the effectiveness of correctional interventions is that well-designed and well-implemented rehabilitative interventions can have a *strong* impact on reducing individual-level recidivism (Andrews et al., 1990; Andrews & Dowden, 2006; Lipsey & Cullen, 2007; McGuire, 2002; Smith, Gendreau, & Swartz, 2009). However, many questions still exist about the key components of effective correctional interventions (Lipsey & Cullen, 2007; Taxman & Belenko, 2012) and the translational process of moving from "bench to trench" (or research to practice) to make effective interventions a routine part of the correctional system (Taxman & Belenko, 2012). An even larger question concerns how to scale up programming to handle the majority of justice-involved individuals that would benefit from

M.S. Caudy, Ph.D. (✉) • S.A. Ainsworth, M.A. • J. Lerch, M.A. • F.S. Taxman, Ph.D.
Department of Criminology, Law and Society, George Mason University,
10900 University Boulevard, Fairfax, VA 20110, USA
e-mail: mcaudy@gmu.edu

L. Tang, Ph.D.
Department of Statistics, George Mason University, Fairfax, VA, USA

F.S. Taxman and A. Pattavina (eds.), *Simulation Strategies to Reduce Recidivism:*   167
Risk Need Responsivity (RNR) Modeling for the Criminal Justice System,
DOI 10.1007/978-1-4614-6188-3_7, © Springer Science+Business Media New York 2013

Box 7.1 The Risk-Need-Responsivity Principles

- The *risk* principle states that the intensity of the intervention should be matched to the offender's risk to reoffend; higher risk offenders should be placed in more intensive services, while lower risk offenders require only minimally intensive interventions or may not be appropriate for interventions at all.
- The *need* principle suggests that correctional interventions should target specific criminogenic needs such as antisocial personality, antisocial values, criminal peers, or substance dependence.
- The *responsivity* principle states that interventions that employ cognitive restructuring techniques (e.g., social learning; cognitive-behavioral therapies; criminal thinking curricula) will have the strongest and most long-term impact on offender behavior.

treatment or correctional interventions. It is believed that only by scaling up, and providing the appropriate services to appropriate offenders, can we have a systematic impact on recidivism. The case for scaling up is strong with the realization that a system-level impact can only occur when well-designed and well-implemented programs are accessible to a large proportion of the individuals in need (Andrews & Dowden, 2005; Austin, 2009; Gendreau, Goggin, & Smith, 1999; Lipsey & Cullen, 2007; Taxman & Belenko, 2012). Two central research questions with roots in the "what works" (i.e., evidence-based practices) movement that are in need of further exploration are: (1) what will it take for correctional programming to have a noticeable impact on recidivism at a system level?; and (2) how can the reach of effective correctional interventions be extended so that more justice-involved individuals can receive them and a greater public health impact can be realized?

The current chapter explores these two questions in the context of the RNR Simulation Tool, an empirically based decision support tool designed to help justice agencies adhere to the principles of effective intervention (Gendreau, Goggin, French, & Smith, 2006; Gendreau, Smith, & French, 2006; Taxman & Belenko, 2012). The principles of effective intervention suggest that correctional programs will be most effective when they: target higher risk offenders (the *risk* principle); address specific criminogenic needs (the *needs* principle); and employ behavioral or cognitive-behavioral treatment approaches (the general *responsivity* principle) (Andrews & Bonta, 2010; Andrews, Bonta, & Hoge, 1990). The chapter begins with a brief review of what we know about "what works" for reducing recidivism at the individual level from existing meta-analyses and continues with an empirically informed discussion of some steps that can be taken by the justice system to reduce recidivism rates at a system level. Finally, the chapter illustrates the process of using findings from meta-analytic research to inform the RNR Simulation Tool and estimate possible recidivism reductions when justice-involved individuals are matched to correctional interventions based on their risk and dynamic needs (Box 7.1).

What Works in Correctional Programming

Correctional programming is a broad term used to describe an array of activities and components that comprise a "program." The notion of a program is that it has some stated goals and objectives; in other words, the program is designed to address particular issues. In the context of correctional programming, the notion is that clinical and control (e.g., drug testing, curfews, area restrictions, electronic monitoring, required daily activities) strategies are used together to address a particular behavior such as substance abuse, criminal thinking and actions, low self-control, social networks that are comprised of antisocial peers, and so on. In other words, the correctional program should address the criminogenic needs or punishment purpose to achieve the desired outcomes. In this section, we discuss the literature on effective correctional programming.

Reviewing What Works: Findings from Existing Meta-Analyses

The findings from existing meta-analyses in corrections generally indicate that rehabilitative correctional interventions (e.g., cognitive-behavioral therapy; drug courts; therapeutic communities) can produce considerable reductions in recidivism at the individual level (Lipsey & Cullen, 2007; McGuire, 2002; Smith et al., 2009). In their extensive review, summarizing over 40 meta-analyses related to correctional interventions, Lipsey and Cullen (2007) provided strong support for rehabilitative programming relative to sanctioning for reducing recidivism. Overall, mean effect sizes indicated modest percent reductions in recidivism associated with community supervision (recidivism reduction ranging from 2 to 8 %) and no effect or increased recidivism for incarceration (recidivism reduction ranging from 0 % to an iatrogenic effect of increasing recidivism by 14 %). The mean effect sizes for treatment programming ranged between a null effect and a 50 % reduction in recidivism with a majority of studies indicating positive effects (~10–30 %) (Lipsey & Cullen, 2007, Table 4). The review of meta-analytic results led the authors to conclude that "treatment is capable of reducing the reoffense rates of convicted offenders and that it has greater capability for doing so than correctional sanctions" (Lipsey & Cullen, 2007, p. 314; see also McGuire, 2002).

Table 7.1 (below) displays a list of existing meta-analytic studies and their findings regarding recidivism reduction. These meta-analyses were identified through a series of systematic searches using the EMTAP process (described later in the chapter) and provide an update to earlier summaries (e.g., Lipsey & Cullen, 2007; McGuire, 2002). The reviews summarized in Table 7.1 affirm the findings of previous "reviews of reviews" that indicate that the effects of rehabilitative correctional interventions on recidivism are generally positive with relatively modest to strong effects. Generally, these findings indicate that treatment programming can produce between a 10 and 30 % reduction in recidivism.

Table 7.1 Percent reductions in recidivism by intervention type from extant meta-analyses

Intervention	Odds ratio	% Reduction
Interventions for general offenders		
Cognitive-behavioral therapy (Lipsey et al., 2007)	1.53 (CI = 1.35–1.73)	25
Moral reconation therapy (Little, 2005; Wilson, Bouffard, & MacKenzie, 2005)	1.38–1.8[a][b]	16[c]–35
Reasoning and rehabilitation (Tong & Farrington, 2006; Wilson, Bouffard, & MacKenzie, 2005)	1.16–1.34[a,b]	14
Restorative justice (Lattimer, Dowden, & Muise, 2005)	–	14[a]
CBT for anger management (Beck & Fernandez, 1998)	0.24[a] (CI = 0.21–0.28)	51
Intensive supervision probation w/treatment (Drake, Aos, & Miller, 2009)	–	17.9
Electronic monitoring (Renzema & Mayo-Wilson, 2005)	0.96 (CI = 0.71–1.31)	2[c]
Interventions for substance using offenders		
General drug treatment (Holloway, Bennett & Farrington, 2006; Prendergast, Podus, Chang, & Urada, 2002)	1.27[b]–1.56 (CI = 1.18–2.07)	12[c]–22[c]
Therapeutic community (Lipton, Pearson, Cleland, & Yee, 2008; Mitchell et al., 2007)	1.38–1.74[b] (CI = 1.17–1.62)	16[c]–27
Therapeutic community (hard drugs) (Holloway et al., 2006)	2.61 (CI = 1.58–4.33)	45
Counseling (general) (Mitchell et al., 2007)	1.50 (CI = 1.25–1.79)	20
Narcotic maintenance (Mitchell et al., 2007)	0.84 (CI = 0.54–1.29)	9 % Increase
Narcotic maintenance (hard drugs) (Holloway et al., 2006)	1.75 (CI = 0.99–3.11)	27[c]
Boot Camp (Mitchell et al., 2007)	1.10 (CI = 0.62–1.96)	5
Intensive Supervision Program (Perry et al., 2006)	1.98 (CI = 1.01–3.87)	33[c]
Post-release supervision (Dowden, Antonowicz, & Andrews, 2003)	–	26[d]
Post-release supervision (Hard Drugs) (Holloway et al., 2006)	1.99 (CI = 0.92–4.31)	33[c]
Interventions for offenders with mental illness		
Mental health treatment (Martin, Dorken, Wamboldt, & Wootten, 2001)	1.41[b]	17[c]
Vocational/educational programs		
General vocation/education (Wilson, Gallagher, & MacKenzie, 2000)	1.52 (CI = 1.37–1.69)	21
Ex-offender employment (Visher, Winterfield, & Coggeshall, 2005)	1.06[b]	3[c]
Academic/Educational (Wilson, Gallagher, & MacKenzie, 2000)	1.44 (CI = 1.15–1.82)	18
Postsecondary correctional education (Wilson, Gallagher, & MacKenzie, 2000)	1.74 (CI = 1.36–2.22)	27

(continued)

Table 7.1 (continued)

Intervention	Odds ratio	% Reduction
Vocational (Wilson, Gallagher, & MacKenzie, 2000)	1.55 (CI = 1.18–1.86)	22
Correctional industries (Wilson, Gallagher, & MacKenzie, 2000)	1.48 (CI = 0.92–2.17)	19
Supervision only interventions for general offenders		
Incarceration (vs. community) (Smith, Goggin, & Gendreau, 2002)	–	14 % Increase
Intermediate sanctions (Smith, Goggin, & Gendreau, 2002)	–	2
Boot Camp (Wilson, MacKenzie, & Mitchell, 2008)	1.02 (CI = 0.90–1.14)	1
Interventions for domestic violence offenders		
General DV treatment (police report)[e] (Babcock, Green, & Robie, 2004; Feder & Wilson, 2005)	1.24[b]–1.60[b]	16–32
General DV treatment (partner report)[e] (Babcock, Green, & Robie, 2004; Feder & Wilson, 2005)	1.18[b]–1.00[b]	0–10
Interventions for sexual offenders		
Sex offender treatment (sexual recidivism) (Gallagher, Wilson, Paul Hirschfield, Coggeshall, & MacKenzie, 1999; Hansen et al., 2002; Hall, 1995; Schmucker & Losel, 2008)	0.81[b]–2.18[b]	16–37
Sex offender treatment (violent recidivism) (Schmucker & Losel, 2008)	1.90[b]	44
Sex offender treatment (general recidivism) (Hanson et al., 2002; Schmucker & Losel, 2008)	0.56[b]–1.67[b]	31–32

[a]Standardized mean difference was converted to odds ratio. Phi coefficient was converted to an odds ratio with an assumed 0.50 control recidivism. Success/failure rates for treatment and control groups were used to calculate odds ratio
[b]Insufficient information to calculate confidence interval
[c]Calculation assumed 0.50 control recidivism baserate
[d]Treatment and control group recidivism rates were converted to percent reduction
[e]Experimental design only

Meta-Analytic Support for the Principles of Effective Intervention

The Risk-Need-Responsivity (RNR) model for offender rehabilitation (see Andrews, Bonta, & Hoge, 1990), as discussed in Chap. 4 of this book, provides an evidence-based framework for improving correctional programming outcomes. The importance of the RNR principles has been well-documented in extant meta-analytic findings (Andrews & Bonta, 2010; Andrews & Dowden, 2006; Andrews,

---

Box 7.2 The Carleton University Databank: Building an Empirical Base for the RNR Model

---

- Most meta-analytic tests of the RNR principles have been conducted using a sample of *230 primary studies* of the effectiveness of correctional interventions housed in the Carleton University Databank (see Andrews & Dowden, 2005, 2006; Dowden, 1998).
- *Inclusion criteria* for studies in the databank:

  - The study compared a group of offenders who received an intervention to a comparison group of offenders who did not receive the primary intervention
  - The study included a follow-up period
  - The study included a measure of recidivism as an outcome
  - The study provided enough information to allow for an effect size estimate to be calculated based on recidivism data

- A total of *374 independent tests of the RNR principles* have been coded from the 230 included studies (Andrews & Bonta, 2006, 2010).

  - 278 of the tests (74 %) indicate adherence to the *risk* principle; adherence to the risk principle is associated with an average 10 % reduction in recidivism.
  - 169 of the tests (45 %) indicate adherence to the *need* principle; adherence to the need principle is associated with an average 19 % reduction in recidivism.
  - 77 of the tests (21 %) indicate adherence to the *responsivity* principle; adherence to the responsivity principle is associated with an average 23 % reduction in recidivism.

Adapted from: Andrews & Bonta (2010); Andrews & Dowden (2005, 2006)

Zinger, et al., 1990; Dowden & Andrews, 1999a, 1999b, 2000). Based on 374 independent tests of the risk principle (from 225 unique studies), Andrews and Dowden (2006) found that programs that adhered to the risk principle by placing higher risk clients into more intensive services produced an average percent reduction in recidivism of about 10 % ($k=278$), while treatment programs for lower risk offenders were only associated with an average 3 % ($k=96$) reduction in recidivism.

In another meta-analysis of the 374 independent effect sizes included in the Carleton University databank (see Box 7.2 above), Andrews and Bonta (2010); and Andrews & Dowden, (2006) found that adherence to the needs principle was associated with an average 19 % ($k=169$) reduction in recidivism while adherence to the responsivity principle was associated with an average 23 % ($k=77$) reduction in recidivism across included studies. They also found that adherence to all three RNR principles produced the greatest recidivism reduction potential (26 %, $k=60$)

particularly for programs delivered in community settings (35 %, $k=30$) (Andrews & Bonta, 2006, 2010; Bonta & Andrews, 2007). Additional work by Lowenkamp and colleagues (Lowenkamp & Latessa, 2005; Lowenkamp, Latessa, & Holsinger, 2006) explored differential recidivism outcomes for high- and low-risk offenders in residential halfway houses and confirmed the importance of using the risk principle for achieving greater gains in recidivism reduction.

In a recent systematic review of extant meta-analyses, Smith et al. (2009) validated the importance of the principles of effective intervention. Using effect sizes from 22 meta-analyses, they found that the general responsivity principle was strongly related to improved recidivism outcomes. When comparing effect sizes for cognitive-behavioral interventions (mean effect sizes ranged from $r=0.02$ to $r=0.63$) to nonbehavioral treatment modalities (mean effect sizes ranged from $r=0.01$ to $r=0.19$), Smith and colleagues found that cognitive-behavioral interventions were more effective in reducing recidivism. Overall, 16 of the 22 effect sizes for cognitive-behavioral interventions were related to at least a 15 % reduction in recidivism for the treatment group relative to the control group. Effect sizes reported from six meta-analyses which provided information on risk indicated that adherence to the risk principle was related to recidivism reductions ranging from 9 to 29 %. Effect sizes for programs that targeted specific criminogenic needs were available from five meta-analyses and ranged from roughly 20–30 % reductions in recidivism. Interventions that targeted noncriminogenic needs produced estimates of less than a 5 % reduction in recidivism (Smith et al., 2009).

Additionally, in their meta-analytic assessments of the effectiveness of cognitive-behavior therapy (CBT) for offenders, Lipsey and colleagues (2005; Lipsey, Landenberger, & Wilson, 2007) confirmed the importance of the RNR model. Based on the analysis of 58 primary studies, Lipsey and colleagues found that CBT was associated with a mean reduction in recidivism of 25 % (OR=1.53) (Landenberger & Lipsey, 2005; Lipsey et al., 2007). Meta-regression moderator analyses revealed that recidivism risk was significantly related to mean effect size. Consistent with the risk principle, this finding indicated that CBT was more effective when it was administered to higher risk offenders. Lipsey and colleagues also estimated the effectiveness of an ideal configuration of CBT and found that when dosage and implementation were appropriate and the risk principle was adhered to, CBT was related to an approximate 52 % (OR=2.86) reduction in recidivism (Lipsey et al., 2007).

Limitations of Relying on Meta-Analyses

Despite these generally promising findings, the extant meta-analytic results regarding the effectiveness of rehabilitative correctional interventions need to be considered in the context of a few caveats. While meta-analysis is a powerful research synthesis technique, it is not without limitations and should not be used to make definitive statements about causality or draw conclusions about the universal effectiveness of an intervention (Austin, 2009). Both Lipsey and Cullen (2007) and McGuire (2002) reported a great deal of variability in effect sizes even when multiple syntheses examined the same interventions with a considerable portion of overlapping primary studies. Across both of these reviews, there were no interventions or intervention types that were found to universally produce positive effects. The consistent finding of variability in outcomes across effectiveness studies and meta-analyses has affirmed that there is "no magic bullet" for correctional programming; in other words, there is no one program or program type that can be identified that will consistently have a large impact on recidivism regardless of how well it is implemented or how appropriate it is for the target population being served (Lipsey & Cullen, 2007; McGuire, 2002, p. 20). The lack of a "magic bullet" has pushed researchers to move beyond simply assessing the effectiveness of specific programs to consider the characteristics/program features that are common to effective correctional interventions (as discussed in Chap. 5 of this book).

Using Meta-Analyses to Inform the RNR Simulation Tool

This section describes the process that was employed to link meta-analytic findings (see Table 7.1) on the effectiveness of correctional interventions to the RNR Simulation Tool. The goal of the systematic process described herein was to identify extant meta-analyses on the effectiveness of correctional interventions and use the findings from these syntheses to estimate the potential recidivism reducing impact of adherence to the RNR principles at the individual level. The utility of meta-analysis for informing simulation model inputs is illustrated in this section.

Meta-analyses are preferred sources for informing simulation model inputs because, by design, they help reduce the impact of several potential threats to the validity of study findings regarding intervention effectiveness. While not immune to validity threats themselves (Borenstein, Hedges, Higgins, & Rothstein, 2009; Lipsey & Wilson, 2001; Wilson, 2009, 2010), well-conducted meta-analyses can help reduce both researcher and publication bias when used to inform simulation model inputs. Additionally, because meta-analyses focus on the magnitude and direction of effect sizes across a wide range of settings, populations, and study designs, they provide perhaps the most objective indicator of intervention effectiveness (or other relevant outcomes) available.

A good research synthesis begins with a systematic search procedure (Hammerstrom, Wade, & Jorgensen, 2010; Lefebvre, Manheimer, & Glanville, 2011). For the RNR Simulation Tool, three separate systematic searches were conducted to identify relevant meta-analytic findings. The primary systematic search was conducted by the fourth study author and a team of researchers from the Center for Advancing Correctional Excellence (ACE!) at George Mason University as part of the Evidence Mapping to Advance Justice Practice (EMTAP) project (see Caudy, Taxman, Tang, & Watson, forthcoming for a discussion of the EMTAP search methodology). This systematic EMTAP search and coding procedure which included a search of the grey literature identified 62 systematic reviews and meta-analyses on the effectiveness of correctional interventions which yielded over 30 meta-analyses that met our inclusion criteria (we eliminated systematic reviews that did not contain meta-analyses and meta-analyses that assessed interventions for juveniles only). This list of extant meta-analyses was then supplemented by an additional systematic search conducted by the study authors and a third search that was conducted by Wilson (2001) on the effectiveness of correctional interventions in secure facilities. As a final check on the exhaustiveness of our search, we cross-referenced the meta-analyses identified by Lipsey and Cullen (2007) on the effectiveness of rehabilitative interventions. Our three-pronged systematic search process identified all of the meta-analyses included in the Lipsey and Cullen review for adult offenders plus several additional meta-analyses that have been conducted since their study was published.

The biggest challenge that we faced in selecting meta-analyses to inform the RNR Simulation Tool was that multiple meta-analyses of the same intervention often reported differing results regarding the mean effect size of the intervention. This variability in effect sizes both within and across meta-analyses is well documented (Lipsey & Cullen, 2007; McGuire, 2002). To avoid objectivity and limit potential researcher bias, we developed a set of decision rules to guide the process of selecting meta-analyses when there are multiple reports of the effectiveness of the same intervention. Similar procedures for dealing with discordant findings from systematic reviews have been discussed in the literature (see e.g., Jadad, Cook, & Browman, 1997). The decision rules were:

1. The meta-analysis must have focused on adult offenders involved in the justice system and reported a mean effect size for general recidivism as an outcome.
2. The meta-analysis must have reported sufficient information to allow for the calculation of a percent reduction in recidivism associated with the intervention. Reporting percent reductions is a preferred metric when disseminating meta-analytic findings to practitioners (Gendreau & Smith, 2007).
3. When multiple meta-analyses examined the same intervention, we favored the most recently published study as long as it did not differ greatly from earlier studies regarding selection criteria or research design. This criterion led to the selection of an updated version of the same meta-analysis conducted by the same group of authors.
4. We prioritized meta-analyses which addressed clearly defined interventions. We selected studies that examined specific interventions (e.g., therapeutic communities for drug-involved offenders, or postsecondary correctional education programs).
5. When available, we prioritized meta-analyses that included assessments of key moderators of intervention effect size.

Calculating Percent Reductions from Meta-Analytic Findings

A potential limitation of meta-analysis is that the statistics (i.e., effect sizes) reported to indicate the magnitude and direction of the effect of a given intervention are not always easily interpretable for the "people who count" (Gendreau & Smith, 2007,

p. 1539). To improve the transportability of meta-analytic findings, it is important to report the findings in a way that makes them meaningful to users such as policy makers and practitioners. Perhaps the best way to achieve this goal is to convert meta-analytic effect sizes to a standardized index of an average percent reduction in recidivism between treatment and comparison conditions (Gendreau & Smith, 2007; Lipsey & Cullen, 2007). Percent reduction can be easily calculated from all of the most commonly reported indicators of treatment effect sizes and can provide a standardized indicator of treatment effectiveness.

Commonly used effect sizes in meta-analysis include the standardized mean difference (Cohen's d), the odds ratio (OR), and the Pearson's r. These effect sizes can be converted to each other (see Borenstein et al., 2009) through fairly simple formulae. However, regardless of what effect size is reported, these results have limited meaning to non-research audiences who have no experience interpreting meta-analyses. It is important to consider converting effect sizes to percent reductions when disseminating meta-analytic findings. In the development of the RNR Simulation Tool, we used the odds ratio (OR) as the basis for our calculations of percentage reduction between treatment and controls across studies included in meta-analyses. The OR of recidivism often refers to the ratio of recidivism odds between a treatment group and a control group. In mathematical terms, suppose that recidivism rates of the treatment and the control are p_1 and p_2, respectively. The odds of the treatment and the control are $p_1/(1-p_1)$ and $p_2/(1-p_2)$, respectively. The OR is calculated using $OR=(p_1/(1-p_1))/(p_2/(1-p_2))$. As can be seen from the formula, if the treatment works effectively to reduce the recidivism rate, the odds of the treatment will be less than that of the control. The resulting OR will also be less than one. And if the treatment has the same effect as the control, we have $p_1=p_2$ and $OR=1$.

Suppose the baseline recidivism rate, p_2, using the control is known, the recidivism rate, p_1, of the treatment is obtained by $p_1 = OR \times p_2/(1-p_2+OR \times p_2)$. The proportion difference in recidivism is given by

$$p_2 - p_1 = (1-OR) \times p_2 \times (1-p_2)/(1-p_2+OR \times p_2),$$

with a positive number indicating an effective treatment. The percentage reduction is defined as the ratio of the proportion difference over the baseline rate, that is, percentage reduction $= (p_2-p_1)/p_2 \times 100\%$.

The OR is often reported as the odds of being a successful non-recidivist in the treatment group relative to the control group. Given the notations for the recidivism rates, the OR is calculated using $OR=((1-p_1)/p_1)/((1-p_2)/p_2)$. This is the inverse of the OR of recidivism. In this case, the OR is greater than one if the treatment is more effective. The calculation of p_1 is different from the one with the OR of recidivism. We can obtain the recidivism rate, p_1, of the treatment using $p_1 = 1-OR \times (1-p_2)/(1+(OR-1) \times (1-p_2))$. The proportion difference in recidivism is given by

$$p_2 - p_1 = (OR-1) \times p_2 \times (1-p_2)/(1+(OR-1) \times (1-p_2)),$$

with a positive number indicating an effective treatment. The percentage reduction can still be calculated by $(p_2-p_1)/p_2 \times 100$ %.

For the purpose of the current project, we converted all effect sizes from the existing meta-analyses that met our inclusion criteria to ORs and then converted these ORs to percent reductions using the formulae provided above (see Table 7.5). While there is always potential for debate regarding which effect size metric is most meaningful (Gendreau & Smith, 2007) and the appropriateness of the effect size is driven by the structure of the data being analyzed (Borenstein et al., 2009; Lipsey & Wilson, 2001; Wilson, 2010), the conversion of effect sizes to percent reductions is an important translational tool for moving research into practice.

Updating the Meta-Analytic Summary of What Works

Despite the publication of several new or updated meta-analyses regarding the effectiveness of specific correctional interventions (see e.g., Mitchell, Wilson, Eggers, & MacKenzie's, 2012 update on the effectiveness of drug courts), the conclusions regarding the effectiveness of rehabilitative correctional interventions have not changed dramatically since the Maryland report (Sherman et al., 1997) or Lipsey and Cullen's (2007) review. Our findings (Table 7.1) reinforce the conclusion that correctional treatment is capable of reducing recidivism (Lipsey & Cullen, 2007; McGuire, 2002). Our summary findings suggest that the treatment effects of correctional programming reported in existing meta-analyses are predominantly positive in favor of treatment and range in magnitude on average from a reduction in recidivism of about 10 to about 30 %. Consistent with prior reviews is the finding of considerable variability in the size of mean treatment effects across existing meta-analyses. These findings suggest the need to dig deeper into what works and to explore the key moderators of correctional treatment effects (Shaffer & Pratt, 2009).

The analysis of moderators of treatment effects is helpful for understanding what works and improving the transportability of meta-analytic findings to everyday correctional practice. Exploring moderators can help reconcile the large variability of effect sizes reported across existing evaluations of the same interventions. If meta-analytic findings are going to be used to inform best practices, it is critical to understand what individual, setting, and program characteristics are related to the effectiveness of the intervention (Shaffer & Pratt, 2009). For the most part, the extant meta-analytic literature base does not contain sufficient information on key moderators of treatment effects to be useful for the RNR Simulation Tool. Thus, we generally relied on aggregate mean effect sizes for a typical configuration of each correctional intervention included in the model.

Lessons Learned in Applying Meta-Analyses to the RNR Simulation Tool

Meta-analyses are now standard research procedures designed to distill and synthesize data from complex studies. Researchers are beginning to explore moderators such as offender demographics, risk level, and different need profiles to understand the question "what works for whom?" This is truly an emerging area of work. For the RNR Simulation Tool, moderator analyses could have assisted in further refining the individual level assessment of recidivism reduction potential. While the extant research on the effectiveness of correctional interventions reviewed throughout this chapter suggests that the effectiveness of interventions is enhanced when risk and needs and responsivity are considered and that program quality impacts program effectiveness (Andrews, Zinger, et al., 1990; Andrews & Bonta, 2006, 2010; Andrews & Dowden, 2006; Bonta & Andrews, 2007; Dowden & Andrews, 1999a, 1999b, 2000; Lowenkamp & Latessa, 2005; Lowenkamp, Latessa, & Holsinger, 2006; Lowenkamp, Latessa, & Smith, 2006; Smith et al., 2009), few studies provided enough information to add more specific parameters to our simulation model.

Table 7.2 displays the percent reductions for each intervention arrayed across the six RNR program groups (see Chap. 6 for a discussion of the RNR program groups). Some interventions may be recommended at multiple program groups because interventions such as cognitive-behavioral therapy and drug courts are potentially effective for reducing recidivism across several risk and need profiles. In order to better guide adherence to the principles of the RNR model, we included moderators of treatment effectiveness in our model when they were available and found to be significantly related to recidivism outcomes in selected meta-analyses. To illustrate, in their analysis of the effectiveness of drug treatment interventions for offenders, Mitchell, Wilson, and MacKenzie (2007) found that in-prison drug treatment programming that included aftercare was more effective than in-prison drug treatment programming alone; an average 20 % reduction in recidivism with aftercare compared to 13 % reduction for in-prison programming alone (Mitchell et al., 2007). When information is available from a participating jurisdiction about whether their drug treatment program includes aftercare or not, the RNR Simulation Tool can adjust the estimated effectiveness of the programming on individual offender recidivism rates accordingly. Considering moderators allows the tool to better match offender profiles to available programs and more accurately estimate effectiveness.

The RNR Simulation Tool model is also built to adjust the estimated effectiveness of an intervention type according to the quality of the programming that is available within a jurisdiction. Based on the findings from two empirical studies which assessed the relationship between program quality and effectiveness using the Correctional Program Assessment Inventory (CPAI) (Lowenkamp & Smith, 2006; Nesovic, 2003), recidivism reduction potential is reduced in the RNR Simulation Tool by half when program quality is low and by 90 % when program quality is assessed as poor using the RNR Program Tool (see Table 7.1, columns 5 and 6). For example, an

Table 7.2 Recidivism reductions associated with RNR Simulation Tool Program categories

Intervention	RNR program category	Mean effect size (OR)	% Reduction (general)	% Reduction (mod quality)	% Reduction (low quality)
Target: substance dependence					
Drug treatment	A	1.27	12	6	1
TC	A	1.38	16	8	2
TC (90+ days)	A	1.45	18	9	2
TC (with aftercare)	A	1.51	20	10	2
TC (no aftercare)	A	1.31	13	7	1
TC (females only)	A	1.65	25	13	3
TC (nonviolent only)	A	1.49	20	10	2
Narcotic maintenance	A	1.40	17	9	2
Boot Camp (w/ treatment)	A	1.10	5	3	1
Drug court	A	1.66	25	13	3
Target: criminal thinking and substance abuse					
CBT	B	1.53	25	13	3
MRT	B	1.80	25	13	3
R&R	B	1.16	14	7	1
ISP with treatment	B	1.40	18	9	2
TC	B	1.38	16	8	2
TC (90+ days)	B	1.45	18	9	2
TC (with aftercare)	B	1.51	20	10	2
TC (no aftercare)	B	1.31	13	7	1
TC (females only)	B	1.65	25	13	3
TC (nonviolent only)	B	1.49	20	10	2
Counseling	B	1.50	20	10	2
Counseling (females only)	B	2.94	49	25	5
Counseling (males only)	B	1.67	25	13	3
Counseling(w/ aftercare)	B	1.82	29	15	3
Counseling (w/out aftercare)	B	1.45	18	9	2
Drug court	B	1.66	25	13	3
Target: social skills					
CBT	C	1.53	25	13	3
MRT	C	1.80	25	13	3
R&R	C	1.16	14	7	1
Counseling	C	1.50	20	10	2
Counseling (females only)	C	2.94	49	25	5
Counseling (males only)	C	1.67	25	13	3

(continued)

Table 7.2 (continued)

Intervention	RNR program category	Mean effect size (OR)	% Reduction (general)	% Reduction (mod quality)	% Reduction (low quality)
Counseling (w/aftercare)	C	1.82	29	15	3
Counseling (w/out aftercare)	C	1.45	18	9	2
Mental health treatment	C	0.19	17	9	2
DV treatment	C	0.12	16	8	2
Sex offender treatment	C	1.67	31	16	3
Drug court	C	1.66	25	13	3
Target: interpersonal skills					
Counseling	D	1.50	20	10	2
Counseling (females only)	D	2.94	49	25	5
Counseling (males only)	D	1.67	25	13	3
Counseling (w/ aftercare)	D	1.82	29	15	3
Counseling (w/out aftercare)	D	1.45	18	9	2
Employment programs	D	1.06	3	2	0
Basic education	D	1.44	18	9	2
Vocational programs	D	1.52	22	11	2
DV treatment	D	0.12	16	8	2
Sex offender treatment	D	1.67	31	16	3
Drug court	D	1.66	25	13	3
Target: life skills					
Restorative justice	E	1.33	14	7	1
Employment programs	E	1.06	3	2	0
Basic education	E	1.44	18	9	2
Vocational programs	E	1.52	22	11	2
Correctional industries	E	1.48	19	10	2
Target: punishment only					
Electronic monitoring	F	0.96	2	1	0
Intermediate sanctions	F	1.04	2	1	0
Boot Camp	F	1.02	1	1	0

intervention type that is found to reduce recidivism by 20 % on average will be assigned a maximum 10 % reduction if program quality and implementation are "low" and a maximum 2 % reduction in recidivism if program quality is "poor". This program quality adjustment adheres to the principles of effective intervention and helps ensure that feedback on potential recidivism impacts is tailored to the specific jurisdiction using the tool.

Improving Recidivism Outcomes at a System Level

The gap between research on the effectiveness of correctional interventions and routine correctional practice remains an unresolved issue. Taxman and colleagues (Taxman, Cropsey, Young, & Wexler, 2007; Taxman, Perdoni, & Caudy, 2013; Chap. 2, this volume) reported that less than 10 % of the offender population can participate in some type of programming and/or treatment on a given day, and that programming is usually not consistent with an offender's dynamic needs. Additionally, Taxman and colleagues (Taxman, Perdoni, & Harrison, 2007) found that only 34 % of 289 criminal justice agencies surveyed during the CJDATS project utilized a validated risk tool to place offenders in appropriate treatment services (Taxman, Cropsey, et al., 2007). While shifts towards using the RNR model are being made, it is important to address these gaps to achieve a greater impact on aggregate recidivism rates.

Increasing Population Impact Through Adherence to the RNR Principles

If the RNR principles were systematically adopted at a population level, what affect could we expect on recidivism? What if treatment matching was a routine practice and not just a recommendation? The RNR simulation tool will help answer these questions. In this section, we explore the potential impact of scaling up the use of effective treatment and achieving widespread adherence to the RNR principles within the field of corrections. To realize recidivism reduction gains at the system level, the reach of effective rehabilitative interventions needs to be extended to a greater proportion of the offender population. This requires improving access to treatment and creating a culture within correctional agencies that embraces treatment as a standard component of the correctional experience (Friedmann, Taxman, & Henderson, 2007; Taxman, Henderson, & Belenko, 2009).

Box 7.3 System-Level Impacts of Scaling Up Effective Interventions

Potential Impact of Offering Treatment to a Greater Percentage of the Offender Population within Justice Settings

1. *Contagion Effect*: Treatment will be normal within the correctional environment. This means that correctional agencies will routinely offer treatment and therefore the activities of the staff and programs will be in sync with the requirements of quality treatment such as cognitive-behavioral therapy. Staff will be integrated into the treatment program to reinforce the treatment potential. And, it means that the "traditional" environment will support the treatment programming. Correctional and probation officers will offer augmented treatment services to reinforce treatment outcomes.
2. *Fidelity Effect*: The more treatment is offered (and to more people), the more the programming will become routine and adhere to treatment principles. The system will be focused on quality—offering treatment that adheres to the principles of effective intervention.
3. *Penetration Effect*: The reach of treatment will be greater because the practice will become routine. The emphasis will be on providing the greatest percentage of offenders with care.
4. *Cultural Effect*: The corrections culture will be altered to embrace treatment as a tool of security. Correctional and probation officers will view their mission as treatment, and the culture will reinforce treatment goals.
5. *Treatment Matching Effect*: With more programs, there is potential to assign offenders to programs based on their risk and need profile. This means there will be more homogeneity within programs, which should provide therapists with a clearer goal for all offenders in the program. This should increase positive outcomes by providing counselors with similar needs to address.

Improving Access to Effective Treatment

In his poignant critique of prison-based treatment programs, Austin (2009, p. 311) argued that the ramping up of correctional treatment would not have the desired effect on recidivism because treatment effects are generally small and often exaggerated in meta-analyses. While we do not disagree with his argument that system level improvement requires policy and contextual change, his critique of treatment effectiveness does not fully consider the potential systematic impact of scaling up effective interventions. Even modest treatment effects can have a considerable impact on outcomes at the system level if two conditions are met: (1) treatment is implemented with fidelity; and (2) treatment is available to a large proportion of the target population. Research from the field of public health supports the claim that reaching a larger proportion of the population in need will yield a greater impact on outcomes even when intervention effects are only modest (see e.g., Frieden, 2010; Heller & Dobson, 2000; Tucker & Roth, 2006). An example of the public health impact

Table 7.3 Population impact of increased treatment utilization in hypothetical population ($N = 10,000$)[a]

Percent in treatment	Recidivism rate[b]	NEPP[c]	Recidivism rate[d] (RNR adjustment)	Recidivism rate[e] (contagion adjustment)
0	60.0	–	60.0	60.0
10	58.8	–	57.6	58.8
20	57.6	118.8	55.2	57.6
30	56.4	237.6	52.8	56.4
40	55.2	356.4	50.4	48.0
50	54.0	475.2	48.0	45.0
60	52.8	594.0	45.6	42.0
70	51.6	712.8	43.2	39.0
80	50.4	831.6	40.8	36.0
90	49.3	950.4	38.4	33.0
100	48.0	1,069.2	36.0	30.0

[a]Assumes equal distribution of all risk levels and a baseline recidivism risk of 0.60
[b]Based on a 0.20 relative risk reduction associated with rehabilitative correctional programming
[c]NEPP = population size × proportion eligible for treatment × proportion with disorder × baseline risk × RRR
[d]Based on a 0.40 relative risk reduction associated with correctional programming adhering to the RNR principles
[e]Based on a 0.50 relative risk reduction once 30 % of the population is involved in treatment

concept is the use of aspirin after a stroke or cardiac event. While the use of aspirin produces only a slight reduction in the risk of subsequent strokes or cardiac events, it has a greater population impact than more effective treatments like surgery because it is available to more patients in need (see Heller & Dobson, 2000 for stroke example). Regardless of effectiveness, when an intervention is able to reach a greater proportion of the population in need, the potential population impact is increased.

Improving access to programming and making treatment a central component of the correctional culture will result in a greater reduction in recidivism than having effective programs that reach a small proportion of the offender population. Box 7.3 summarizes the potential systematic effects of expanding treatment utilization in the justice system. When treatment is offered routinely, it impacts referrals, access, retention, and outcomes. That is, the correctional system will not consider treatment a foreign entity, which has been found to undermine many of the collateral benefits of having a system of effective and responsive treatment services.

To illustrate the potential population impact of improved adherence to the RNR principles and a cultural change within corrections agencies, we estimated a series of hypothetical models depicting changes in aggregate recidivism rates associated with increasing the proportion of offenders who have access to treatment (Table 7.3). Additionally, we calculated the number needed to treat (NNT) (Cook & Sackett, 1995) and the number of events prevented in your population (NEPP) (Heller, Edwards, & McElduff, 2003). The NNT statistic provides an indicator of the number of individuals who need to be treated with an intervention to prevent one negative event (i.e., one recidivist). The NEPP provides an estimate of the number of events (e.g., incidences of recidivism) that will be prevented by offering an intervention to a specific population (Heller et al., 2003).

The results depicted in Table 7.3 are based on a hypothetical offender population of 10,000 individuals in one jurisdiction. A conservative estimate of baseline recidivism risk of 0.60 was used in this model based on the Bureau of Justice Statistics (BJS) 1994 recidivism study which found that over two thirds (67.5 %) of all released prisoners were rearrested within 3 years (Langan & Levin, 2002). The relative risk reduction (RRR) associated with correctional treatment was specified as 0.20 (a modest effect) consistent with the meta-analytic findings reviewed by Lipsey and Cullen (2007). The proportion of offenders currently receiving treatment was specified as 0.10 based on the findings of Taxman and colleagues (Taxman, Cropsey, et al., 2007; Taxman et al., 2013 (and in Chap. 2 of this book)).

Using the model parameters described above we explored the potential population impact of increasing the proportion of offenders receiving treatment without assuming any collective impact on the correctional culture stemming from treating more offenders. As displayed in Table 7.3, each 10 % increase in the proportion of the population receiving treatment results in only a small improvement (1.2 % absolute rate reduction) in the aggregate population recidivism rate. While the impact of moderately effective treatment on population level recidivism rates is not large, it is still meaningful. By moving from the 10 % of the population that is currently receiving treatment to 50 % receiving treatment, the population recidivism rate can be reduced by 8 % (RRR); a modest but meaningful impact. The NNT to prevent one recidivism event in this population is eight. This suggests that for every eight offenders who receive treatment, one will be prevented from further offending. The NNT of eight is much lower than many medical interventions used in preventing death from cardiovascular disease and stroke (see e.g., Chamnan, Simmons, Khaw, Wareham, & Griffin, 2010; Heller & Dobson, 2000; Heller et al., 2003) and is a marked improvement over the NNT of 33 * needed for punishment sanctions to prevent one recidivism event. If we increase the rate of providing treatment from 10 to 50 %, this would prevent approximately 475 (NEPP) recidivism events over the course of 1 year for a population of 10,000 offenders. That results in about 475 less victims of crime.

The analysis described above did not consider the potential collateral consequences of expanding treatment to a greater percentage of the population. Based on the literature reviewed above which indicates that programs that adhere to the RNR principles are capable of producing significantly larger recidivism reduction effects, we recalculated the potential impact of treatment on the aggregate recidivism rate using a relative risk reduction of 0.40 (Table 7.3, Column 4). The model also takes into consideration the potential impact of a contagion or public health impact (Table 7.3, Column 5) wherein as the proportion of the population receiving care increases, cultural shifts occur making treatment utilization a more routine part of correctional practices which subsequently increases the effectiveness of interventions (Frieden, 2010; Tucker & Roth, 2006).

---

*This estimate assumes a .05 relative risk reduction for punishment sanctions.

Table 7.4 Example treatment dosage matching strategy from Jurisdiction A

	Low	Moderate	Moderate/High	High
Risk score (male)	0–14	15–23	24–33	34+
Risk score (female)	0–14	15–21	22–28	29+
Dosage	N/A	100 h of CBT-qualified interaction and programming	200 h of CBT-qualified interaction and programming	300+ hours of CBT-qualified interaction and programming
Supervision length	Minimal length of supervision	3–6 months	6–9 months	9–18 months
Supervision intensity	Not required to meet with staff	Meet with CM once per week	Meet with CM twice per week	Meet with CM thrice per week

In the RNR adjusted model we see that increasing the effectiveness of treatment through widespread adherence to the principles of effective intervention and changing the culture of corrections produces a greater impact at the population level: providing treatment to 30 % of the population would reduce the aggregate recidivism rate by 12 % (RRR) in the RNR model, while providing treatment to 50 % of the population would yield a 20 % (RRR) reduction in recidivism. Based on the RRR of 0.40, the NNT for this example is four. This suggests that when the principles of effective intervention are adhered to, treating four offenders will prevent one recidivism event. These scenarios could produce even larger population-level impacts if the results were configured to include age, gender, and other key determinants of recidivism rates.

Treatment Matching as a Policy

As detailed by Crites and Taxman (in Chap. 6 of this book), implementing strategies for matching offenders to levels of care and assessing programs to ensure that they possess the appropriate RNR components is needed to maximize effectiveness. Table 7.4 illustrates an evidence-based strategy that one community corrections agency has implemented to match treatment dosage to offender risk and needs using a validated offender risk assessment tool. The strategy detailed in Table 7.4 varies treatment dosage by risk level and adheres to the general responsivity principle of the RNR model by requiring that all treatment programs be CBT-based. While this approach does adhere to the risk and responsivity principles, it *does not* specify which programs are appropriate for addressing specific criminogenic needs and therefore neglects the need principle, a key component of the of the RNR model. Despite this limitation, strategies like this one and the more complete algorithm-based strategy employed by the RNR Simulation Tool (see Chap. 5) are an important step toward maximizing the effectiveness of correctional interventions.

Table 7.5 Estimated impact of treatment matching strategy in Jurisdiction A

Risk level[a]	Current 3-year reconviction rate	Relative risk reduction[b]	3-Year reconviction rate w/matching[c]
High	52.1	0.35	33.9
Moderate	38.7	0.32	26.3
Low	27.1	0.57	11.7

[a]Risk level was identified from the criminal history subscale of the LSI-R
[b]Relative risk reductions (RRR) calculated from Bourgon & Armstrong, 2005 (Table 2, p. 17)
[c]Estimated 3-year reconviction rate for correctly matched offenders who complete recommended dosage

Translational research that has tested the principles of effective interventions has identified treatment dosage and treatment matching as key elements of program effectiveness (Bourgon & Armstrong, 2005). In their evaluation of a prison-based CBT program guided by the RNR principles, Bourgon and Armstrong (2005) found support for a linear relationship between treatment dosage and effectiveness. They found that each additional week of treatment (~20 h of group-based CBT) was related to about a 1.5 % absolute recidivism risk reduction for the average offender. This impact was even greater when dosage was matched to offender risk and need profiles. For moderate risk offenders with few criminogenic needs, completion of low-dosage CBT programming (100 h over 5 weeks) was associated with a 16 % absolute risk reduction relative to a untreated control group. For high-risk offenders with multiple criminogenic needs, completion of high-dosage CBT (300 h over 15 weeks) was associated with a 20 % absolute risk reduction. When high-risk, high-need offenders completed the low-dosage treatment, no significant recidivism reduction was observed (Bourgon & Armstrong, 2005). These findings suggest that treatment matching strategies like the one described in Table 7.4 can have a significant impact on recidivism and that recidivism reductions are maximized when treatment matching strategies are guided by the RNR principles. These strategies improve treatment outcomes by tailoring key intervention components to offender risk and need profiles.

Applying the findings of Bourgon and Armstrong (2005) to data from the community corrections agency employing the proposed treatment matching strategy depicted in Table 7.4 provides an illustration of the potential recidivism reduction impact of making treatment matching a routine correctional practice. As displayed in Table 7.5, applying relative risk reductions (0.35 for high risk, 0.31 for moderate risk, and 0.57 for lower risk) from Bourgon and Armstrong to Jurisdiction A data results in a considerable drop in 3-year reconviction rates for correctly matched offenders who complete their recommended dosage of CBT programming. This example suggests that the proposed matching strategy could result in as much as a 35 % drop in the 3-year reconviction rate (52.1–33.9 %) for high-risk offenders in Jurisdiction A (Table 7.5). This example provides strong support for the potential impact of adhering to the RNR model and making treatment matching a policy.

Discussion

Understanding what works for whom is an important and still emerging area of science. We know that responding to the dynamic criminogenic needs of offenders is a complex process. Existing meta-analysis have identified "what works" but more information is needed about tailoring interventions to specific individual risk and need profiles. Since meta-analyses have the ability to provide standardized metrics of the robustness and consistency of empirical findings across time, settings, and target populations, this body of knowledge is particularly well-suited for informing what works and is subsequently a good source of data for informing simulation model inputs. As explicated by Lipsey and Cullen (2007) following their review of treatment effectiveness from over 40 meta-analyses, "the greatest obstacle to using rehabilitation treatment effectively is not a nothing-works research literature with nothing to offer but, rather, a correctional system that does not use the research available" (Lipsey & Cullen, 2007, p. 315).

The literature reviewed in the current chapter demonstrates that well-designed and well-implemented rehabilitative correctional programming can have an appreciable impact on recidivism at the individual level. The findings summarized here support the principles of effective intervention as the primary evidence-based framework for moving research into practice in the area of correctional treatment. While risk and need information is collected in many correctional agencies, it is not routinely used in making treatment matching decisions and an overwhelming majority of offenders who are in need of treatment do not receive it (Taxman, Cropsey et al., 2007; Taxman, Perdoni, et al., 2007; Taxman, Perdoni, et al., 2013). Without question, what we know about what works does not currently constitute routine correctional practice (Lipsey & Cullen, 2007).

Knowledge translation is perhaps the most pressing challenge facing the evidence-based corrections movement. While we are in need of considerably more research on what interventions work for whom and in what setting, moving what we already know from research into practice can have a profound impact on recidivism rates at the system level. The argument advanced throughout this volume is that simulation modeling can be used to help the field better apply what we know about what works and subsequently help foster adherence to the RNR principles and facilitate a system-level impact on recidivism. The RNR Simulation Tool, which is informed by existing meta-analyses on the effectiveness of correctional interventions, provides a mechanism for helping link offenders to interventions to maximize potential recidivism reduction outcomes. Through simulation, the tool provides decision support to assist justice agencies in building up the capacity to achieve responsivity to the risk and needs of their offender population. The tool bridges the gap between research and practice through translation and application of the principles of effective intervention.

Potential Impact of a Limited Evidence Base

"You cannot develop evidence-based policies and practices without first conducting high quality research on the problem under review" (Byrne & Lurigio, 2009, p. 304). Unfortunately, this observation remains a problem and the current effort to build a simulation tool that provides decision support for matching offenders to effective interventions and estimates potential recidivism reductions associated with adherence to the RNR principles is not immune to some limitations. The primary limitation of the current effort is that there are insufficient primary studies to conduct meta-analysis on the effectiveness of some interventions and there are few moderator analyses of the impact of demographic factors within existing meta-analyses. While we took considerable measures to ensure that our search for existing meta-analyses was exhaustive and that we identified the most valid findings available regarding treatment effectiveness, there are still some interventions for which we do not have effectiveness estimates and to the degree that treatment effects may be overestimated in existing meta-analyses (Austin, 2009) they may be overestimated in our tool as well. That caveat aside, empirical evaluations of the overestimation of treatment effects in meta-analyses that include nonexperimental primary studies suggest that the evidence in support of this claim is mixed at best and that well-designed meta-analyses provide an accurate indication of treatment effects even when they include nonexperimental designs (Lipsey & Cullen, 2007).

A final limitation of the existing evidence base that makes moving research into practice difficult is the lack of clearly defined interventions within existing meta-analyses. Extant meta-analyses, and primary studies, of treatment effectiveness often do a poor job of defining the key components of the interventions being assessed (Caudy et al., forthcoming; Glasziou et al., 2010; Michie, Fixsen, Grimshaw, & Eccles, 2009; Shaffer & Pratt, 2009). This ambiguity surrounding key intervention components represents a major barrier to knowledge translation. The ability of meta-analyses to inform practice will be greatly enhanced by better attention to the functional components of effective correctional interventions.

While these limitations are beyond the control of our model, they do potentially limit the ability of our tool to facilitate evidence-driven treatment matching. Future meta-analytic research should examine key individual-level moderators of treatment effectiveness including risk, problem severity, and demographics as well as contextual and study-level moderators. Additionally, both primary and meta-analytic studies of treatment effects need to accurately define the key functional components of interventions being tested to improve knowledge translation potential (Glasziou et al., 2010; Michie et al., 2009 Shaffer & Pratt, 2009).

Conclusions

The process of linking meta-analytic findings regarding the effectiveness of correctional interventions to the RNR Simulation Tool was demonstrated in the preceding pages. The use of simulation modeling techniques like those employed in the RNR Simulation Tool can help practitioners and policy makers see the potential impact of scaling up evidence-based practices and can serve as a valuable asset in the effort to translate research into everyday correctional practice to maximize the return on our correctional investment. The development of decision support tools like the RNR Simulation Tool will play an important role in influencing evidence-based decision making and researchers should continue to develop knowledge translation strategies to improve both public safety and public health impacts of correctional interventions.

If the US correctional system is going to have an appreciable impact on aggregate recidivism rates, the overarching correctional culture must change and embrace a human services approach (Andrews & Bonta, 2006, 2010; Bonta & Andrews, 2007). While empirical evidence regarding the effectiveness of rehabilitative correctional interventions has expanded rapidly over the past 30 years, this research has largely not translated into practice (Lipsey & Cullen, 2007). The US correctional system still lacks the capacity to provide treatment and when treatment is provided it seldom constitutes what the evidence base has identified as best practices (Taxman, Cropsey, et al., 2007; Perdoni, et al., 2007; Taxman, Perdoni, et al., 2013). Impacting recidivism at the population level requires improving access to evidence-based interventions and aligning justice system goals with the principles of effective intervention. Extensive empirical research has clearly established that America's get tough policies and incarceration boom have failed miserably (Austin, 2009; Lipsey & Cullen, 2007). Only through improved availability of quality correctional programming and improved access to rehabilitative interventions can the justice system expect to have an effect on aggregate recidivism rates and improve offender outcomes.

The examples provided in this chapter clearly show that even moderately effective correctional interventions can have a considerable impact on aggregate recidivism rates if they are a part of a justice system that embraces treatment and rehabilitation as primary goals. Based on available estimates of treatment effectiveness we calculated that the number of offenders needed to treat (NNT) to prevent one recidivism event was eight for rehabilitation treatment compared to 33 for correctional sanctions without treatment. This suggests that even current programming that does not reflect best practices can have a noticeable impact on recidivism. When treatment adhered with the principles of effective intervention, it was estimated that the NNT was reduced to four. These findings illustrate the vital need to move research into practice and develop an evidence-based justice system that is guided by the principles of effective intervention and what we know about what works. The population impact of a responsive justice system on recidivism can be substantial.

References

Andrews, D. A., & Bonta, J. (2010). *The psychology of criminal conduct* (5th ed.). Cincinnati, OH: Anderson.

Andrews, D. A., Bonta, J., & Hoge, R. D. (1990). Classification for effective rehabilitation: Rediscovering psychology. *Criminal Justice and Behavior, 17,* 19–52.

Andrews, D. A., & Dowden, C. (2005). Managing correctional treatment for reduced recidivism: A meta-analytic review of programme integrity. *Legal and Criminological Psychology, 10,* 173–87.

Andrews, D. A., & Dowden, C. (2006). Risk principle of case classification in correctional treatment. *Int J Offender Ther Comp Criminol, 50*(1), 88–100.

Andrews, D. A., Zinger, I., Hoge, R. D., Bonta, J., Gendreau, P., & Cullen, F. T. (1990). Does correctional treatment work? A clinically relevant and psychologically informed meta-analysis. *Criminology, 28*(3), 369–404.

Austin, J. (2009). The limits of prison based treatment. *Victims and Offenders, 4,* 311–20.

Babcock, J. C., Green, C. E., & Robie, C. (2004). Does batterers' treatment work? A meta-analytic review of domestic violence treatment. *Clinical Psychology Review, 23*(8), 1023–53.

Beck, R., & Fernandez, E. (1998). Cognitive-behavioral therapy in the treatment of anger: a meta-analysis. *Cognitive Therapy and Research, 22*(1), 63–74.

Bonta, J., & Andrews, D. A. (2007). *Risk-need-responsivity model for offender assessment and treatment.* Ottawa, ON: Public Safety Canada (User Report 2007–06).

Borenstein, M., Hedges, L. V., Higgins, J. P. T., & Rothstein, H. R. (2009). *Introduction to meta-analysis.* Hoboken, NJ: Wiley.

Bourgon, G., & Armstrong, B. (2005). Transferring the principles of effective treatment into a "real world" prison setting. *Criminal Justice and Behavior, 32*(1), 3–25.

Byrne, J. M., & Lurigio, A. J. (2009). Separating science from nonsense: Evidence-based research, policy, and practice in criminal justice and juvenile justice settings. *Victims and Offenders, 4,* 303–10.

Caudy, M.S., Taxman, F.S., Tang, L., & Watson, C. (forthcoming). Using systematic reviews and meta-analyses to advance knowledge translation and dissemination. In: D. Weisburd, D. F. Farrington (eds.). *Systematic reviews in criminology: What have we learned?* New York: Springer.

Chamnan, P., Simmons, R. J., Khaw, K. T., Wareham, N. J., & Griffin, S. J. (2010). Estimating the population impact of screening strategies for identifying and treating people at high risk of cardiovascular disease: Modeling study. *Br Med J, 340,* 1–11.

Cook, R. J., & Sackett, D. L. (1995). The number needed to treat: A clinically useful measure of treatment effect. *Br Med J, 310,* 452–4.

Dowden, C. (1998). A meta-analytic examination of the risk, need and responsivity principles and their importance within the rehabilitation debate. Unpublished M.A. Thesis, Carleton University, Ottawa, ON.

Dowden, C., & Andrews, D. A. (1999a). What works for female offenders: A meta-analytic review. *Crime and Delinquency, 45,* 438–51.

Dowden, C., & Andrews, D. A. (1999b). What works in young offender treatment: A meta-analysis. *Forum on Corrections Research, 45,* 438–52.

Dowden, C., & Andrews, D. A. (2000). Effective correctional treatment and violent reoffending: A meta-analysis. *Canadian Journal of Criminology, 42,* 449–67.

Dowden, C., Antonowicz, D., & Andrews, D. A. (2003). The effectiveness of relapse prevention with offenders: a meta-analysis. International Journal of Offender Therapy and Comparative Criminology, 47(5), 516–28.

Drake, E., Aos, S., & Miller, M. (2009). *Evidence-based public policy options to reduce crime and criminal justice costs: Implications in Washington State* (No. 09-00-1201). Washington State Institute for Public Policy.

Feder, L., & Wilson, D. B. (2005). A meta-analytic review of court-mandated batterer intervention programs: Can courts affect abusers' behavior? *Journal of Experimental Criminology, 1*(2), 239–62.

Frieden, T. R. (2010). A framework for public health action: The health impact pyramid. *Am J Public Health, 100*(4), 590–5.

Friedmann, P. D., Taxman, F. S., & Henderson, C. (2007). Evidence-based treatment practices for drug-involved adults in the criminal justice system. *J Subst Abuse Treat, 32*(3), 267–77.

Gallagher, C. A., Wilson, D. B., Paul Hirschfield, M. A., Coggeshall, M. B., & MacKenzie, D. L. (1999). Quantitative review of the effects of sex offender treatment on sexual reoffending. *Corrections Management Quarterly, 3*(4), 11.

Gendreau, P., Goggin, C., French, S. A., & Smith, P. (2006). Practicing psychology in correctional settings: "What works in reducing criminal behavior". In A. K. Hess & I. B. Weiner (Eds.), *The handbook of forensic psychology* (3rd ed., pp. 722–50). New York: Wiley.

Gendreau, P., Goggin, C., & Smith, P. (1999). The forgotten issue in effective correctional treatment: Program implementation. *Int J Offender Ther Comp Criminol, 43*(3), 180–7.

Gendreau, P., & Smith, P. (2007). Influencing the "people who count": Some perspectives on the reporting of meta-analytic results for prediction and treatment options with offenders. *Criminal Justice and Behavior, 34*(12), 1536–59.

Gendreau, P., Smith, P., & French, S. A. (2006). The theory of effective correctional intervention: Empirical status and future directions. In F. T. Cullen, J. P. Wright, & K. R. Blevins (Eds.), *Taking stock: The status of criminological theory* (Advances in criminological theory, Vol. 15, pp. 419–46). Transaction: New Brunswick, NJ.

Glasziou, P., Chalmers, I., Altman, D. G., Bastian, H., Boutron, I., Brice, A., et al. (2010). Taking healthcare interventions from trial to practice. *Br Med J, 341*, 384–7.

Hall, G. C. N. (1995). Sexual offender recidivism revisited: a meta-analysis of recent treatment studies. *J Consult Clin Psychol, 63*(5), 802–9.

Hammerstrom, K., Wade, E. & Jorgensen, A. K. (2010). *Searching for studies: A guide to information retrieval for Campbell systematic reviews*. Retrieved from http://www.campbellcollaboration.org/resources/research/new_information_retrieval_guide.php

Hanson, R. K., Gordon, A., Harris, A. J. R., Marques, J. K., Murphy, W., Quinsey, V. L., et al. (2002). First report of the collaborative outcome data project on the effectiveness of psychological treatment for sex offenders. *Sexual Abuse: A Journal of Research and Treatment, 14*(2), 169–94.

Heller, R. F., & Dobson, A. J. (2000). Disease impact number and population impact number: Population perspectives to measures of risk and benefit. *Br Med J, 321*, 950–3.

Heller, R. F., Edwards, R., & McElduff, P. (2003). Implementing guidelines in primary care: Can population impact measures help? *BMC Public Health, 3*(7).

Holloway, K. R., Bennett, T. H., & Farrington, D. P. (2006). The effectiveness of drug treatment programs in reducing criminal behavior: a meta-analysis. *Psicothema, 18*(3), 620–9.

Jadad, A. R., Cook, D. J., & Browman, G. P. (1997). A guide to interpreting discordant systematic reviews. *Can Med Assoc J, 156*(10), 411–6.

Landenberger, N. A., & Lipsey, M. W. (2005). The positive effects of cognitive-behavioral programs for offenders: A meta-analysis of factors associated with effective treatment. *Journal of Experimental Criminology, 1*, 451–76.

Latimer, J., Dowden, C., & Muise, D. (2005). The effectiveness of restorative justice practices: a meta-analysis. *The Prison Journal, 85*(2), 127–44. doi:10.1177/0032885505276969.

Langan, P. A., & Levin, D. J. (2002). *Recidivism of prisoners released in 1994*. Washington DC: Bureau of Justice Statistics.

Lefebvre, C., Manheimer, E., & Glanville, J. (2011). Searching for studies. In J.P.T. Higgins & S. Green (Eds.). *Cochrane Handbook for Systematic Reviews of Interventions* Version 5.1.0 (Chapter 6). Retrieved from www.cochrane-handbook.org

Lipsey, M. W. & Landenberger, N. A. (2005). Congitive-behavioral interventions: A meta-analysis of randomized controlled studies. In B. C. Welsh & D. P. Farrington (Eds.), *Preventing crime: What works for children, offenders, victims, and places*. Springer, Berlin Heidelberg New York.

Lipsey, M. W., & Cullen, F. T. (2007). The effectiveness of correctional rehabilitation: A review of systematic reviews. *Annual Review of Law and Social Science, 3*, 297–320.

Lipsey, M. W., Landenberger, N. A. & Wilson, S. J. (2007). Effects of cognitive-behavioral programs for criminal offenders. *Campbell Systematic Reviews*. Retrieved from http://campbellcollaboration.org/lib/project/29/

Lipsey, M. W., & Wilson, D. B. (2001). *Practical meta-analysis*. Thousand Oaks, CA: Sage.

Lipton, D. S., Pearson, F. S., Cleland, C. M., & Yee, D. (2008). The Effects of therapeutic communities and milieu therapy on recidivism: Meta-analytic findings from the Correctional Drug Abuse Treatment Effectiveness (CDATE) Study. In J. McGuire (Ed.), *Offender rehabilitation and treatment* (pp. 39–77). Chichester: Wiley.

Little, G. L. (2005). Meta-analysis of moral reconation therapy recidivism results from probation and parole implementations. *Cognitive Behavioral Treatment Review, 14*(1/2), 14–6.

Lowenkamp, C. T., & Latessa, E. J. (2005). Increasing the effectiveness of correctional programming through the risk principle: Identifying offenders for residential placement. *Criminology and Public Policy, 4*(2), 263–90.

Lowenkamp, C. T., Latessa, E. J., & Holsinger, A. M. (2006). The risk principle in action: What have we learned from 13,676 offenders and 97 correctional programs? *Crime & Delinquency, 52*(1), 77–93.

Lowenkamp, C. T., Latessa, E. J., & Smith, P. (2006). Does correctional program quality really matter? The impact of adhering to the principles of effective intervention. *Criminology and Public Policy, 5*(3), 575–94.

McGuire, J. (2002). Integrating findings from research reviews. In J. McGuire (Ed.), *Offender rehabilitation and treatment: Effective programmes and policies to reduce re-offending* (pp. 3–38). Hobokon, NJ: Wiley.

Michie, S., Fixsen, D., Grimshaw, J. W., & Eccles, M. P. (2009). Specifying and reporting complex behaviour change interventions: The need for a scientific method. *Implementation Science, 4*(1), 40–5.

Mitchell, O., Wilson, D. B., Eggers, A., & MacKenzie, D. L. (2012). Assessing the effectiveness of drug courts on recidivism: A meta-analytic review of traditional and non-traditional drug courts. *Journal of Criminal Justice, 40*(1), 60–71.

Mitchell, O., Wilson, D. B., & MacKenzie, D. L. (2007). Does incarceration-based drug treatment reduce recidivism? A meta-analytic synthesis of the research. *Journal of Experimental Criminology, 3*(4), 353–75.

Nesovic, A. (2003). *Psychometric evaluation of the Correctional Program Assessment Inventory (CPAI)*. Unpublished doctoral dissertation, Carleton University, Ottawa, ON.

Perry, A., Coulton, S., Glanville, J., Godfrey, C., Lunn, J., McDougall, C., et al. (2006). Interventions for drug-using offenders in the courts, secure establishments and the community. *Cochrane Database of Systematic Reviews*, (3):CD005193.

Prendergast, M. L., Podus, D., Chang, E., & Urada, D. (2002). The effectiveness of drug abuse treatment: a meta-analysis of comparison group studies. *Drug Alcohol Depend, 67*(1), 53–72.

Renzema, M., & Mayo-Wilson, E. (2005). Can electronic monitoring reduce crime for moderate to high-risk offenders? *Journal of Experimental Criminology, 1*, 215–37.

Schmucker, M., & Lösel, F. (2008). Does sexual offender treatment work? A systematic review of outcome evaluations. *Psicothema, 20*(1), 10–19.

Shaffer, D. K., & Pratt, T. C. (2009). Meta-analysis, moderators, and treatment effectiveness: The importance of digging deeper for evidence of program integrity. *Journal of Offender Rehabilitation, 48*(2), 101–19.

Sherman, LW, Gottfredson, D., C., MacKenzie, D.L, Eck, J., Reuter, P., & Bushway, S.D. (1997) Preventing crime: What works, what doesn't, what's promising. Washington DC: National Institute of Justice. http://www.ojp.usdoj.gov/nijJustice Information Center

Smith, P., Goggin, C., & Gendreau, P. (2002). *The effects of prison and intermediate sanctions on recidivism: General effects and individual differences*. Ottawa, Canada: Department of Solicitor General Canada, Ottawa.

Smith, P., Gendreau, P., & Swartz, K. (2009). Validating the principles of effective intervention: A systematic review of the contributions of meta-analysis in the field of corrections. *Victims and Offenders, 4*, 148–69.

Taxman, F. S., & Belenko, S. (2012). *Implementing evidence-based practices in community corrections and addiction treatment*. New York: Springer.

Taxman, F. S., Cropsey, K. L., Young, D. W., & Wexler, H. (2007). Screening, assessment, and referral practices in adult correctional settings. *Criminal Justice and Behavior, 34*(9), 1216–34.

Taxman, F. S., Henderson, C. E., & Belenko, S. (2009). Organizational context, system change, and adopting treatment delivery systems in the criminal justice system. *Drug Alcohol Depend, 103S*, S1–6.

Taxman, F. S., Perdoni, M. L., & Caudy, M. S. (2013). The plight of providing appropriate substance abuse treatment services to offenders: Modeling the gaps in service delivery. *Victims & Offenders, 8*, 70–93.

Taxman, F. S., Perdoni, M. L., & Harrison, L. D. (2007). Drug treatment services for adult offenders: The state of the state. *J Subst Abuse Treat, 32*(3), 239–54.

Tong, J., & Farrington, D. P. (2006). How effective is the "Reasoning and Rehabilitation" programme in reducing reoffending? A meta-analysis of evaluations in four countries. *Psychology, Crime & Law, 12*(1), 3–24. doi:10.1080/10683160512331316253.

Tucker, J. A., & Roth, D. L. (2006). Extending the evidence hierarchy to enhance evidence-based practice for substance use disorders. *Addiction, 101*, 918–32.

Visher, C., Winterfield, L., & Coggeshall, M. (2005). Ex-offender employment programs and recidivism: A meta-analysis. *Journal of Experimental Criminology, 1*(3), 295–316.

Wilson, D. B., Gallagher, C. A., & MacKenzie, D. L. (2000). A meta-analysis of corrections-based education, vocation, and work programs for adult offenders. *Journal of Research in Crime and Delinquency, 37*(4), 347–368. doi:10.1177/0022427800037004001.

Wilson, D. B., Bouffard, L. A., & MacKenzie, D. L. (2005). A quantitative review of structured, group-oriented, cognitive-behavioral programs for offenders. *Criminal Justice and Behavior, 32*(2), 172–204. doi:10.1177/0093854804272889.

Wilson, D. B., MacKenzie, D. L., & Mitchell, F. N. (2008). Effects of correctional boot camps on offending (Campbell Systematic Review). Oslo, Norway: The Campbell Collaboration.

Wilson, D. B. (2001). Meta-analytic methods for criminology. *The Annals of the American Academy, 587*, 71–89.

Wilson, D. B. (2009). Missing a critical piece of the pie: Simple document search strategies inadequate for systematic reviews. *Journal of Experimental Criminology, 5*(4), 429–40.

Wilson, D. B. (2010). Meta-analysis. In A. R. Piquero & D. Weisburd (Eds.), *Handbook of quantitative criminology* (pp. 181–208). New York: Springer.

Wilson, D. B. (forthcoming). Correctional interventions. In D. Weisburd & D. F. Farrington (Eds.). *Systematic reviews in criminology: What have we learned?* New York: Springer.

Part III
Simulation Applications

Chapter 8
RNR Simulation Tool: A Synthetic Datasets and Its Uses for Policy Simulations

Avinash Bhati, Erin L. Crites, and Faye S. Taxman

Introduction

Consider, as a point of departure, that analysts are interested in studying a population of offenders in prison or on community supervision in order to assist policy-makers or practitioners make evidence-based decisions. This might require several pieces of information. What are the available choices? Who are the offenders? What types of offenders would benefit most (or least) from a specific policy intervention? What is the expected base rate—i.e., the failure rate in the absence of this intervention? And so on. To answer all of these questions, one might require a considerable data collection and analysis effort. A specialized data collection effort—typically very expensive—might need to be commissioned. But suppose there already existed relevant information on different parts of the problem. Can this information—possibly from diverse sources—be combined in some way to obtain a dataset that can be analyzed as if a sample had actually been obtained from the population? The information may exist in other datasets, published results, aggregate numbers from annual reports, etc. Can this information be synthesized in some way to allow the analyst to proceed? This chapter describes and applies an information-theoretic framework to this problem.

The next section describes the structure of the synthetic dataset and the methodology used to populate it. Following that, we present application of the approach to develop a national-level synthetic dataset that is then customized to reflect the

A. Bhati, Ph.D. (✉)
Maxarth, LLC, Gaithersburg, MD, USA
e-mail: abhati@maxarth.com

E.L. Crites, M.A. • F.S. Taxman, Ph.D.
Department of Criminology, Law and Society, George Mason University, 10900 University, Boulevard, Fairfax, VA 20110, USA

F.S. Taxman and A. Pattavina (eds.), *Simulation Strategies to Reduce Recidivism:*   197
Risk Need Responsivity (RNR) Modeling for the Criminal Justice System,
DOI 10.1007/978-1-4614-6188-3_8, © Springer Science+Business Media New York 2013

aggregate attributes from two jurisdictions as well as to develop estimates of key variables that a jurisdiction is missing information on.

Finally, the chapter will describe how this synthetic data is used in the full web-based RNR Simulation Tool. This web-based model includes three portals—assess an individual, rate a program, and assess your jurisdiction. The assess an individual and assess a jurisdiction portals draw directly from the synthetic data described in the early parts of this chapter, while the rate a program portal uses the RNR Program Tool described in Chap. 6. This final section will provide a brief discussion of how each of these three portals function and present an example of how the portals use the synthetic data to provide feedback to a jurisdiction.

Methodology

To make things concrete, let's consider a simple situation. Suppose we wish to offer a community correction agency guidance on how to decide on program placement. Let there be a set of $k = 1 \ldots K$ different attributes they can observe as the offender comes onto supervision. This may include standard demographic characteristics, current offense, risk measures, needs, and lifestyle measures. The practitioner has a choice of programs he can offer to the client. At an operational level, the practitioner needs to know what is the best choice to offer to the next offender that walks through the door. More broadly, at a tactical level, agency management needs to allocate staff to various program or monitoring teams to keep track of the program activities. At a strategic level, the agency or jurisdiction policy-makers need to plan for future programming needs based on projections. The goal here is to create a tool that can offer this diverse group of policy-makers and practitioners assistance in making these decisions.

In order to provide this assistance, we would need to gather relevant information from a relevant population. The traditional approach is to obtain this data from a sample from this population. This can be both time-consuming and expensive for most agencies or jurisdictions. Generating synthetic data is an appealing alternative approach that offers several benefits.

The notion of synthetic data is not new. A great deal of attention has recently been paid to the potential of using synthetic data as a way of archiving data regarding individual behavior without actually providing sensitive micro-data (Abowd & Woodcock, 2001). In that literature, the main purpose of synthetic data is to replace the actual micro-data with a scientifically valid replacement, allowing for the robust estimation of outcomes without violating the confidentiality of individuals represented by the data. One valuable extension of this approach is the use of many synthetic datasets to generate expected outcomes, which allows for the modeling of statistical uncertainty and the generation of standard errors and confidence intervals around outcomes. The method has been in general use in statistics for more than 25 years for handling missing data and has recently been formalized for the synthetic data problem (Raghunathan, Reiter, & Rubin, 2003; Reiter, 2002, 2003).

The motivation for using synthetic data here is slightly different. We use this design for the purpose of bringing together information from several disparate data sources. A similar strategy was used in Bhati, Roman, and Chalfin (2008). In that study, the authors developed a synthetic dataset to analyze whether and to what extent expanding the drug court model to cover more drug-involved offenders would be cost beneficial. Despite a growing consensus among scholars that substance abuse treatment is effective in reducing offending, strict eligibility rules have generally limited the impact of current models of therapeutic jurisprudence on public safety.

Since data needed for providing evidence-based analysis of that issue are not readily available, microlevel data from three nationally representative sources were used to construct a synthetic dataset—defined using population profiles rather than sampled observation—that was used to analyze this issue. Data from the National Survey on Drug Use and Health (NSDUH) and the Arrestee Drug Abuse Monitoring (ADAM) program were used to develop profile prevalence estimates. Data from the Drug Abuse Treatment Outcome Study (DATOS) were used to compute expected crime reduction benefits of treating clients with particular profiles. The resulting synthetic dataset—comprising of over 40,000 distinct profiles—permitted the benefit–cost analysis of a limited number of simulated policy options.

This chapter describes the development of the synthetic dataset for providing correctional agencies assistance in making programming decision. The first step in the process is to design a synthetic database that allows all possible attribute combinations to exist. This would mean designing a dataset with rows constructed with all possible cross-combinations of the attributes deemed relevant. Each of the rows in the database would be considered a profile. Figure 8.1 provides a graphical depiction of a synthetic dataset. Rather than collect a sample of real microlevel data from real sample members (denoted with stars in Fig. 8.1), one defines a set of possible profiles. The crucial column in this dataset is labeled "prevalence." This column reflects the relative size of this profile in the population of interest. Once estimated, this column allows us to compute a host of interesting quantities. For example, as shown in the extreme right of Fig. 8.1, the recidivism rate of the first type of individual can be computed as

$$\Pr(\text{Recid} \mid X_1) = \frac{\text{Prev}(X_1 + \text{Recid})}{\text{Prev}(X_1)}$$

$$= \frac{630}{630 + 220} = 74\%$$

where $\Pr(\text{Prev} \mid X_1)$ is the probability of recidivism for the profile X_1 (i.e., with attribute combination represented by X_1), $\text{Prev}(X_1 + \text{Recid})$ is the prevalence of recidivists with profile X_1, and $\text{Prev}(X_1)$ is the prevalence of profile X_1 (i.e., the prevalence of recidivists and non-recidivists with profile X_1).

In a similar manner, other implications of the prevalence column can be computed (e.g., recidivism rates across groups of profiles). When collecting samples of data, the prevalence of profiles *is* what we collect. However, in this setting we need some way to generate (compute, develop, or estimate) that column. How do we do that?

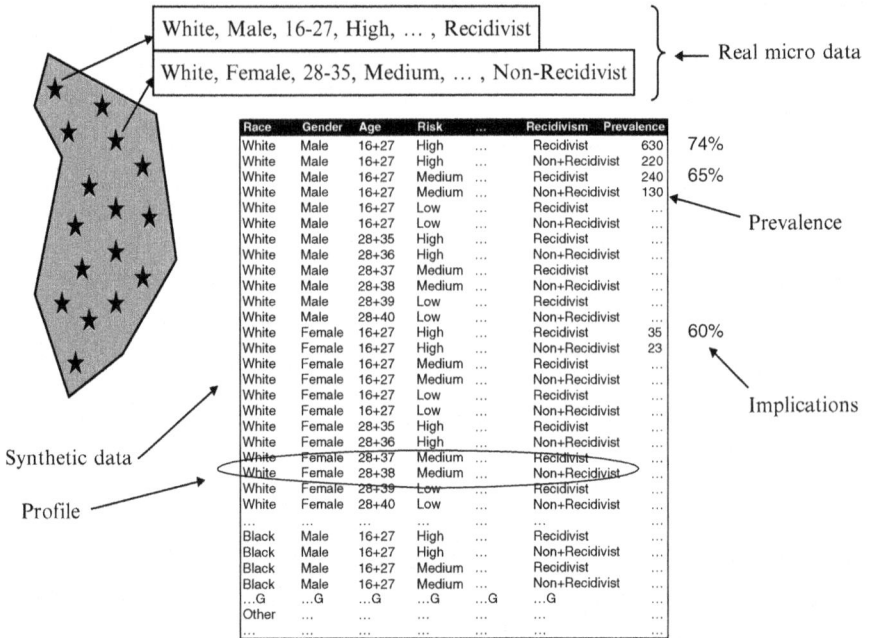

Fig. 8.1 Graphical depiction of a synthetic dataset

Let the prevalence of each of these profiles be denoted by the weight w_i. Clearly, in any real-world dataset, some of the possible profiles will not exist. However, unless a particular profile is theoretically not possible, it is part of the synthetic dataset—only its prevalence, once estimated, may be close to 0. Assume we have information about various moments or proportions of the attributes in the sample and a population size (also called control totals). For example, we may know that the total population is 50,000. This would mean that the sum of the weights should equal 50,000. In other words, we'd have the following requirement:

$$\sum_i w_i = 50,000$$

We might have other moment information such as the proportion of the relevant population that is male (say 70 % of the sample) which would imply the following requirement:

$$\sum_i w_i \text{Male}_i = 0.7 \times 50,000$$

where Male$_i$ is an indicator function (dummy variable) set to 1 if that profile includes the condition Gender = Male, and 0 otherwise.

Clearly, any other pieces of information we wish to build into this data could be incorporated in this fashion. This may include quantities like the means and variance of specific variables but also bivariate relationships among variables (covariances or conditional probabilities). The possibilities are endless. The only limit is that there are fewer requirements than the total number of profiles, as that would render the problem unsolvable.[1]

Let's say there are $j \in J$ such conditions we wish to impose on the dataset. J may be much greater than K because we can impose interactions between variables as constraints. For example, if we know the proportion of males who were young, we'd like the synthetic data to reflect that as well. We want the weights w_i to be such that they satisfy all of these requirements.

Let us generically write the functions based on the x data as $f_j(X_i)$. X_i now represents a vector of attributes as the same function could involve multiple attributes—i.e., $X_i = (x_{i1}, x_{i2}, \ldots, x_{iK})$. Let μ correspond to the evidence we wish the data to be consistent with (this is the available evidence). For example, in the discussion above, we required the sample to sum up to 50,000 and the males to account for 70 % of that sample. These two pieces of information would constitute μ_1 and μ_2. We are generically permitting J such conditions. These $j \in J$ requirements we wish to impose on the dataset can be written as

$$\sum_i w_i f_j(X_i) = \mu_j \ \forall j \in J$$

There are typically far more unknowns (the w_i) than there are constraints linking them (J). For example, even if we included all two-way covariances between $K = 10$ continuous attributes, this would only imply $J = 55$ constraints which are much smaller than the size of the synthetic dataset. This sort of a problem is referred to as an ill-posed problem—there are possibly an infinite number of solutions (configurations of w_i) that might satisfy the constraints. So how can we select one set of values for the w_i?

Information Theory—a branch of statistics—provides one way to solve ill-posed inversion problems (Jaynes, 1957; Kullback, 1959). The approach has philosophical roots in the principle of indifference. The principle of indifference (maximum entropy or minimum cross-entropy) says that the weights should be as uniform as possible (or, in the case of cross-entropy, as close to the priors as possible) while being just consistent with the evidence (Golan, Judge, & Miller, 1996). In the current context, the divergence between the weights w_i and a set of prior weights (u_i) can be defined as

$$I = \sum_i \left[w_i \log\left(\frac{w_i}{u_i}\right) - w_i + u_i \right]$$

---

[1] In actual application, the number of requirements needs to be much smaller because the higher the number of constraints we impose, the harder it is to find a set of weights that will satisfy all of them simultaneously.

The information-theoretic solution to this problem is therefore to minimize this function subject to all the constraints. This can be solved using the method of Lagrange. The primal Lagrange function for this constrained optimization problem is set up

$$L = \sum_i \left[w_i \log\left(\frac{w_i}{u_i} \right) - w_i + u_i \right] + \sum_j \theta_j \left[\mu_j - \sum_i w_i f_j(X_i) \right]$$

where L represents the Lagrange function and θ_j are the Lagrange multipliers for each of the $j \in J$ constraints. Solving the first-order conditions (i.e., setting $dL / dw_i = 0$), we get the solution:

$$w_i = u_i \exp\left(\sum_j f_j(X_i)\theta_j \right)$$

We can simplify the problem further by plugging the solution back into the primal Lagrange function to derive a dual *un*constrained maximization problem. The dual for this problem is

$$F = \sum_j \theta_j \mu_j - \sum_i u_i \exp\left(\sum_j f_j(X_i)\theta_j \right)$$

This is an unconstrained maximization problem that can be solved using any standard software that is capable of doing maximum likelihood estimation.

Once the dual is maximized, the estimated parameters θ can be inserted into the solution for w_i to compute the weights of the various profiles. If a solution exists, then it is guaranteed to be consistent with all the pieces of evidence introduced into the problem at the onset.

Synthetic Datasets

This section describes the various synthetic datasets that were designed and populated using the methods described above. These datasets were populated to be nationally representative. They provide the baseline prevalence and can be reweighted to reflect jurisdiction-specific attribute distributions.

Designing the Synthetic Datasets

Because most jurisdictions may not have detailed data on the various attributes relating to offenders under their supervision (whether in prison, jail, or community), synthetic datasets were developed using national-level estimates. These national-level synthetic datasets were defined to have all possible combinations of attributes of interest. In the current analysis, the attributes of interest include risk factors, criminogenic need factors, as well as standard demographic attributes (a total of nine profile elements). Each of these profile elements are summarized below. In-depth discussion of the creation and validation of these variables can be found in Chap. 5.

V1: Substance use programming increase is defined as abuse of a criminogenic drug (opiates, methamphetamines, amphetamines, crack, cocaine, and heroin) or clinical dependence on alcohol. In the synthetic data, we tried to replicate the clinical abuse on drugs or dependence on alcohol and also included risky behaviors associated with substance use, such as use at the time of an offense or use of a needle to inject a drug. Drug use scale scores ranged from 0 to 12 with an average score of 9 (SD = 1.84). Offenders who scored less than three on the drug use scale or who had not regularly or ever used a substance were classified as no/low substance users. Offenders who had a score of 3 or greater or reported regular use of a drug associated with offending behavior (crack, cocaine, heroin, other opiates, methamphetamine, or other amphetamines) were classified as abusers of a criminogenic drug or dependent on a non-criminogenic drug (68.0 %). In the synthetic data, a value of 0 represents no substance use and 1 represents indication of a substance abuse problem requiring an increase in programming.

V2: Stabilizers are defined as having family support, at least a high school diploma, full-time employment, lack of criminal peers, and stable housing (see Chap. 5 for more detail on how these variables were defined). Family support is defined as the offender having had family visit while he or she was incarcerated. Having a high school diploma, but not a GED, is considered a stabilizer. Full-time employment is defined as working 30 or more hours per week. The lack of association with peers who engaged in crime is another stabilizer. Stable housing was scored if an individual had not been homeless at any point in the previous year. Stabilizers were grouped into three categories: none to two (0), three to five stabilizers (1),[2] and six or more (2).

V3: Sex is defined using a standard demographic variable asking about gender. The values were restricted to male and female. Males were given a value of 1 and females a value of 2. In all included datasets, males were the modal group.

V4: Race is defined by self-reported race or ethnic identification. Six race ethnicity categories are used: White (1), Black (2), American Indian/Alaskan Native (3), Asian/Hawaiian/Pacific Islander (4), Hispanic (5), and other (5).

V5: Age was condensed into four categories: 16–27 years (1), 28–35 years (2), 36–42 years (3), and 43 years and older (4).

V6: Criminal justice risk (5 levels) was created using a number of criminal history variables available in the BJS datasets described above and in Chap. 5. The cutoffs below reflect the creation of risk scores in the synthetic data using a number of criminal history variables including age at first arrest, number of prior arrests, on probation as a juvenile, on probation as an adult, revocation of probation, parole as an adult,

---

[2] Number in parenthesis is the value given to the category in the dataset.

guilty of escape, and number of times found guilty of escape. The creation of this measure is discussed in detail in Chap. 5. Criminal justice risk can be grouped into three, four, or five levels. The synthetic data discussed in this chapter uses five risk levels: very low (1–5), low (6–17), moderate (18–19), high (20), and very high (21–27).

V7: Mental health concerns are considered in the model. Programming increase for mental health (1) is defined as having had treatment for a mental health disorder at any point in one's lifetime. No history of mental health treatment is scored as 0.

V8: Population type is based on the individual's history of offending behaviors. Individuals who accumulated three or more of any of the following offenses were considered a special population: violent (1), sexual (2), drug (5), and other (9).

V9 (*Primary Criminogenic Need*): Two primary needs are identified—criminal thinking and substance dependence on a criminogenic drug. Because criminal thinking information was not available in the datasets used to create the synthetic dataset, individuals having fewer than two stabilizing factors were considered to have an unstable lifestyle. Having an unstable lifestyle was used as a proxy for criminal thinking. Individuals with criminal thinking are reflected by a value of 2 in the synthetic data. Substance dependence is defined as clinical dependence on a substance for which there is clear drug–crime nexus such as cocaine, crack, heroin, methamphetamine, other amphetamines, and other opiates. Individuals with substance dependence are given a value of 2 in the data. Individuals with neither of these needs are denoted with a 0.

Given this set of nine features and the number of categories among them, the total number of possible attribute combinations is 34,560. Each unique combination of the values that each of the nine variables can take is considered a unique profile. Therefore, there are 34,560 possible profiles in the synthetic data.

Each synthetic data contains variables identifying each of the nine features (total of nine variables), a set of outcome probabilities (total of six outcomes), profile prevalence weights (one weight variable), and a profile ID (one ID field uniquely identifying each profile). The outcomes are the base rates that a specific profile can be expected to fail at if not assigned to any programming level. The profile prevalence quantifies how common (or prevalent) the profile is in the population.

Six alternate recidivism definitions are provided because different jurisdictions might have different preferred outcomes that they consider in their decision-making. These include measures of rearrest, reconviction, and re-incarceration, each measured at a follow-up period of 1 year and 3 years.

Populating the Synthetic Datasets

Three different synthetic datasets—pertaining to three different criminal justice populations—were designed and populated. These include prison-based, jail-based, and community-based. Each of the synthetic datasets was constructed using data collected by the Bureau of Justice Statistics (BJS)—the statistical arm of the Office of Justice Programs, US Department of Justice. The data sources include:

- Survey of Inmates in Local Jails (2002): These data are used for the jail-based synthetic dataset. The survey provides detailed information on the demographic and risk/needs measures for creating profiles and estimating their prevalence. However, the data does not contain recidivism measures.
- Survey of Inmates in State Correctional Facilities (2004): These data are used for the prison-based synthetic dataset. The survey contains information on demographic and risk/need measures for creating profiles and their prevalence. The data do not include recidivism measures.
- State Court Processing Statistics (SCPS) 1990–2006: These data are used for the community-based synthetic dataset. This data collection provides information on risk/needs and demographic and minimal information on recidivism.
- BJS National Recidivism Study of Released Prisoners 1994: This study was used as the primary source for recidivism information in all the synthetic datasets.

Because the datasets provide much of the information at the microlevel, two different approaches were used to compute the prevalence of the various profiles that were then combined. These approaches included:

- Cross-Tabulation: In this approach, actual prevalence estimates are taken by cross-tabulating the raw data when available, i.e., the number of occurrences of a particular profile is calculated from the raw data. Note these are cross-tabulations involving nine categorical variables. Unfortunately, in any finite sample, there are rare profiles that might have no realizations. For example, it is possible that the raw data contain no instance of an Asian, male, violent offender, with very low criminal justice risk level and a mental health indicator (even though it is theoretically possible to observe such an individual). Indeed, in the empirical analysis, we found that only between 10 and 20 % of the synthetic dataset could be populated using this method.
- Modeled Estimates: In this approach, we used the information-theoretic estimation of weights consistent with marginal features from the raw data to fill in the cells with zero actual prevalence (as explained in the previous section).

The final weight is a combination of the actual and estimated prevalence based on the number of observations in the actual cross-tabulation cell. Let the estimate prevalence weights be $w_i(e)$ for any profile and the actual prevalence weights be $w_i(a)$. If the number of cells in the actual cross-tabulation was c_i, the final weight was computed as

$$w_i = w_i(a) \times \left[1 - \exp(-c_i / 5)\right] + w_i(e) \times \exp(-c_i / 5)$$

so that for larger cells, more of the final prevalence w_i comes from the actual and less from the estimate. Conversely, for smaller cells, more of the final weight comes from the estimated number and not from the actual. The parameters were chosen so that when the cell count is 30 or more, the final weight is almost identical to the actual weight. When the cell count is as low as 2 or 3, then the final weight is almost identical to the information-theoretic estimate.

Tables 8.1–8.3 present summary statistics from synthetic datasets representing prison, jail, and community populations, respectively. Because the synthetic

Table 8.1 Summary statistics from synthetic dataset representing prison-based offender populations

		Recidivism rates					
		Rearrest		Reconviction		Re-incarceration	
		1 year	3 years	1 year	3 years	1 year	3 years
	(%)	(%)	(%)	(%)	(%)	(%)	(%)
	100.0	35.9	60.2	20.6	39.7	17.2	30.6
Substance use programming increase							
0 No increase	24.4	32.1	55.4	17.3	34.6	14.5	27.1
1 Increase for drug abuse and alcohol dependence	75.6	37.1	61.7	21.6	41.3	18.0	31.8
Stabilizers							
0 0–2 stabilizers	65.6	35.2	59.7	20.2	39.3	16.8	30.3
1 3–5 stabilizers	27.9	37.4	60.9	21.5	40.5	18.2	31.5
2 6+ stabilizers	6.5	36.8	61.3	20.2	40.2	16.0	30.1
Sex							
1 Male	93.2	36.5	60.8	21.0	40.2	17.6	31.0
2 Female	6.8	28.0	51.1	14.5	32.7	11.4	25.3
Race recode							
1 White	34.2	31.8	53.2	18.2	34.7	14.9	26.2
2 Black	41.0	39.5	66.6	22.5	45.1	18.9	34.5
3 American Indian/Alaskan Native	3.9	36.4	59.1	18.7	32.9	16.2	28.5
4 Asian/Hawaiian/Pacific Islander	1.4	13.2	46.8	4.9	23.0	4.5	13.7
5 Hispanic	18.8	37.7	60.4	22.4	39.9	18.8	32.4
6 Other	0.8	23.1	45.7	16.7	29.4	15.4	25.0
Age categories							
1 16–27	27.3	50.8	77.6	31.1	54.1	26.4	43.0
2 28–35	25.8	37.3	62.9	20.9	41.3	17.5	31.8
3 36–42	22.2	32.5	56.4	17.8	36.4	14.6	27.9
4 43+	24.7	21.1	41.3	11.1	24.9	8.9	18.2
Criminal justice risk (5 levels)							
1 Very low (1–5)	11.5	27.0	47.5	13.4	26.6	11.7	20.8
2 Low (6–17)	33.8	29.6	53.3	15.5	33.2	13.2	26.2
3 Moderate (18–19)	24.7	34.8	60.4	20.4	40.6	17.1	31.6
4 High (20)	13.9	42.0	68.4	25.8	47.2	21.0	35.4
5 Very high (21–27)	16.2	51.9	76.0	32.1	54.4	26.2	41.4
Programming increase for mental health							
0 No MH program increase	75.5	35.7	60.5	20.4	39.8	17.1	30.6
1 MH program increase	24.5	36.7	59.0	20.9	39.3	17.4	30.6
Population type to match SKIPS data							
1 Violent	56.5	38.6	63.1	21.9	42.0	18.2	31.7
3 Sex offender	16.7	25.7	47.5	13.0	26.4	10.1	19.7
5 Drug offender	17.2	34.3	61.5	21.2	41.3	18.5	34.5
9 General	9.6	40.7	62.4	24.9	45.8	20.8	36.4
Primary criminogenic need							
0 No primary need indicated	65.5	32.2	56.3	17.8	35.7	15.0	27.5
1 Criminal thinking	24.4	44.7	69.5	27.1	49.4	22.5	38.6
2 Substance dependence on a criminogenic drug	10.0	38.6	62.4	22.6	42.1	18.1	32.0

Table 8.2 Summary statistics from synthetic dataset representing jail-based offender populations

		Recidivism rates					
		Rearrest		Reconviction		Re-incarceration	
		1 year	3 years	1 years	3 years	1 years	3 years
	(%)	(%)	(%)	(%)	(%)	(%)	(%)
	100.0	33.4	55.5	19.8	36.7	17.0	30.8
Substance use programming increase							
0 No increase	33.3	28.1	43.4	18.1	29.2	15.6	24.8
1 Increase for drug abuse and alcohol dependence	66.7	36.1	61.5	20.6	40.5	17.7	33.7
Stabilizers							
0 0–2 stabilizers	49.4	35.1	59.0	19.7	38.4	16.7	31.6
1 3–5 stabilizers	37.0	31.7	52.7	19.5	35.1	17.0	29.9
2 6+ stabilizers	13.6	31.9	50.2	20.9	35.3	17.7	30.1
Sex							
1 Male	71.0	34.8	56.9	20.6	37.7	18.1	31.8
2 Female	29.0	30.0	51.9	17.6	34.4	14.2	28.3
Race recode							
1 White	34.0	33.3	53.7	19.7	35.6	16.5	29.4
2 Black	36.6	37.4	65.6	20.4	43.3	17.2	35.5
3 American Indian/Alaskan Native	3.3	33.1	54.1	11.4	28.4	11.3	26.9
4 Asian/Hawaiian/Pacific Islander	0.9	18.1	51.1	7.4	18.6	4.2	13.3
5 Hispanic	15.6	35.6	57.9	22.9	38.3	20.0	32.7
6 Other	9.6	16.9	19.8	16.8	17.8	16.1	17.1
Age categories							
1 16–27	45.8	38.3	60.2	25.2	41.2	21.8	35.1
2 28–35	22.2	32.6	55.9	17.0	37.6	15.4	32.0
3 36–42	17.9	32.7	54.6	16.2	34.7	14.0	27.8
4 43+	14.2	19.8	40.5	11.0	23.7	7.7	18.5
Criminal justice risk (5 levels)							
1 Very low (1–5)	12.5	23.9	45.1	10.6	26.6	7.9	21.0
2 Low (6–17)	42.2	31.1	56.7	17.6	35.2	15.1	29.5
3 Moderate (18–19)	29.8	33.3	50.1	21.6	35.9	19.4	30.6
4 High (20)	8.3	45.6	68.6	27.5	48.7	23.1	40.4
5 Very high (21–27)	7.1	50.2	73.1	31.9	53.0	26.5	44.2
Programming increase for mental health							
0 No MH program increase	73.5	33.6	54.6	20.2	36.4	17.4	30.3
1 MH program increase	26.5	33.1	57.9	18.6	37.8	15.8	32.0
Population type to match SKIPS data							
1 Violent	20.6	40.5	63.9	23.3	41.2	20.2	34.7
3 Sex offender	2.8	26.3	49.0	16.4	28.0	13.2	22.5
5 Drug offender	51.7	34.7	60.1	19.1	39.7	16.1	32.9
9 General	24.9	25.7	39.5	18.5	27.9	16.5	23.9
Primary criminogenic need							
0 No primary need indicated	60.9	30.7	51.9	17.3	32.8	15.0	27.3
1 Criminal thinking	32.7	37.4	61.5	23.6	43.2	20.2	36.3
2 Substance dependence on a criminogenic drug	6.4	38.9	59.1	23.3	41.5	19.7	34.7

Table 8.3 Summary statistics from synthetic dataset representing community-based offender populations

| | | Recidivism rates | | | | | |
| | | Rearrest | | Reconviction | | Re-incarceration | |
	(%)	1 year (%)	3 years (%)	1 year (%)	3 years (%)	1 year (%)	3 years (%)
	100.0	37.7	60.4	23.4	41.1	19.4	32.7
Substance use programming increase							
0 No increase	31.1	32.9	53.8	18.6	34.2	14.6	26.5
1 Increase for drug abuse and alcohol dependence	68.9	39.8	63.4	25.5	44.3	21.5	35.4
Stabilizers							
0 0–2 stabilizers	58.6	37.1	60.2	22.9	39.8	18.8	31.2
1 3–5 stabilizers	31.6	39.4	61.2	24.1	43.6	20.8	35.7
2 6+ stabilizers	9.8	35.8	59.7	23.7	41.3	18.3	31.8
Sex							
1 Male	81.1	39.0	61.1	24.2	41.7	20.2	33.0
2 Female	19.0	32.2	57.6	20.1	38.8	15.7	31.4
Race recode							
1 White	28.2	34.1	55.0	21.0	36.5	17.3	28.6
2 Black	48.2	40.3	63.8	25.1	44.2	21.0	35.4
3 American Indian/Alaskan Native	0.3	38.9	62.5	23.8	38.0	22.4	34.9
4 Asian/Hawaiian/Pacific Islander	1.3	27.6	59.5	13.2	32.2	11.5	23.6
5 Hispanic	19.1	37.5	60.3	23.6	41.0	19.2	32.4
6 Other	2.9	35.1	58.0	21.1	40.7	17.2	32.3
Age categories							
1 16–27	47.4	43.8	65.7	29.5	46.7	24.1	37.0
2 28–35	24.5	38.2	62.8	21.7	40.7	18.7	33.0
3 36–42	15.6	32.2	56.2	17.5	37.5	14.7	29.4
4 43+	12.4	20.1	41.1	10.8	25.5	8.4	19.4
Criminal justice risk (5 levels)							
1 Very low (1–5)	20.0	30.8	52.5	15.2	30.6	13.5	25.1
2 Low (6–17)	31.4	31.0	53.9	19.3	34.3	16.8	28.7
3 Moderate (18–19)	18.4	41.1	64.6	26.3	45.1	22.3	35.7
4 High (20)	12.6	40.0	65.3	24.4	45.1	19.0	35.2
5 Very high (21–27)	17.7	52.2	73.3	36.2	58.2	27.9	43.2
Programming increase for mental health							
0 No MH program increase	94.2	37.6	60.6	23.4	41.2	19.5	32.8
1 MH program increase	5.8	39.3	57.6	23.1	39.1	18.0	31.0
Population type to match SKIPS data							
1 Violent	1.4	49.9	72.8	27.0	49.6	22.8	37.4
3 Sex offender	1.7	32.8	54.4	17.7	34.8	14.0	25.6
5 Drug offender	35.4	40.0	64.0	25.2	44.5	19.9	35.1
9 General	61.6	36.2	58.3	22.4	39.2	19.1	31.3
Primary criminogenic need							
0 No primary need indicated	62.6	33.4	56.7	19.6	36.4	16.2	28.9
1 Criminal thinking	29.0	45.9	67.0	30.3	48.9	25.5	39.6
2 Substance dependence on a criminogenic drug	8.4	41.3	66.0	27.4	50.0	22.1	37.1

datasets are based on real microlevel data, it is not surprising that they produce distributions that seem reasonable.

In each of the tables, the first column reflects the aggregate distribution of the attribute in the population of interest, while the remaining six columns reflect the average recidivism rates within the attribute categories. The frequencies of each of the nine attributes as well as the recidivism rates are computed using the prevalence weights for the profiles.

For example, 93.2 % of the prison-based offenders are males (Table 8.1). The recidivism rate among males is typically higher than among females, irrespective of the recidivism measure considered. As expected, recidivism rates are lower among older offenders (irrespective of the measure considered). Recidivism rates among higher-risk offenders are typically higher than the lower-risk categories. Recidivism rates among Blacks and Hispanics are typically higher than among Whites. Recidivism rates appear not to vary very much with respect to stabilizers, mental health programming increases, and primary criminogenic needs.

The distributions of the various attributes are different for each of the populations described. For example, the proportion of females is lowest in the prison-based offender population (6.8 %, Table 8.1), higher in the community-based population (19 %, Table 8.3), and higher still in the jail-based offender population (29 %, Table 8.2). Similarly, the community-based offender population is the youngest, followed by the jail-based population, while the prison-based population is the oldest.

As expected, irrespective of the population being described, recidivism rates are much higher for the 3-year follow-up period relative to the 1-year follow-up period. Similarly, in all three types of synthetic dataset, re-incarceration rates are lower than the reconviction rates. Rearrest rates are typically the highest.

These tables (Tables 8.1–8.3) provide basic descriptive statistics on the three populations using the baseline synthetic datasets. They appear to provide prevalence and recidivism rate estimates that are consistent with expectation—at least of the national population—as they reflect the data from nationally representative BJS surveys.

Re-weighting the Synthetic Datasets

As noted in the introductory sections, the synthetic dataset can be re-weighted to reflect local jurisdiction characteristics so that smaller jurisdiction might be able to utilize the available national-level datasets for making decisions. Since the final weight w_i is an adjustment to a prior weight u_i, a local jurisdiction that might have a slightly different offender population than the nation as a whole would use the current national prevalence estimates provided with the nationally representative synthetic dataset (those described in Tables 8.1–8.3) as its priors u_i. Subsequently, the jurisdiction would need to reestimate the final weights imposing its own information constraints. This would result in a set of revised prevalence weights that would reflect the local jurisdiction's population but would be as close as possible to the nationally representative weights. Moreover, if the jurisdiction was missing information on some aspect of its population, it could impute the national-level estimates (as closely as possible).

To demonstrate this feature of the synthetic dataset, we conducted validation exercises. We obtained detailed microlevel data from two jurisdictions that were willing to share the information. We computed aggregate characteristics from this data and imposed a minimal set of constraints on the national-level synthetic datasets. The constraints imposed include just the aggregate attribute distributions. We then computed the recidivism rates within each attribute groups using the synthetic datasets as well as the actual microlevel data provided by the jurisdiction. Because only a minimal amount of information is used by the strategy, we expect the two sets of estimates to not be identical. However, we do expect that the recidivism rates computed from the synthetic datasets will provide similar inferences as those computed from the actual microlevel data.

Table 8.4 provides these comparisons for two jurisdictions using the 3-year reconviction rate. The first three columns of numbers in Table 8.4 are for jurisdiction A and the remaining three for jurisdiction B. Within a jurisdiction, the column labeled "Prev." reflects the frequency distribution of each of the nine attributes. The next two columns, under each jurisdiction, reflect the "actual" and "synthetic" versions of the 3-year reconviction rates.

Recall that this is a validation exercise. We do have detailed microlevel data from these jurisdictions. So the "actual" columns reflect the actual recidivism rates from the raw data in each of these jurisdictions. However, if we were to ignore this detailed microlevel data and were to re-weight the nationally representative synthetic data using just the frequency distributions provided in the first column (under each jurisdiction), then we would obtain different estimates. The validation exercise is to assess how close the re-weighted synthetic estimates are to the actual estimates.

Although the specific recidivism rates are somewhat different, there appears to be general concordance between the inferences one would derive from the two methods (actual and synthetic).

With few exceptions, the synthetic estimates almost replicate the recidivism rates in the actual dataset (compare column 2–3 and column 5–6). Where there are large divergences, typically, it is the case that the number of cases available in the microlevel data was extremely small. As a result, it is unclear whether the divergence is because the synthetic data is not producing a sensible estimate or if the microlevel data is unreliable (at least for that category). For example, the proportion of actual offenders within the American Indian and Asian categories in the real microlevel data was only 1.4 % and 0.7 %, respectively, in jurisdiction A but only 0.1 % and 0.8 %, respectively, in jurisdiction B. Similarly, the proportion of the population

Table 8.4 A comparison of recidivism rates from the re-weighted synthetic data and jurisdiction-specific microlevel data

	Jurisdiction A			Jurisdiction B		
	Prev. (%)	Reconviction (3 years)		Prev. (%)	Reconviction (3 years)	
		Actual (%)	Synthetic (%)		Actual (%)	Synthetic (%)
	100.0	20.6	20.6	100.0	35.4	35.4
Substance use programming increase						
0 No increase	73.5	19.7	19.8	9.3	29.1	28.6
1 Increase for drug abuse and alcohol dependence	26.5	23.2	22.9	90.8	36.1	36.1
Stabilizers						
0 0–2 stabilizers	52.7	18.6	19.8	70.0	32.7	33.9
1 3–5 stabilizers	47.0	22.8	21.4	29.8	42.0	39.1
2 6+ stabilizers	0.3	12.9	21.0	0.2	25.0	38.5
Sex						
1 Male	81.8	20.5	20.9	79.1	37.1	35.8
2 Female	18.2	21.0	19.0	20.9	29.2	34.1
Race recode						
1 White	66.8	20.9	20.0	51.7	28.1	33.1
2 Black	21.8	20.6	22.3	41.0	44.5	38.3
3 American Indian/Alaskan Native	1.4	21.0	17.9	0.1	50.0	32.4
4 Asian/Hawaiian/Pacific Islander	0.7	18.1	14.6	0.8	20.0	31.5
5 Hispanic	9.1	18.5	21.7	6.2	36.9	37.0
6 Other	0.3	20.8	20.2	0.1	100.0	36.1
Age categories						
1 16–27	37.6	24.5	24.9	40.3	42.8	42.1
2 28–35	21.8	20.9	22.0	25.3	33.3	35.7
3 36–42	16.5	19.6	19.1	16.0	33.0	31.9
4 43+	24.0	14.9	13.6	18.4	24.5	23.5
Criminal justice risk (5 levels)						
1 Very low (1–5)	9.1	18.3	13.2			
2 Low (6–17)	15.1	19.3	15.3	39.4	27.1	30.3
3 Moderate (18–19)	48.7	19.5	21.0	50.7	38.7	36.5
4 High (20)	14.7	23.5	23.1	9.9	52.1	50.6
5 Very high (21–27)	12.4	24.7	28.0			
Programming increase for mental health						
0 No MH program increase	89.4	20.5	20.7	80.8	37.1	36.2
1 MH program increase	10.6	21.5	19.8	19.2	28.4	32.2
Population type to match SKIPS data						
1 Violent	12.1	16.1	23.2	11.0	37.1	42.2
3 Sex offender	5.5	8.3	16.1	2.6	12.9	26.6
5 Drug offender	27.9	23.2	20.9	39.6	32.8	36.7
9 General	54.5	21.5	20.3	46.9	38.5	33.3

(continued)

Table 8.4 (continued)

	Jurisdiction A			Jurisdiction B		
	Prev. (%)	Reconviction (3 years)		Prev. (%)	Reconviction (3 years)	
		Actual (%)	Synthetic (%)		Actual (%)	Synthetic (%)
Primary criminogenic need						
0 No primary need indicated	57.3	18.3	17.8	64.4	35.5	30.9
1 Criminal thinking	16.8	21.9	23.6	21.5	34.6	43.3
2 Substance dependence on a criminogenic drug	25.9	24.9	24.7	14.1	36.5	44.2

Note: Jurisdiction B provided criminal justice risk information only in three categories (low, moderate, and high). As such, an appropriate nationally representative synthetic dataset was re-weighted to produce the reconviction rate and prevalence distributions for jurisdiction B

Table 8.5 Comparison of actual prevalence estimates of primary criminogenic need with estimates imputed from synthetic data

		Jurisdiction A		Jurisdiction B	
	National (%)	Actual (%)	Synthetic (%)	Actual (%)	Synthetic (%)
Primary criminogenic need					
0 No primary need indicated	62.6	57.3	59.9	64.4	68.8
1 Criminal thinking	29.0	16.8	24.0	21.5	24.9
2 Substance dependence on a criminogenic drug	8.4	25.9	16.1	14.1	6.3

with six or more stabilizers is 0.3 % in jurisdiction A and 0.2 % in jurisdiction B. It is typically these cases where the "actual" and "synthetic" estimates diverge.

It is possible that the synthetic data provides more accurate estimates in some cases. For example, in jurisdiction A, the actual raw data does not provide sufficient separation between the five risk categories. However, when using the synthetic dataset, we find that there is a much more pronounced gradation in the recidivism rates as one goes from the very low-risk category to the very high-risk category.

A second validation test was conducted to test whether jurisdictions missing information about a specific attribute could use the estimates provided by the synthetic dataset for decision-making. As a test, we ignored the information about the primary criminogenic need (V9) measure when re-weighting the synthetic datasets for both jurisdictions. Once re-weighted, we computed the estimated prevalence of the primary criminogenic need categories using the re-weighted and the real microlevel datasets. These estimates are provided in Table 8.5.

In both jurisdictions, the re-weighted synthetic data is able to impute the missing information for some of the categories well. For example, it is able to provide good estimates of the proportion of the respective populations having no primary need. In jurisdiction A, 57.3 % of the population had no primary criminogenic need, while value imputed from the synthetic data is 59.9 %. In jurisdiction B, the actual value is 64.4 %, while the value imputed from the synthetic data is 68.8 %.

The imputation procedure does not provide very good estimates of the prevalence of the two need categories (criminal thinking and substance dependence on criminogenic drug). For example, in jurisdiction A, the criminal thinking need has a prevalence of 16.8 % while the synthetic data imputes a value of 24 %; substance dependence need has a prevalence of 25.9 % while the imputed value is 16.1 %. On the other hand, in jurisdiction B, the criminal thinking need prevalence is fairly well approximated (21.5 % actual vs. 24.9 % synthetic) but the substance dependence need is not (14.1 % actual vs. 6.3 % synthetic).

Although there appears to be some large discrepancy, recall that the re-weighting algorithm uses less information now—only on the remaining eight attributes. Moreover, it ignores all interactions among these attributes. Given the minimal amount of information provided to the algorithm, it is still able to produce credible estimates of the prevalence of many of the primary criminogenic need categories in these jurisdictions. At the very least, the synthetic data provides the practitioners a ballpark figure on the prevalence of these needs in their jurisdictions.

Using the Synthetic Data: The RNR Simulation Tool

The RNR Simulation Tool is a web-based decision support tool that can be used by a variety of stakeholders: (1) case managers who want to know what are the best options for a particular offender, (2) local or state correctional agency(s) that wants to know what configuration of programming/controls are useful to reduce recidivism and to reduce costs, and (3) policy-makers, budget offices, and a variety of other stakeholders that desire to engage in a series of "what if" analyses. A series of assumptions drive the individual-, program-, and jurisdiction-level models, as described in Chap. 4 (the RNR framework), Chap. 5 (the supporting data), Chap. 6 (the program measurement tool), and Chap. 7 (the use of meta-analyses findings). Additionally, the individual-level portal relies heavily on the synthetic data discussed above in this chapter. The flexibility of the tool is that the simulation can be populated by various levels and types of data, depending on the jurisdiction. Below we are only providing a brief summary of the tool. In the supporting chapters, we provide examples of the application of each component.

Assess an Individual Portal

For jurisdictions that do not have an assessment tool or that do not have a clear procedure for transforming risk and needs into program placement level, the assess the individual portal is designed to support that decision. It helps the case manager (or probation/parole officer, correctional officer, or counselor) identify the

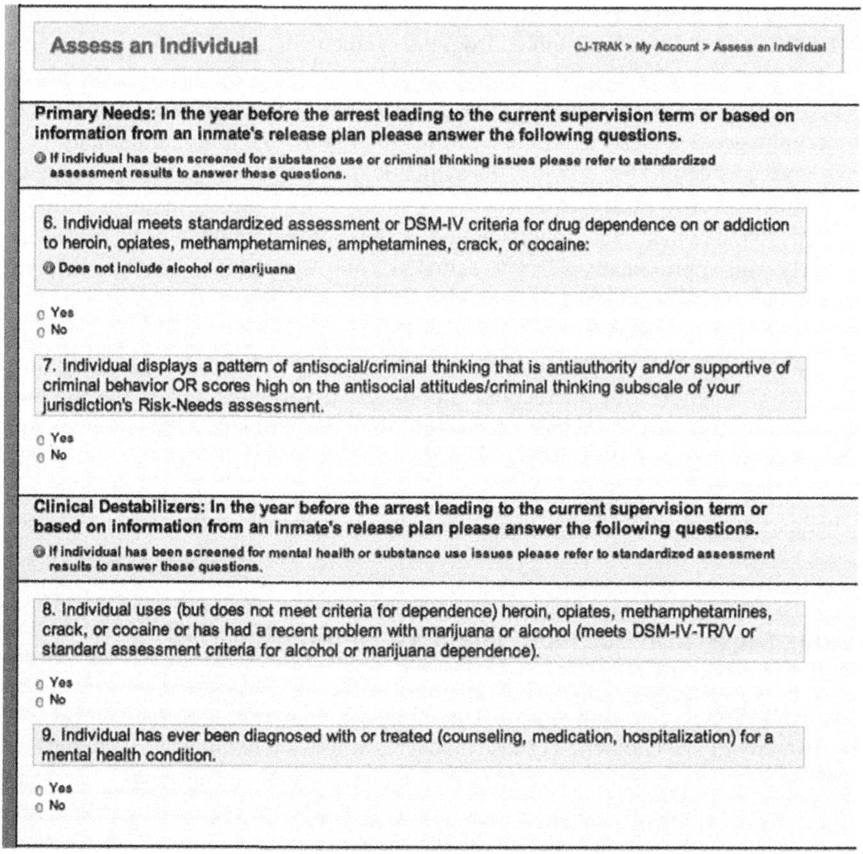

Assess an Individual   CJ-TRAK > My Account > Assess an Individual

---

Primary Needs: In the year before the arrest leading to the current supervision term or based on information from an inmate's release plan please answer the following questions.

ⓘ If individual has been screened for substance use or criminal thinking issues please refer to standardized assessment results to answer these questions.

---

6. Individual meets standardized assessment or DSM-IV criteria for drug dependence on or addiction to heroin, opiates, methamphetamines, amphetamines, crack, or cocaine:

ⓘ Does not include alcohol or marijuana

○ Yes
○ No

---

7. Individual displays a pattern of antisocial/criminal thinking that is antiauthority and/or supportive of criminal behavior OR scores high on the antisocial attitudes/criminal thinking subscale of your jurisdiction's Risk-Needs assessment.

○ Yes
○ No

Clinical Destabilizers: In the year before the arrest leading to the current supervision term or based on information from an inmate's release plan please answer the following questions.

ⓘ If individual has been screened for mental health or substance use issues please refer to standardized assessment results to answer these questions.

---

8. Individual uses (but does not meet criteria for dependence) heroin, opiates, methamphetamines, crack, or cocaine or has had a recent problem with marijuana or alcohol (meets DSM-IV-TR/V or standard assessment criteria for alcohol or marijuana dependence).

○ Yes
○ No

---

9. Individual has ever been diagnosed with or treated (counseling, medication, hospitalization) for a mental health condition.

○ Yes
○ No

Fig. 8.2 RNR Simulation Tool: asses an individual screenshot

differential impact of placing offenders of different profiles in different types of programs/services.

What's in the portal? The assess an individual portal is specifically designed for line officers and case managers to put in pertinent information about an offender and get a recommended programming level. That is, the user will supply the following: age, gender, ethnic background, risk level, primary criminogenic needs (substance dependence or criminal thinking), clinically relevant factors (mental illness, substance abuse), destabilizers, and stabilizers. These types of information are discussed in-depth in Chap. 5. Once an individual's information is put into the tool, a program level is recommended (levels A–F) based on the individual's risk and needs profile (using a number of decision rules also discussed in Chap. 5). In addition, predicted recidivism rates are provided to the user. These recidivism values are based on those identified in synthetic dataset, which associates a program level and recidivism rate to each individual profile (or combination of individual demographic factors, risk, and needs). Figure 8.2 (above) provides a screenshot from the assess an individual portal of the web tool.

Example: As above let's continue with the example of assisting a community correction's agency in determining what type of programming a hypothetical client should receive and what the expected recidivism rate and recidivism reduction value for this profile are. The expected recidivism and recidivism reductions will come from the synthetic data. The individual is a 23-year-old incarcerated white male who has been assessed as moderate risk. His primary criminogenic need is criminal thinking, he also reports abusing alcohol, being unemployed, and having friends and family who engage in criminal activity. He does have a GED and stable housing with his parents. He therefore has more than two destabilizers and only two stabilizers. This combination of characteristics creates a string of numerical values (101112091) that can be found in the synthetic dataset. This profile represents about 2 % of the male prison population and 0.046 % of the synthetic data population. His estimated risk for recidivism is 52 % for 1-year rearrest and 28 % for 1-year re-incarceration. As discussed in the earlier sections of this chapter, while our hypothetical individual does not come from a jurisdiction that know exactly what the recidivism rate for him would be, we can get an idea based on the average recidivism rate identified for an individual with the same demographic and risk–need profile from the synthetic database. Based on his combination of risk and need factors, we would recommend this individual be placed in program category B. This category focuses on criminal thinking. In our hypothetical jurisdiction, we have a number of programs available. However, in order to identify which programs are appropriate for this individual needing category B programming, we must complete the RNR Program Tool in the *rate a program* portal (see next section) (Fig. 8.3).

Categorize and Rate a Program Portal

Information on Portal: Given that labels of programs tend to be misleading, it is important for jurisdictions to be able to understand the nature of their current programs. A program inventory tool—the RNR Program Tool discussed in-depth in Chap. 6—is being validated that allows user agencies to assess their current programs for the ability to achieve gains in reducing recidivism. This portal will require users to complete an online survey that asks questions in each of the four areas of target, content, dosage, and fidelity. The "rate your program" portal using the RNR Program Tool will (1) categorize the program in terms of the category of services and (2) provide an estimate of the recidivism reduction potential of the program based on the research literature. The goal will be to categorize programs into one of the following six categories:

Level A: Focuses on drug dependence and targets individuals of any risk level with a clinical dependence on a criminogenic drug. Programs have a dosage

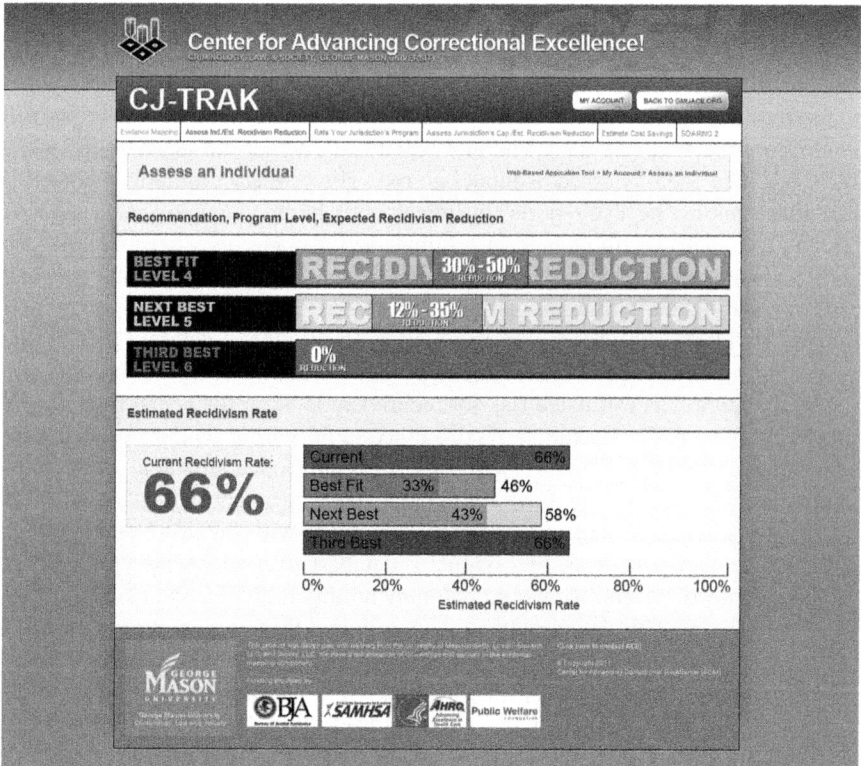

Fig. 8.3 RNR Simulation Tool: screenshot of placement decisions

of approximately 300 clinical hours and are implemented by staff with advanced degrees using and evidence-based treatment manual.

Level B: Focuses on criminal thinking and targets predominately high-risk offenders. These programs have a dosage of approximately 300 clinical hours and are implemented by staff with college degrees in related fields using an evidence-based treatment manual.

Level C: Focuses on developing interpersonal and social skills to reduce criminal activity but also includes some cognitive restructuring work to address developing criminal thinking patterns. These programs target predominately moderate-risk offenders, have a dosage of approximately 200 clinical hours, and are implemented by staff who are certified in the programs' evidence-based curriculum.

Level D: Focuses on social skills and interpersonal skills targeting multiple destabilizing issues. These programs target moderate- and low-risk offenders, should target a dosage of less than 200 h, and are implemented by staff possessing

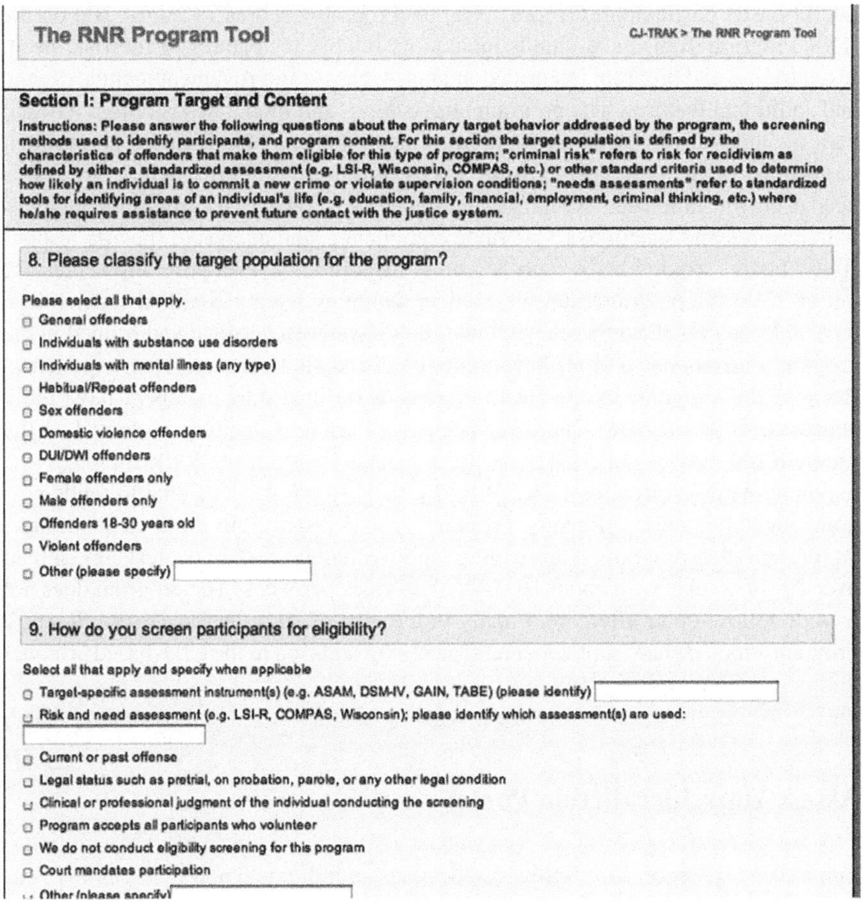

The RNR Program Tool

CJ-TRAK > The RNR Program Tool

Section I: Program Target and Content

Instructions: Please answer the following questions about the primary target behavior addressed by the program, the screening methods used to identify participants, and program content. For this section the target population is defined by the characteristics of offenders that make them eligible for this type of program; "criminal risk" refers to risk for recidivism as defined by either a standardized assessment (e.g. LSI-R, Wisconsin, COMPAS, etc.) or other standard criteria used to determine how likely an individual is to commit a new crime or violate supervision conditions; "needs assessments" refer to standardized tools for identifying areas of an individual's life (e.g. education, family, financial, employment, criminal thinking, etc.) where he/she requires assistance to prevent future contact with the justice system.

8. Please classify the target population for the program?

Please select all that apply.

- General offenders
- Individuals with substance use disorders
- Individuals with mental illness (any type)
- Habitual/Repeat offenders
- Sex offenders
- Domestic violence offenders
- DUI/DWI offenders
- Female offenders only
- Male offenders only
- Offenders 18-30 years old
- Violent offenders
- Other (please specify)

9. How do you screen participants for eligibility?

Select all that apply and specify when applicable

- Target-specific assessment instrument(s) (e.g. ASAM, DSM-IV, GAIN, TABE) (please identify)
- Risk and need assessment (e.g. LSI-R, COMPAS, Wisconsin); please identify which assessment(s) are used:
- Current or past offense
- Legal status such as pretrial, on probation, parole, or any other legal condition
- Clinical or professional judgment of the individual conducting the screening
- Program accepts all participants who volunteer
- We do not conduct eligibility screening for this program
- Court mandates participation
- Other (please specify)

Fig. 8.4 RNR Simulation Tool: rate your program screenshot

generic certifications (e.g., PO, CO) using an internally generated treatment manual.

Level E: Targets predominately low-risk individuals, has a dosage of about 100 h, and is implemented by staff with relevant experience using an internally generated treatment manual.

Level F: Few to no restrictions on behavior, punishment only.

Figure 8.4 provides a screenshot of the RNR Program Tool housed within the "rate your program" portal.

Example: For our hypothetical individual above who would be placed into program category B, we have the "Reentry Program" available. The "Reentry Program" is assigned to program category B because it focuses on criminal thinking using a CBT-based curriculum. Overall "Reentry Program" scores 79 out of 100 on the RNR Program Tool. Its strengths include its fidelity to features of the risk, need, responsivity, and program integrity domains. Areas for improvement include dosage and additional features. The program targets high- and moderate-risk offenders and uses a validated risk–need assessment to identify these individuals. The program targets one of the core criminogenic needs (criminal thinking) and identifies this need using the validated. "Reentry Program" uses a CBT-based curriculum, includes both rewards and sanctions to facilitate compliance, and attends to specific responsivity factors (individuals with learning disabilities). The program scores 21 out of 25 on the program integrity domain including having qualified staff: 100 % have at least a BA/BS in a relevant field, half have been certified and trained in the program curriculum, and all have generic certifications and relevant experience. Because the program is operated in a prison setting, staff members have daily contact with participants. There has been an external evaluation conducted on the program and they use internal measures for quality assurance. While the programs list brand-name curriculum (T4C, WRAP), they do not use a manual to guide program implementation. "Reentry Program" could improve in the area of dosage. At present the program is providing approximately 200 h of services spread over 13–17 weeks, with about 10–14 h of services per week. The program does not include follow-up or aftercare. Finally in the area of additional features, "Reentry Program" does include supplemental features (in addition to the CBT-based criminal thinking curriculum).

Assess Your Jurisdiction Portal

Information on the Portal: Similar to the asses an individual portal, the assess your jurisdiction portal collects information on demographics, risk level, needs, stabilizers, and destabilizers. However, instead of collecting this at an individual level, the portal asks the user to input the proportion of individuals within the jurisdiction's population who have these characteristics as well as the average recidivism rate for the jurisdiction. The jurisdiction can also input how they define recidivism (i.e., rearrest, reconviction, re-incarceration, technical violation). This information is then used to re-weight the synthetic data (see section "Re-weighting the Synthetic Datasets" above). This is an important feature of the RNR Simulation Tool because many jurisdictions do not have the resources to collect and analyze data to the same degree the data used to define the synthetic database was. Figure 8.5 presents a screenshot of the "assess your jurisdiction" portal in the RNR Simulation Tool and Fig. 8.6 presents the results of looking at the gaps in services. Providing the capacity at each level and the needs for each level can then provide a clear indication of where a jurisdiction should enhance service delivery.

Example: Section "Re-weighting the Synthetic Datasets" provides examples of how the re-weighting of the synthetic data is conducted using information

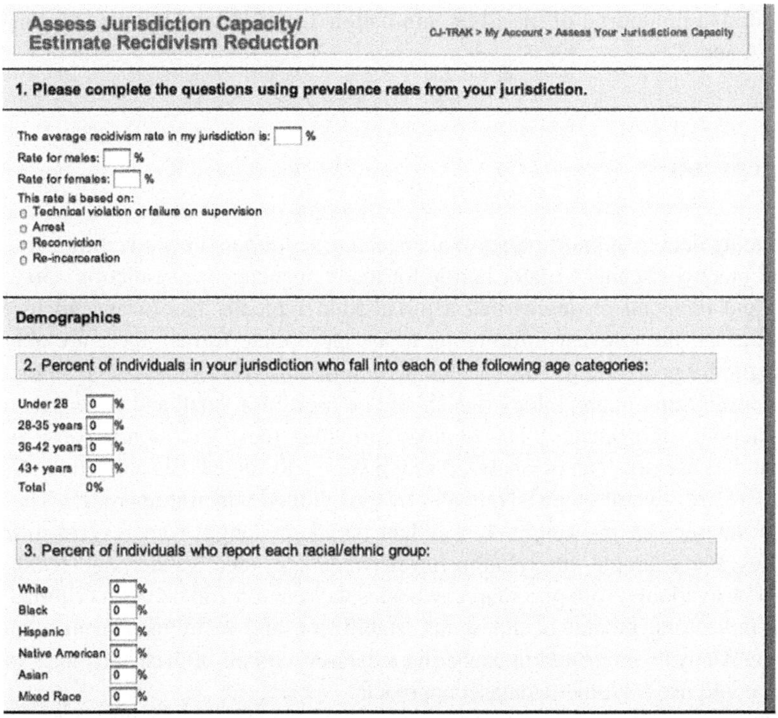

Fig. 8.5 RNR Program Tool: describe your jurisdiction screenshot

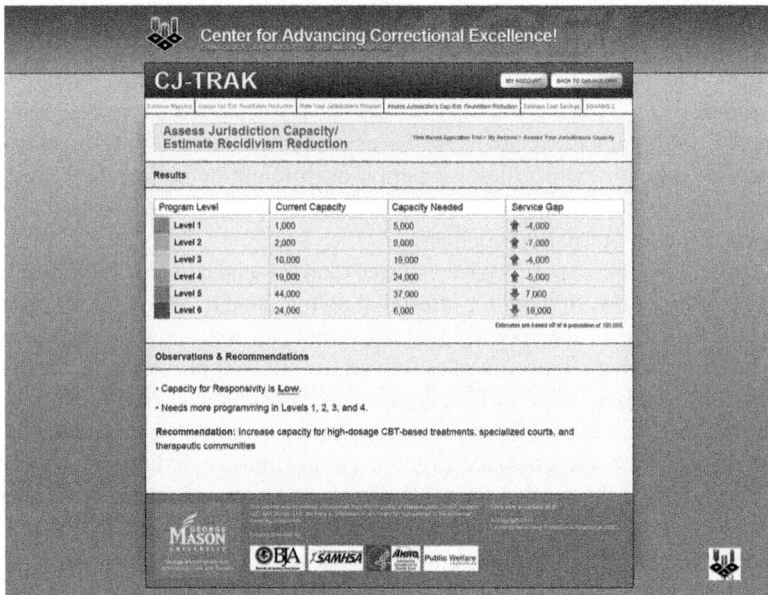

Fig. 8.6 RNR Program Tool: gap analysis screenshot

provided in this portal of the RNR Simulation Tool for two hypothetical jurisdictions A and B.

Conclusion

Producing data to support policy-makers and practitioners in conducting evidence-based practices seems a major hurdle for many agencies or jurisdiction. Given the time and financial resources that a jurisdiction typically has to expend in order to gather all the relevant information in an appropriate format, there are alternate strategies for proceeding. This chapter has described a synthetic dataset development and re-weighting methodology that should be useful for small and under-resourced jurisdictions or agencies. The strategy provides these decision-makers with a ynthetic dataset that can be analyzed as *if* it were a localized data collection effort.

There are several other advantages of the synthetic dataset approach. The synthetic dataset can be weighted to reflect data/knowledge from several different sources. For example, the relationship between age, race, gender, and recidivism might be available from one source, whereas the relationship between criminal lifestyle indicators, mental health status, stabilizers, and recidivism from a second source. Analysts interested in analyzing a dataset with all of these attributes in one place could use a synthetic dataset approach.

The ability to re-weight synthetic datasets means that it can be revised to reflect more current information and different jurisdictions. For example, the Bureau of Justice Statistics is currently undertaking a recidivism study reflecting a population of prisoners released from state prisons in 2005. The latest national-level recidivism available (and the data used in this chapter) was from a similar study describing recidivism for a 1994 prison release cohort. As soon as the 2005 recidivism data are published, the synthetic data can be re-weighted to reflect this more current information.

Finally, the synthetic dataset can be expanded to include several outcomes of interest—e.g., different follow-up period or different definitions (e.g., rearrest, reconviction, re-incarceration, drug test failure, supervision revocation, failure to appear, or other pretrial misconduct).

The two validation experiments conducted (and presented in this chapter) support the plausibility of this strategy for informing policy-makers. When sufficient information is provided to the re-weighting algorithm, it produces very credible estimates of the expected recidivism rates for various offender profiles (and categories) thereby not requiring local jurisdictions to conduct detailed recidivism studies. Similarly, when missing information on some specific attributes, the synthetic data is able to provide some guidance to these jurisdictions regarding the prevalence of the categories in their jurisdictions. However, as fewer and fewer constraints are imposed on the re-weighting algorithm, one can expect the synthetic data to provide estimates more closely resembling the national-level baseline synthetic datasets.

After all, if no jurisdiction-specific constraints were imposed, then the localized weights would be identical to the nationally representative ones.

References

Abowd, J. M., & Woodcock, S. (2001). Disclosure limitation in longitudinal linked data. In P. Doyle, J. Lane, J. Theeuwes, & L. Zayatz (Eds.), *Confidentiality, disclosure and data access: Theory and practical applications for statistical agencies* (pp. 215–277). Amsterdam: North Holland.

Bhati, A., Roman, J., & Chalfin, A. (2008). *To treat or not to treat: Evidence on the prospects of expanding treatment to drug-involved offenders*. Washington, DC: The Urban Institute.

Golan, A., Judge, G., & Miller, D. (1996). *Maximum entropy econometrics: Robust estimation with limited data*. Chichester: Wiley.

Jaynes, E. T. (1957). Information theory and statistical mechanics. *Physics Review, 106,* 620–630.

Kullback, S. (1959). *Information theory and statistics*. New York, NY: Wiley.

Raghunathan, T. E., Reiter, J. P., & Rubin, D. B. (2003). Multiple imputation for statistical disclosure limitation. *Journal of Official Statistics., 19,* 1–19.

Reiter, J. (2002). Satisfying disclosure restrictions with synthetic data sets. *Journal of Official Statistics, 18,* 531–544.

Reiter, J. (2003). Releasing multiply-imputed, synthetic public use microdata: An illustration and empirical study. *Journal of the Royal Statistical Society, Series A, 168,* 185–205.

Chapter 9
A Simulation Modeling Approach for Planning and Costing Jail Diversion Programs for Persons with Mental Illness

David Hughes

Introduction

The Surgeon General's report in 1999 highlighted that effective mental health interventions for people with mental illness reached only a small proportion of those who could benefit (U.S. Department of Health and Human Services, 1999). Years later, the President's New Freedom Commission on Mental Health echoed this same theme, envisioning a transformed mental health system in which science-driven interventions are widely available (New Freedom Commission on Mental Health, 2003). The Criminal Justice Subcommittee emphasized this situation specifically regarding people with mental illness accessing services in the criminal justice system, where high rates of co-occurring substance use disorders further complicate the issue (Abram & Teplin, 1991; Abram, Teplin, & McClelland, 2003). Unfortunately, the programs to date have not included measurable evidence-based practices (EBP) (Case, Steadman, Dupuis, & Morris, 2009), so studies on the outcomes and costs of jail diversion programs have yielded limited and equivocal results (Clark, Ricketts, & McHugo, 1999; Cowell, Broner, & Dupont, 2004; Solnit, 2004).

The seriousness of this problem becomes especially apparent in jails. Jails bear the public health burden for many who cannot access, or who are ineffectively treated by, community-based mental health service providers. Annually, approximately 13 million people are arrested in the United States and best estimates are that 16 % (2.1 million) of these have current acute symptoms of serious mental illness (Steadman, Osher, Robbins, Case, & Samuels, 2009). An estimated 75–80 % of those 2.1 million people also have co-occurring substance use disorders (Abram et al., 2003; Abram & Teplin, 1991).

D. Hughes, Ph.d.(✉)
Human Services Research Institute, Cambridge, MA, USA
e-mail: dhughes@hsri.org

F.S. Taxman and A. Pattavina (eds.), *Simulation Strategies to Reduce Recidivism:*
Risk Need Responsivity (RNR) Modeling for the Criminal Justice System,
DOI 10.1007/978-1-4614-6188-3_9, © Springer Science+Business Media New York 2013

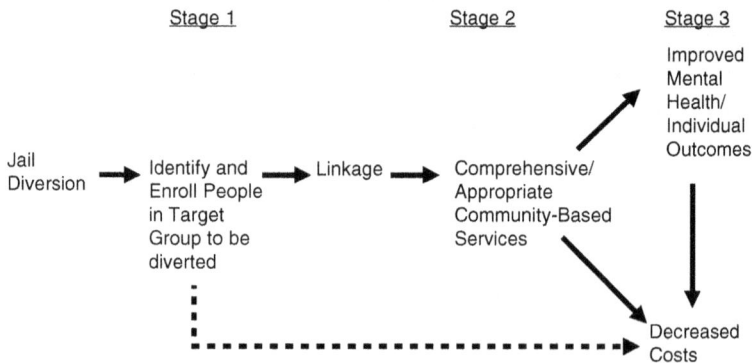

Fig. 9.1 Jail diversion logic model

Jail diversion is one of the strategies recommended by the Subcommittee on Criminal Justice of the President's New Freedom Commission on Mental Health in order to connect justice-involved people who have serious mental illness with comprehensive and effective community-based mental health treatment (New Freedom Commission on Mental Health, 2004). Jail diversion is a strategy by which jail time is reduced or avoided and community-based treatment is used as an alternative. Figure 9.1 presents the logic model of the jail diversion strategy, and intended results are based on the Technical Assistance and Policy Analysis (TAPA) Center for Jail Diversion's Logic Model for diversion and reentry programs.

Stage 1 is to identify and divert (or enroll) eligible individuals into the program. The second and most important stage is to link individuals to the correct mix of mental health and substance abuse services in the community. As discussed in more detail below, this stage has been a problem for jail diversion programs to date. The goal of jail diversion programs is stage 3—to improve mental health outcomes (e.g., improved functioning) and to spend fewer resources than if individuals were not-diverted.

In general, diversion programs follow similar models to enroll, treat, and supervise offenders. Programs first screen and assess offenders for program eligibility and needed treatment and services. For those eligible, enrollment in a diversion program is offered in lieu of incarceration. Program participants are supervised through intensive or regular probation or monitoring to ensure they comply with program requirements; some programs, especially those with a criminal justice focus, provide sanctions such as incarceration for participants who do not comply. Successful program participants can often have their cases dismissed.

Recent research on the outcomes of jail diversion programs has found that these programs do increase access to community-based mental health treatment but also that the services to which divertees are linked often employ practices that are insufficiently evidence-based or comprehensive (Steadman & Naples, 2005). However, research on the effectiveness of jail diversion has shown that the strategy improves mental health and public health outcomes (Case et al., 2009). Specifically, recent research on mental health courts has shown effectiveness in connecting people to services, in reducing episodes of inpatient treatment, and in reducing subsequent involvement with the criminal justice system in terms of jail time, new arrests, and technical violations (Cosden, Ellens, Schnell, & Yamini-Diouf, 2005; Herinckx, Swart, Ama, Dolezal, & King, 2005; Hiday & Ray, 2010).

A core public policy question for jail diversion programs is whether, and to what extent, they generate cost savings. This question is particularly relevant in difficult economic times. There is a general pattern of cost shifting from the criminal justice system to the community mental health system, but otherwise studies on the costs of jail diversion programs have yielded limited and equivocal results (Clark et al., 1999; Cowell et al., 2004; Solnit, 2004).

In response to mixed results on cost findings and the lack of evaluations of implemented comprehensive mental health service packages, Human Services Research Institute (HSRI), along with the GAINS Center (2007), developed and tested a simulation model that projects client outcomes and the fiscal impact of jail diversion programs that represent evidence-based practices. The model uses data inputs from actual criminal justice and mental health systems and, when such data are not available, employs expert judgment and reviews of the most current literature. The model also takes into account knowledge of requirements for capacity and staffing of systems.

This simulation approach simplifies the comparison of systems of care through mathematical modeling of consumer services and provides comparative data on costs, consumer outcomes, and cost-effectiveness. This is important—better planning for mental health interventions can result in safer and more effective mental health systems and can prepare stakeholders for the risks and limitations of system change in addition to its opportunities. Inadequate planning for any of the interventions being conducted systemwide, including jail diversion programs, poses risks to both program participants and the public.

Framework and Description of the Mental Health/Jail Diversion Simulation Model

The Mental Health/Jail Diversion Cost Simulation Model was developed to help communities plan and budget for jail diversion programs. This computerized model projects the costs, effectiveness, and potential cost offsets of implementing a jail diversion program for people with mental illnesses. The model is a research and strategic planning tool intended to provide program staff and stakeholder groups with information for planning resource allocation strategies and for prioritizing and choosing options for jail diversion programs.

The Mental Health/Jail Diversion Simulation Model draws from the Mental Health Simulation Model (Leff, Graves, Natkins, & Bryan, 1985; Leff, Hughes, Chow, Noyes, & Ostrow, 2010) that has been in place for over two decades and was developed for use in planning mental health service systems. The Mental Health

Simulation Model uses information collected from data systems (i.e., claims and encounter data, outcomes) with gaps filled in by expert panels composed of system stakeholders. The model seeks to describe an entire system of programs taken together and as such provides planners with the tools to explore the global effects of local changes and possible policy implications. A robustly developed simulation performs numerous tasks—including forecasting of capacity and budgets, conducting cost-effectiveness analysis, and estimating the effects of initiating new or replacement services such as evidence-based practices—and serves as a pilot study for an actual experiment or quasi-experiment (Berk, Bond, Lu, Turco, & Weiss, 2000). Simulation simplifies the comparison of systems of care through mathematical modeling of consumer services, providing comparative data on budgets, consumer outcomes, cost-effectiveness, and requirements for capacity. The ability to anticipate and model such side effects via simulation and sensitivity analysis is a particular strength of the model.

The model takes the service utilization, consumer outcomes, and costs for a group of individuals who could potentially be diverted into community-based supports and services and compares them to the costs for the same group of individuals in the absence of a jail diversion program. The goal is for researchers and program planners to be able to use the model to explore the fiscal implications of implementing different jail diversion strategies, providing different services, and choosing different target populations.

The approach takes advantage of operations research (OR) methods, sometimes called the "science of decision-making," by using information technology, data, and other relevant sources to produce informed decisions (Pierskalla & Brailer, 1994). In the case of the intersection of the mental health and criminal justice systems, an OR approach considers what happens when groups of consumers with differing disorders receive multiple services that vary in type and duration in response to the nature of the disorders, effectiveness of treatments, and social environment (Leff, Dada, & Graves, 1986). These data are combined with the expected impacts of criminal justice involvement for a group that is not-diverted (police, interaction with courts, pretrial detention, jail time, probation, parole, etc.), and from this the model provides a comprehensive assessment of the service capacity needs for the desired system and of the fiscal implications for different service systems.

The inputs to the model include the mental health functional level of potential participants, mental health and substance abuse service needs, available or planned treatment services, mental health and criminal justice service unit costs, and probable service impacts. Data needed to construct inputs can be obtained from published reports, state and local mental health systems, criminal justice databases, instruments developed specifically for the model, and published literature for modeling functional changes in client progress over time.

The simulation model is outcome based. The heart of the model is the impact of diverse service configurations on client transition at a functional level on a month-to-month basis. The simulation model weighs needs and resources according to system constraints and stakeholder recommendations. Clients, providers, and administrators may offer unique perspectives regarding the impact of reducing certain services in support of others. The model can calculate the impact of those considerations

Table 9.1 Example RAFLS consumer health states and categories

Functional level	FL category	FL name
1	Low functioning	Dangerous
2	Low functioning	Acute
3	Low functioning	Residual
4	Mid-functioning	Dependent
5	Mid-functioning	Vulnerable
6	High functioning	Recovering
7	High functioning	Independent

simultaneously and therefore arrives at a system-level, cost-effectiveness analysis. The simulation estimates service utilization and expenditures for a comprehensive, full-capacity service system, taking into account projected service needs and outcomes. The remainder of this chapter describes the stages and steps followed in the jail diversion modeling process and presents examples from the latest testing model.

Stage 1: Defining Functional Level Groups and Estimating Individuals in Need of Service

An intrinsic aspect of the model is the need for a consumer[1] (i.e., any individual who has been booked) classification system that groups consumers with similar characteristics into a well-defined scale of functional levels (FLs). An example of an FL scale, known as the Resource Associated Functional Level Scale (RAFLS), is summarized in Table 9.1 (see Leff et al., 2010 for a more complete account of the RAFLS). RAFLS is a global level functioning scale that aggregates consumers based on the types and amounts of service and resource needs. For the purposes of analysis, it may also be convenient to group certain FLs together into broader conceptual categories. Examples of such categories can be seen in Table 9.1.

It is assumed that consumers will transition between health states during their participation in the system. In general, consumers may move freely from one FL to any other, with a few exceptions. In the case of RAFLS functional level 7, consumers essentially graduate from the system by achieving system independence. Functional level 7 is therefore defined as an absorbing state because it disallows movement from itself to other states, in essence trapping or collecting all consumers who enter it.

Achieving system independence is one way for consumers to exit the system entirely. However, in order to more fully account for how consumers enter and exit the system, it is necessary to augment the FL system with two additional absorbing states: death and disappearance. In the real world, consumers also exit the overall system by dying, or by simply disappearing. Consumers who "disappear" are those who are present for services at a given time interval but not in the following interval,

---

[1] Consumer refers to person in need of or currently receiving services. This is often referred to as client in other systems, but consumer is used more often and used throughout the chapter.

or who disappear from the system completely. Many persons with serious mental illnesses leave the mental health system in unplanned ways; under such circumstances, often little is known about the levels of functioning of these individuals at the time. Some may have become system independent (i.e., reached FL7), while others may have regressed and become homeless, and still others may have been hospitalized in some other system or become involved with the criminal justice system (Leff et al., 2010).

The program is mathematically formulated using a deterministic first-order Markov simulation model. Markov models have been used in mental health planning for years and continue to be used today (Bala and Mauskopf 2006; Hargreaves, 1986; Heeg et al., 2008; James, Sugar, Desai, & Rosenheck, 2006; Korte, 1990; Miller, Brown, Pilon, Scheffler, & Davis, 2009; Norton, Yoon, Domino, & Morrissey, 2006; Patten, 2005; Perry, Lavori, & Hoke, 1987; Shumway et al., 1994; Sweillam & Tardiff, 1978). Two great advantages of Markov models are their basis in states for which services can be planned and their ability—given that even the most effective services do not have favorable outcomes for all persons at all times—to describe backward as well as forward movement or change from one state to another.

In this model, planners assign service packages to functional level groups. Functional level groups describe states through which individuals pass (although not necessarily linearly) in the course of mental illness. Service packages start as menus of multiple services. For each functional level, planners "prescribe" the services that individuals in the functional level group should receive, the percentage of individuals in the functional group who should receive those services, and the average amount of service that individuals in need should receive. The percentage of individuals prescribed a service multiplied by the average amount of service prescribed is the utilization rate for that service. Each service is associated with a unit cost (or any other resource requirement, e.g., staffing) and can also be associated with revenues realized. In addition, for each service package, planners estimate a set of monthly Markov transition probabilities that reflect the effectiveness of the service package in improving the functioning of service recipients. Simply put, for each month in the planning time frame, the model multiplies the number of new and arriving individuals in each functional level group by the service utilization rates and uses these numbers to estimate service costs and revenues. The model also uses the Markov transition probabilities to distribute individuals to functional levels, in order to set the stage for the next month.

Stage 2: Service Packages in the Community

Service packages are built by selecting from the given service taxonomy and services available to and planned for consumers in each consumer health state or functional level. Service packages are determined by an expert panel of mental health professionals and other relevant system stakeholders. Data provided to the expert panel include current utilization for services and service prescriptions made by other panels. The panel is also provided information on the scientific evidence for the effectiveness of candidate services. This group then specifies the percentage of consumers to receive a given service (prescription rate) and the amount (or number of units) of service prescribed (prescription amount) for each functional level. Once the percentage of consumers at each functional level is assigned, a recommended monthly penetration rate per service is made—this rate represents the average percentage of consumers at each functional level expected to receive a given service. After a penetration rate has been assigned, a monthly average amount of service per person is assigned to individuals in each functional level.

Expected penetration rates multiplied by recommended amounts of service yield average utilization amounts for each service and functional level. Taken together, the calculated utilizations for a functional level constitute a service package. The calculated service utilization multiplied by the estimated unit cost yields an estimated monthly cost for each service, which results in the estimated monthly cost for the service package for each functional level. Both the expected penetration rate and calculated amounts of service constitute critical inputs into the model.

The amount of each service utilized is expressed in a 4D array with axes of time period, consumer health state (CHS), service package, and service. This service utilization array stems from the 3D census array multiplied by the appropriate amount of each service for the service package being calculated. The values in the array represent the amount of each service used by the particular group of consumers (service package $(SP)_k$, CHS_i, $time_t$, $service_r$) expressed in the same units as the service taxonomy and service recommendations. With service utilization array $R_{i,t,k,r}$ for each CHS_i, $time_t$, SP_k, and $service_r$, let S_{ikr} be the average amount of $service_r$ being administered to CHS_i consumers assigned SP_k.

$$R_{i,t,k,r} = C_{i,t,k} \times S_{i,k,r}$$

Current service utilization can be obtained from state Medicaid or other claims and encounter records. The model applies an algorithm to estimate the amount of services received by each individual in the planning population. Designed to arrive at the average service received on a monthly basis, this algorithm can compare services that individuals actually receive to the expected or prescribed services as recommended by the expert panel. This information can then be used by expert panels as a starting point for creating an enhanced service package, as well as for understanding the probable service utilization of a group that is not-diverted and so receives this more basic service package in the community upon release from jail.

This formula also answers the "diversion to what" question that has been an issue in many diversion programs by showing the services consumers are actually needing to receive on a monthly basis to achieve the intended outcomes. Diverting people from the criminal justice system is easier than having comprehensive, appropriate mental health and substance abuse services in place for consumers who are diverted to the service system from the criminal justice system. This kind of modeling allows stakeholders to plan for and put services in place.

Table 9.2 Sample population characteristics for planned jail diversion intervention and two tracks for the nonresidential treatment for defendants with housing stability and co-occurring disorders

Diversion alternative	Nonresidential treatment for defendants with housing stability and co-occurring disorders
Criminal justice status	Track 1: misdemeanors
	Release on bond
	Divert within 5 days of jail admission
	Pretrial services, 6 months
	Dismiss charges at 6 months
	Track 2: felonies
	Condition of probation
	Consider waiving indictment as a requirement of admission
	Probation, 2–5 years
Clinical criteria	Priority population: diagnosis of bipolar disorder, major depression, schizoaffective disorder, or schizophrenia
	Co-occurring substance use disorder
Legal criteria	Eligible offense categories: misdemeanor A, misdemeanor B, state jail felony, felony 3
Other criteria	Not homeless or do not have major housing instability

Setting Up a Diversion Alternative or Scenario

Table 9.2 provides the summary of a simulation scenario that a planning committee wanted to explore. The table identifies the characteristics of the population to be targeted for a diversion program. One problem that has been identified in communities that do not engage in this kind of planning is overestimating the number of persons that would actually be divertible. For example, in the first test of the model, it was estimated that approximately 40 individuals per month could be diverted; however, when detailed criteria similar to the ones below were applied, only ten individuals on average per month could be diverted. The criteria included as Table 9.2 above helps communities clearly articulate the charges to be included and any conditions tied to the diversion, the clinical criteria, and any other important factors. This articulation helps avoid any ambiguity for the diverse set of stakeholders that would be responsible for implementing any agreed-upon interventions.

Diverted and Not-Diverted Groups Defined

This model also produces treatment utilization and cost, criminal justice cost, and outcome estimates for a group that is not-diverted. Members of this group receive typical criminal justice sanctions and services in the community and in jail. The model produces the same outputs for a group that is diverted to appropriate community-based mental health and substance abuse services. Below is a summary of the assumptions for each group:

Diverted group assumptions

- Receives current basic service package with enhanced service package (e.g., assertive community treatment, supported housing), depending on diversion option.
- Criminal justice costs are calculated up to the diversion point, and recidivism rates from a model program are included in subsequent criminal justice costs.
- Recidivism rates can be calculated for each period based on model program results in the literature, or on assumptions about program impact on recidivism.

Not-diverted group assumptions

- Receives basic service package when not in jail.
- Receives mental health services in jail.
- Criminal justice costs include original sentence plus current recidivism rates in the community.
- Recidivism rates are based on cohort's current recidivism rates by offense when available, or the literature when not available.

Stage 3: Assigning Costs and Outcomes

Service packages in jail. Mental health services in jail or prison can be a major expense for counties and states. As such, it is important to collect and analyze data to determine how much service occurs in prison and how to cost these services. A county jail may also be responsible for the cost of referrals to a hospital during an inmate's sentence, adding to the total cost of the sentence.

Assigning unit costs and revenues. Each service has a unit cost that is derived during data collection. The expenditures array is 4D, with the same axes as the service utilization array. Let $E_{i,t,k,r}$ be the total expenditure for service r, given to CHS_i, at time t, assigned SP_k; also, let u_r be the unit cost for service r as determined in the service taxonomy.

$$E_{i,t,k,r} = R_{i,t,k,r} \times u_r$$

Estimating outcomes. This dynamic model uses information about population functional levels, service recommendations for each functional level, consumer outcomes (expressed as functional level improvement), and unit costs to provide decision-makers with estimates of service utilization, costs, and system effectiveness for a diverted vs. non diverted population. Consumer outcomes are expressed as monthly transition probabilities estimated from current literature on treating mental health service consumers and—when available—from the more specific criminal-justice-involved population. The literature and previous model projects demonstrate that individuals receiving appropriate and comprehensive community-based mental

Table 9.3 Example of transition rate matrix

		Following FL						
		FL1	FL2	FL3	FL4	FL5	FL6	FL7
Current FL	FL1	0.624	0.118	0.050	0.154	0.007	0.005	0.000
	FL2	0.099	0.624	0.129	0.037	0.068	0.002	0.000
	FL3	0.006	0.031	0.716	0.184	0.022	0.001	0.000
	FL4	0.014	0.019	0.069	0.734	0.111	0.013	0.000
	FL5	0.004	0.007	0.015	0.073	0.747	0.103	0.013
	FL6	0.000	0.008	0.000	0.008	0.050	0.879	0.017
	FL7	0.000	0.000	0.000	0.000	0.000	0.000	1.000

health service packages improve in functioning at a faster rate than individuals receiving a more basic service package that does not include the appropriate best and evidence-based practices (Leff et al., 2010).

Calculating Transition Probabilities

This section describes the method used to calculate FL transition probabilities for a given population of consumers of mental health services. Transition probabilities are derived via a first-order Markov process that describes the likelihood a given consumer will transition from one functional level to another after receiving service for a given period (usually a month). Table 9.3 is an example of a transition rate matrix based on the RAFLS for a fictional population.

Table 9.3 shows a 6.9 % chance that a given consumer at FL4 will transition to FL3 in the following month and a 73.4 % chance that a given consumer at FL4 will remain there in the following month (i.e., transition from FL4 to FL4). This method for deriving transition probabilities assumes that complete monthly functional level scores are known.

Entering Model Parameters

The previous section described the conceptual and mathematical assumptions included in the model and expanded on the research objectives. This section includes additional technical detail about how the inputs are entered into the model at each step.

Time Period

The first decision is to determine for how long a period the model will be run. As shown in Fig. 9.2, this step includes entering the number of periods and the

Fig. 9.2 Step 1 model setup

Number of Periods	24
Period Unit	Month
Number of Units in Period	1

Fig. 9.3 Number diverted
and insurance coverage table

	Snapshot	Arrivals	Coverage
Death	0	0	0
Disappearance	0	0	0
Functional Level 1	0	6	0
Functional Level 2	0	25	0
Functional Level 3	0	40	0
Functional Level 4	0	36	0
Functional Level 5	0	12	0
Functional Level 6	0	4	0
Functional Level 7	0	0	0

period unit (typically measured in months). Most models are run in periods of years
(1-, 2-, 3-, and 5-year models seem to be the most common).

Clients by Functional Level

For each FL, the model needs an initial number of consumers as well as an estimate
of the number of new arrivals to that FL from outside the system in the chosen time
interval. Arrivals to the system consist of consumers who are just becoming mentally ill (new incidence), those who have been ill for some time but are new to the
service system (latent demand), and those persons who are reentering the system
after "dropping out" or "disappearing" from care at an earlier point. The sources
of these population estimates typically consist of management information systems,
jail mental health screens, and/or expert opinion. For new jail diversion programs,
all the program participants would be considered arrivals. If a program wanted
to implement the model on an existing jail diversion program, the snapshot or existing numbers would be entered as in Fig. 9.3. In this example, existing numbers
would be entered in the snapshot column; new arrivals to the program would be
entered in the arrivals column. The table shows that there are 0 in the snapshot column and a different mix of numbers in the arrival column, representing the different
functioning levels of the system. The coverage column is optional and can be helpful in determining reimbursement from federal and state programs (e.g., Medicaid
or Block Grant). Death and disappearance are options when rates are known in large
systems (typically not).

+	Category Name					
1	Treatment					U D X

	Service Name	Unit	Units	Cost	Coverage	
A	Outpatient Assessment	event	events	99.01	0	U D X
B	Diagnosis Evlauation / Consultation	event	events	114.87	0	U D X
C	Individual Therapy	hour	hours	134.82	0	U D X
D	Individual SA Therapy	hour	hours	109.15	0	U D X
E	Group SA Therapy	hour	hours	13.27	0	U D X
F	Family SA Therapy	hour	hours	101.92	0	U D X
G	COPSD Services	hour	hours	83.31	0	U D X
H	Medication Management	15 min.	15 min.s	27.35	0	U D X
I	Medication Review	15 min	15 mins	55.95	0	U D X
J	Methadone Maintenance Clinic	Units	Unitses	0	0	U D X
K	Injection Administration	event	events	100.55	0	U D X
L	FICM	hour	hours	119.74	0	U D X
M	Treatment Plan Review	hour	hours	99.78	0	U D X

Fig. 9.4 Services, units, and costs table

How to Use the Mental Health/Substance Abuse Service Options

After the setup of the population, the planner enters category names for the different types of services that will be used. Categories include "emergency care," "rehabilitation services," "treatment," "support," and "inpatient." The user enters these categories into the text boxes under "category name." For each category, the planner can enter as many or as few services as needed.

At this point planners can also add service units, costs, and the proportion of people covered. In the "unit" column, planners add the units used to measure service. Units of service vary and may be 15 min, 1 hour, 1 day, or 1 mile, depending on the service (e.g., medication management, group therapy, inpatient residential treatment, or transportation, respectively). "Cost" refers to the cost per unit of service (e.g., "15 min of medication management bills at $79.68"). "Coverage" refers to the proportion of the cost that is covered by funding mechanisms like Medicaid. Figure 9.4 illustrates the treatment domain; the above process would be repeated for the inpatient, emergency, support, and rehabilitation domains.

Service definition and selection of services: A population of persons with serious mental illnesses can require between 20 and 40 services in the service domains of medical inpatient and outpatient treatment, mental health inpatient and outpatient treatment, case management, housing, rehabilitation, and social support (Leff et al., 2010). Table 9.4 contains an illustrative list of services and their definitions. Depending on the level of functioning and other considerations (e.g., family supports), individuals with serious mental illness typically receive four to six services. For a desired service system, planners prescribe the percentage of persons in each functional level group who are in need of a service and the average amount of service persons in need should receive. If the purpose of a plan is to change the service system, the services available and the percentages and amounts prescribed for the desired system will always differ from those available and utilized in the current system. In some cases, services will be added or increased. However, an important part of service planning is removing or reducing ineffective or inefficient services, a process Frank and Glied (2006) have described as *exnovation*. The multiservice prescriptions for each functional group are called "service packages."

Assigning unit costs and revenues: A unit cost and revenue generated are also assigned to each service. Units differ as a function of service. For example, hospital units of service are typically days and outpatient therapy units of treatment are typically hours. Unit cost and revenue data for existing services are usually available from system financial divisions. It should be noted that the unit cost data available is more accurately unit price data. If new services are being planned, unit cost data may have to be obtained from outside the system. In some cases, data may have to be estimated based on staffing and other resource requirements. Revenues in this context are typically the amounts that will be reimbursed by insurance agencies at the federal level (e.g., Medicaid), state (e.g., Block Grant), or other funding source (e.g., foundation).

Criminal Justice Service Options and Costs

In order to prepare the comparison data of persons that are not-diverted, it is necessary to compute the costs of a group that is not-diverted. This process involves collecting costs for the following criminal justice related costs:

Jail sentence length estimates by charge type: For each major charge category, the model will need the average jail sentence estimates. It is best to use actual sentence length, computed after the sentence is complete, rather than using the official sentence as individuals often serve much less time than their original sentence, for a variety of reasons (e.g., time off for good behavior or sentence reduction because of overcrowding). If possible, a planner could also collect sentencing guidelines that factor in number of prior convictions and other possible criteria.

Probation length estimates by charge type: For low-level offenders, probation time can be much longer than jail days and will need to be factored in to costs for the not-diverted group and the diverted group (when factoring in recidivism). The easiest way to calculate this cost is to come up with an average daily cost and sentence estimates for probation time.

Table 9.4 Service variables, component services, and service definitions

Service domain	Component services	Definition
Inpatient	Specialty inpatient	Provides continuous treatment that includes general psychiatric care, medical detoxification, and/or forensic services in a general hospital, a general hospital with a distinct part or a freestanding psychiatric facility
Emergency	Crisis intervention services	Crisis intervention services for the purpose of stabilizing or preventing a sudden episode or behavior
	Crisis respite	24-hr services for individuals in crisis in homelike settings
Residential treatment	Short-term and long-term residential	Residential services that are provided by a behavioral health agency. These agencies provide a structured treatment setting with 24-hr supervision and counseling or other therapeutic activities for persons who do not require on-site medical care
Community treatment	Assessment	Evaluation for the purposes of intake, treatment planning, eligibility determination
	Individual counseling	Scheduled outpatient mental health services provided on an individual basis in a clinic or similar facility
	Group counseling	Psychotherapy to multiple clients in same session
	Family counseling	Psychotherapy to a family or couples to improve insight, decision-making, reduce stress
	Medication evaluation/ management	Services provided by physician or other qualified medical provider to evaluate, prescribe, and monitor psychiatric medications
	Substance abuse treatment	Programs for persons with both mental illness and substance abuse
	Assertive community treatment (ACT)	ACT is a multidisciplinary approach to providing an inclusive array of community-based rehabilitation services following SAMHSA evidence-based practices (EBP) guidelines
Rehabilitation	Supported employment	Job finding/retention services following SAMHSA EBP guidelines
	Skills training	Individual or group training in activities of daily and community living skills

(continued)

Table 9.4 (continued)

Service domain	Component services	Definition
Support	Case management	Assistance in accessing services and making choices about opportunities and services in the community
	Peer support	Self-help/peer services are provided by persons or family members who are or have been consumers of the behavioral health system. This may involve assistance with more effectively utilizing the service delivery system or understanding and coping with the stressors coaching, role modeling, and mentoring
	Supported housing	Supported housing services are provided to assist individuals or families to obtain and maintain housing in an independent community setting including the person's own home or homes that are owned or leased by a subcontracted provider

It is preferable, if possible, to compute average costs per day on a specialized caseload, as the costs can be much higher for persons with mental health needs. The more contacts and more referrals required may result in lower case loads, but it increases costs. For example, in Travis County, TX, the regular caseload costs are $2.27 per day and for persons on the mental health caseload the cost was almost twice as high at $4.53.

Police costs: Calculating police time generally involves using police wages and benefits to come up with an average cost of arrest. In Travis County the average arrest time was 2.5 hr; based on an hourly wage and benefit cost of $38, the total cost was $95. This estimate is much lower than in other communities, but in keeping with the conservative cost estimate approach, this amount was selected for Travis County.

Pretrial services costs: A service that is often overlooked when factoring in criminal justice costs is pretrial services. Pretrial services programs perform two functions in the administration of criminal justice: First, they gather and present information about newly arrested defendants and available release options, for use by judicial officers in deciding what (if any) conditions are to be set for defendants' release prior to trial; second, they supervise the defendants released from custody during the pretrial period by monitoring their compliance with release conditions and helping to ensure they appear for scheduled court events. In Travis County the average number of days used for pretrial services was 145, and the average cost was $5.15 per day.

Court costs: Collecting court costs can be complicated, as numerous offices may be involved. Offices that may provide data include criminal courts, the district attorney, the county attorney, the district clerk, and the county clerk. It is important to distinguish costs by type of charge or levels of intensity that might be involved in processing a case. In Travis County the important distinction was felony vs. misdemeanor. In 2008, felony costs were more than double misdemeanor costs, with felonies at $1,663 per case and misdemeanors at $702 per case as calculated by the court.

Jail costs: When multiplied across the full sentence and factored in to any recidivism rates, the cost of a jail day can end up being a large portion of the overall costs. There are several ways to calculate the jail bed cost. One way is to take the entire budget of the jail and divide it by the number of jail beds and then divide by 365 (the number of days in the year) or you can try to account for special populations. For Travis County, this cost was $45 per inmate per day in the general population and $55 for the mental health population, with increased staff-to-inmate ratio accounting for the increase. This calculation includes all county overhead costs, support services, and booking operations.

Mental health services in jail: Mental health and substance abuse services provided to incarcerated persons generally need to be calculated based on the wages and benefits of professional staff. Unlike community services, which often have unit costs tied to billing systems, these services in jail generally do not have unit costs and need to be computed. Another approach is to compute average daily or monthly costs for an individual requiring mental health services while in jail. For the Travis County case study, a combination of these approaches was used. Based on staffing patterns, unit costs were created for each service and applied to individuals in the sample.

Calculating number of refusers/rejections: Not everyone who is provided an option to be diverted from the criminal justice system will select this option nor will the courts allow every case recommended for diversion to occur. It may be that a short jail stay is preferable to having to adhere to conditions applied by the court for what might be, in many cases, a longer time than a jail sentence. If the local community has encountered this situation in previous jail diversion programs, this number could be estimated and entered. The number refusing can also be estimated by type of charge if desired. A planner could anticipate that individuals with a more severe charge category would be more likely to accept a diversion option than someone with the lowest level of offense. A planner could also anticipate that the most severe charge category would also be the cases most often rejected by the courts. Figure 9.5 shows an example of a form to enter all the criminal justice data, including the assumptions on the percent that would refuse to participate or be rejected from the courts. It is also known that people leave the program after starting it; however, this fact is taken into consideration and included as part of the disappearance rate in the transition probabilities.

Mental Health / Jail Diversion Cost Simulation Cost Calculations

Scenario 1

1. Choose a diversion type: | Court Based   ▼ |

2. Choose a time period for this scenario: | 12 months   ▼ |

3. Enter the charge names, the number of people to be diverted over a month for each charge type and whther or not it is a felony or misdemeanor charge.

	Charge Name	% Refused	# of people diverted	Charge Type	
1.	Felony 1	35%	0	☑ Felony	☐ Midemeanor
2.	Felony 2	35%	6	☑ Felony	☐ Midemeanor
3.	Felony 3	35%	9	☑ Felony	☐ Midemeanor
4.	State Jail	35%	19	☑ Felony	☐ Midemeanor
5.	Misdemeanor 1	35%	16	☐ Felony	☑ Midemeanor
6.	Misdemeanor 2	35%	0	☐ Felony	☑ Midemeanor
7.	Misdemeanor 3	35%	0	☐ Felony	☑ Midemeanor
8.				☐ Felony	☐ Midemeanor
9.				☐ Felony	☐ Midemeanor
10.				☐ Felony	☐ Midemeanor

4. Enter the number of days for each expense; based on charge type.

	Charge Type	# of Jail days	# of Probation days	# of Transportation days	# of pretrial days
1.	Felony 1	118	2,800	4	120
2.	Felony 2	60	2,545	4	120
3.	Felony 3	41	2,072	4	120
4.	State Jail	36	1,456	4	120
5.	Misdemeanor 1	10	604	3	120
6.	Misdemeanor 2	5	534	3	120
7.	Misdemeanor 3	2	0	2	120
8.					
9.					
10.					

5. Enter the costs for the following services

1.	Police	$	98.00	per event
2.	Court Costs - Misdemeanor	$	702.20	total over the course of the case
3.	Court Costs - Felony	$	1,663.00	total over the course of the case
4.	Pretrial services	$	5.15	daily average
5.	Transportation Cost	$	30.00	per trip
6.	Jail Day	$	45.00	daily average
7.	Probation Day	$	3.35	daily average
8.	Average Mental Health Services Cost	$	80.00	per Jail Day

Fig. 9.5 Criminal justice sentence and cost estimate form

Service Packages by Functional Level

The service package feature is designed to allow for different service delivery patterns for different groups of people. With this feature, a planner can create service delivery packages that correspond to the "ideal" scenario, as well as to those that just maintain services as usual. For each functional group, the planning workgroup generated service prescriptions (percentages of persons to receive a service and average amount of service per recipient) for each of the services in Fig. 9.6. These

	Death	Disap	FL1	FL2	FL3	FL4	FL5	FL6	FL7
Outpatient Assessment	0	0	0.5	0.6	0.79	0.49	0.61	0.61	0
Diagnosis Evaluation / Consultation	0	0	0.32	0.28	0.26	0.28	0.24	0.24	0
Individual Therapy	0	0	0.01	0.01	0.14	0.01	0.02	0.02	0
Individual SA Therapy	0	0	0.02	0.04	0.03	0.02	0	0	0
Group SA Therapy	0	0	0	0	0	0	0	0	0
Family SA Therapy	0	0	0	0	0	0	0	0	0
COPSD Services	0	0	0.01	0.03	0.02	0.09	0.06	0.06	0
Medication Management	0	0	0.01	0.55	0.65	0.36	0.42	0.42	0
Medication Review	0	0	0.28	0.64	0.72	0.64	0.82	0.82	0
Methadone Maintenance Clinic	0	0	0.02	0.13	0.01	0.02	0	0	0

Fig. 9.6 Services by functional level (percent who receive)

	Death	Disap	FL1	FL2	FL3	FL4	FL5	FL6	FL7
Case Management	0	0	2	12	6	20	2	2	0
Intensive Case Management	0	0	0	0	0	0	0	0	0
Peer Support	0	0	0	0	0	0	0	0	0
Supported Housing	0	0	0	12	7	30	3	3	0
Supported Housing Skills Training	0	0	0	12	7	0	0	0	0
Transportation	0	0	0	0	0	0	0	0	0
Advocacy	0	0	0	0	0	0	0	0	0

Fig. 9.7 Services by functional level (average number of units per month)

prescriptions were based on prescriptions from a previous study conducted in Chester County, PA, other state prescriptions from earlier studies, expert judgment, information about the current Chester County system, and the scientific literature. The model allows planners to enter the proportion of individuals within a given functional level that receive a given service over the course of the periods under study (typically over one month). For each service and functional level, a planner would first enter the proportion of people receiving services (1 = 100 %, 0.05 = 5 %, etc.) in the corresponding box (see Fig. 9.6). The service amounts table in Fig. 9.7 includes the number of service units that individuals in a given functional level are expected to receive during the unit of time under study (usually one month).

Transitional Probabilities

Figure 9.8 shows how transition probabilities are entered. As noted above, transition probabilities are matrices that show the probability of individuals in a given

		Resulting States								
		Death	Disap	FL1	FL2	FL3	FL4	FL5	FL6	FL7	Total
Starting States	Death	1	0	0	0	0	0	0	0	0	1
	Disap	0	1	0	0	0	0	0	0	0	1
	FL1	0	0.2	0.1	0.1	0.01	0.3	0.25	0.03	0.01	1
	FL2	0	0.1	0	0.05	0	0.1	0.5	0.23	0.02	1
	FL3	0	0.1	0	0.03	0.55	0.25	0.07	0	0	1
	FL4	0	0.15	0	0.01	0	0.45	0.2	0.15	0.04	1
	FL5	0	0.1	0	0.01	0	0.02	0.55	0.22	0.1	1
	FL6	0	0.1	0	0.01	0	0.02	0.04	0.63	0.2	1
	FL7	0	0	0	0	0	0	0	0	1	1

Fig. 9.8 Transitional probabilities table

functional level ("starting states") moving to another functional level and dying or disappearing ("resulting states") during the period being studied (typically one month). Starting states are represented by rows and resulting states by columns. The far right column represents the totals of the starting states, which should always be 1 (i.e., 100 %), because all individuals in a given functional level are accounted for by movement through functional levels, death, and disappearance. The model is programmed to display an error message if the proportions entered in any given starting state (row) do not equal 1.

Model Outputs Defined

The model produces a number of mental health system utilization outputs, clinical outcomes, criminal justice costs and total system costs, and costs to the mental health and criminal justice systems, which are described in greater detail below.

Mental Health System Costs

It is important to be able to break down to the service level of costs for any intervention for planning and implementation preparation. The outputs for the mental health system include different ways to summarize service utilization, cost, and outcomes.

Total cost and population by time period: There are a number of different concepts that can be used to evaluate the total costs and the census over a selected time period. Choosing a specific time period that has clinical and policy relevance is an important step, which must be completed prior to beginning to evaluate costs and counting population figures. In computer-based simulations, the time period can be expanded or reduced to whatever is needed to suit the simulation. In real-world test simulations however, the simulation often utilizes a one-month time interval. Using the one-month time period, the concepts that can be used to take a census of the population are the initial consumer census, the arriving consumer, the continuing consumer, and the departing consumer. For the initial consumer census, the population counted is the number of consumers already determined to be in the simulation from the beginning. Within a prison simulation, this group would include inmates already within the jail facility. The second concept is the arriving consumer. This concept takes into account the consumers arriving from outside of the simulation— for example, new inmates who arrive at each interval throughout the time period of the simulation.

Continuing consumers are those consumers who are present from the beginning to the end of the simulation and are often included in the simulation outcome. Departing consumers are participants who for some reason exit the simulation half-way through, perhaps because of death, disappearance, or achieving independence with no reliance on the system.

The cost of these simulated models or scenarios can be estimated using the data from the census to get an accurate figure of who is using what service packages throughout the simulation. Each service has its own unit cost which can then be integrated into the simulation. Calculating the cost for each unit using the census information then gives the accurate cost of the simulated models or scenarios over the determined time period of one month or more.

Comparison of costs over time: The census information is used to determine the number of each type of consumer/inmate in the simulation; once that determination is made, costs can then be compared with the difference between the number of services being used and the type of consumer/inmate who is using them. Once this comparison has been done, an accurate comparison of each unit cost can follow, correlating the information among the census data, consumer/inmate type, and the service packages used.

Units and cost by service: The model produces units and costs on a monthly basis for all services. This provides figures for yearly cost per service, monthly cost, and the units associated with the dollars. This can be helpful when thinking about the capacity to provide the services. For example, assumptions about case manager ratios could be used to project how many case management teams you would need for the first six months, the first year, or any given time frame.

Reimbursement by category of service and by individual service: The model will produce for each different service type (e.g., individual therapy, crisis respite, medication management, peer specialist) the total amount of reimbursement and the total reimbursement over all services. For example, results may show that individual therapy is covered by Medicaid and at least some of the costs are paid by the federal government, while a service like peer specialist may not be covered and not have any reimbursement. If a planner needs to make decisions based on cost, the model outputs show the cost for each individual service.

Total monthly cost by time period: If a planner needs to think about budget implications across months or years, the outputs to the model can be used to calculate how much must be available each month or year under each scenario. This information can be particularly helpful when starting an intervention with a small number of participants and a goal to increase that number.

Functional level steps (forward, backward, net steps): One of the outcomes the model tracks is change in functional levels. This outcome is tracked by number of "forward" and "backward" steps in measured levels of functioning. For example, if a person improves from FL2 to FL4 in a given time period, then the result is two positive steps. If a person moves from FL4 to FL3, then the result is one negative step. The model sums these forward and backward steps to provide an indicator of the quality of the system. It is important to note that even in the best practice systems, consumers will experience both forward and backward steps. People in systems will recover or relapse, sometimes as a result of treatment and sometimes as a result of other factors, and their functional level will improve or decline. The model makes it possible to compare forward movement, backward movement, and the resulting net steps (forward steps–backward steps) for different scenarios.

Number "disappearing" from system: The number of people disappearing can be viewed as a program outcome. It is possible that some individuals may have become system independent (i.e., reached **FL7** using a scale such as the RAFLS) or that some have moved away from the service area. Unfortunately, the other reasons people leave the system include regressing and becoming homeless, having become hospitalized in some other system, becoming involved with the criminal justice system, or dying. Different scenarios and assumptions about the impact of the program can provide key data for a system. If a number of people are disappearing from higher functional levels, they might be achieving system independence. However, high disappearance rates from the lowest functional levels might indicate problems with the system.

Mental health costs in jail: As is the case with mental health system costs, this output will be able to provide monthly costs for these services. The diversion program is meant to minimize the amount of mental health treatment costs in jail; therefore, this number is presented as just a summary cost number for the purposes of calculating total costs to the criminal justice system.

Criminal Justice Costs

The model produces detailed cost data from the criminal justice system, including:

- Total cost for charge category from arrest through sentence completion (jail and/ or probation or parole)

- Total cost per offense over the complete time period selected (e.g., 24 months)
- Total cost for each criminal justice cost center including:

    - Police
    - Pretrial services
    - Court
    - Transportation
    - Jail or prison
    - Probation or parole

Test Model: Inputs

This section provides details on the latest test of the model in order to illustrate the process using a real-world example. Details of the inputs to the model are provided first, followed by model results.

Study Site Description

In 2007, the Mayor's Mental Health Taskforce Monitoring Committee submitted a request on behalf of Travis County, Texas, to the Substance Abuse and Mental Health Services Administration (SAMSHA) for technical assistance related to setting up a jail diversion program. At the time, their key policy question was how to build affordable residential services with a jail diversion program. Also of concern was whether cost savings would accrue under different scenarios of diversion.

Over the next two years, the Travis County Criminal Justice Planning Committee helped collect cost, jail admission, criminal charges, criminal history, and service utilization data. Many county agencies participated in the data collection process: notably the Sheriff's Office, the Criminal Justice Planning Department, the Community Supervision and Corrections Department, the Criminal Court Administration, the Indigent Care Commission, and the Planning and Budget Office. The assessment data collected from the sample were then matched with current criminal charge data.

Study Participants

The initial sample included all individuals who were booked and administered a Texas Recommended Assessment Guidelines (TRAG) assessment in a six-month period in 2007 at the Travis County Jail ($n = 878$). The TRAG was designed to be used face-to-face by a Qualified Mental Health Professional Community Services

Table 9.5 Clinical disorders
($n = 592$)

Clinical disorders	Number in sample	Percentage in sample
Bipolar	160	27
Depression	130	22
Mood/anxiety	30	5
Rule outs	124	21
Schizophrenia	118	20
Unknown/other	30	5

Table 9.6 Current functioning level ($n = 586$)

Current functioning level	Number in sample	Percentage in sample*
RAFL 1	47	8
RAFL 2	65	11
RAFL 3	71	12
RAFL 4	154	26
RAFL 5	142	24
RAFL 6	71	12
RAFL 7	36	6
Total	586	100

Note: Due to rounding errors, percentages do not equal 100 %

Provider at each Local Mental Health Authority and their providers, to assess the service needs and recommend a level of care for adults in the public mental health system. First, the goal of the TRAG is to develop a systematic assessment process for measuring mental health service needs among adults, based on their most recent diagnosis and nine dimensions. Second, the aim is to propose a methodology for quantifying the assessment of service needs to allow reliable recommendations into the various levels of care or service packages with specified types and amounts of services. The TRAG was a very good match to data needed to assess level of functioning and level of service need and was one of the main reasons Travis County was selected as a test site.

Diagnosis

Although no one clinical disorder was diagnosed in the majority of the sample, over a quarter of the population was diagnosed with bipolar disorder. In addition, depression, rule outs (could not be determined), and Schizophrenia ranked high among the clinical disorders most commonly diagnosed among the sample population, as displayed in Table 9.5.

In Table 9.6, only 6 % of the sample group was determined to be at the highest Resource Associated Functional Level (RAFL), with the largest percentages of the sample group at RAFL 4 and 5, respectively. Thus, the majority of the sample group is at a mid-functioning level, although they are still vulnerable and have not yet achieved independence.

The study also showed that the risk of harm was low among the majority of inmates, with a total of 124 inmates determined to be at significant or high levels of risk of harm, while 85 % of inmates are in the "moderate" to "none" categories. Of this number, 49 % were in the "none" category. In comparison to the risk of harm, the differences in needs of the inmates were more evenly spread out, with the majority being in the "moderate" category. Although 49 % were determined to be at no risk of harm, over one quarter (28 %) of the inmates needed a "moderate" level of support and 17 % needed significant support. Only 24 % had no support needs. Additionally, for a majority of the inmates (53 %), criminal justice involvement was low and only 17 % had significant involvement.

In accordance with the TRAG scores, only 9.0 % of the inmates studied required hospitalization as a result of mental illness. In addition, the degree of functional impairment among the majority of inmates was determined to be moderate to none, with only 5 % of inmates having a high level of functional impairment. The TRAG scores show that a high number of inmates abused substances. The total number of inmates considered to be moderate to frequent users totaled 44 %, with 21 % considered to be low users. While 36 % of inmates were determined not to be abusing substances, the number of abusers was higher, totaling 64 %, including the low users.

In regard to housing stability, 47 % of the inmates had no housing stability, 26 % had low stability, and 9 % had moderate stability. Only 18 % were deemed to have significant housing stability, with no inmates having a high amount of stability.

Behavioral Health Services in the Community

Data on current services utilization was obtained from the Texas Mental Health and Mental Retardation (MHMR) service utilization database. The most common services provided over the previous 18-month period included:

- Residential bed days
- Psychosocial rehabilitation living skills
- Substance abuse counseling
- Crisis intervention-rehabilitation
- Medication review
- Medication management
- Substance abuse case management
- Group substance abuse counseling
- Routine case management
- Office-based medication services

Table 9.7 Basic service package (based on 100 persons transitioning per level)

FL	Stay the same (%)	Move forward[a] (%)	Move backward[b] (%)	Total (%)
FL1	78	21	1	100
FL2	76	21	3	100
FL3	78	13	9	100
FL4	79	5	16	100
FL5	80	4	16	100
FL6	84	9	7	100
Average	79	12	9	100

TPs total numbers by transition type
[a]Includes disappear percentage for FL6
[b]Includes disappear percentage for FL1–FL5

Table 9.8 Enhanced or Evidence Based service package (based on 100 persons transitioning)

FL	Stay the same (%)	Move forward[a] (%)	Move backward[b] (%)	Total (%)
FL1	80	20	0	100
FL2	78	21	1	100
FL3	73	20	7	100
FL4	76	15	9	100
FL5	78	7	15	100
FL6	76	16	8	100
Average	77	17	7	100

TPs total numbers by transition type
[a]Includes disappear percentage for FL6
[b]Includes disappear percentage for FL1–FL5

Functional Level Improvement (Transition Probabilities)

Transition probabilities were calculated for Travis County service recipients. The data for calculating transition probabilities came from service utilization and functional level outcome data from Travis County MHMR on 606 consumers (during the same time period as this study) that had enough functional level assessments (minimum of six each) over an 18-month period to compute transitions. Of this group, 408 had not received a comprehensive service package including an evidence-based practice (EBP; aka basic service package), and 198 had received a comprehensive service package that included an EBP (aka enhanced service package). This sample included 145 of the persons included in the jail sample, and the rest of the sample included persons who had some involvement with the criminal justice system (most were in a specialized mental health probation program). The calculated transition probabilities (TP) are included in Tables 9.7 and 9.8, summarizing the transition probabilities as the number that stayed the same, moved forward, and moved backward in a typical month. In the example below we also make assumptions on whether disappearance numbers are positive or negative. For persons who are lower functioning and up to mid-functioning (FL1–FL5), we generally consider someone in these categories who disappear from the system as a negative outcome. It is possible that some are leaving for good or neutral reasons (e.g., at a state of functioning that does not require services or moved out of the area), but generally the assumption is that the person left because he/she was dissatisfied with the services and no longer engaged, was in another system (e.g., criminal justice), or experienced other negative outcomes. For persons in FL6 who only need services under extremely stressful situations, disappearance can often be a positive step as it is likely the person is at a FL7 and no longer needs services at this time. So, the table above includes the disappearance rates as backward steps for FL1–FL5 and forward steps for FL6.

Comparing Table 9.7 (basic package) above with Table 9.8 (enhanced package), we can quickly see the package containing EBPs led to more forward movement (on average 5 % more) and fewer people moving backward (on average 2 % less).

Behavioral Health Services in Jail

Mental health service options in jail were tracked and included the following service options:

- Screening
- Counselor follow-up
- Group counseling for stable mental health inmates
- Individual follow-up for mental health inmates on disciplinary status
- Individual follow-up for unstable mental health inmates
- Treatment team review and jail management team review
- Full mental health assessment and report
- Psychiatrist/nurse practitioner appointment

As shown in Table 9.9, the study found that 671 inmates utilized the mental health services available in jail. This service usage cost a total of $126,572 over a three-month period, an average of $188 per person per month.

Current and Past Criminal Charges

Of the 685 individuals with current charges, 77 % were males and 67 % were white. Current criminal charges were defined by selecting the highest level charge. The breakdown of inmates for each category is shown in Table 9.10.

Criminal Justice History

As is apparent in Table 9.11, the majority of individuals (55 %) had been previously booked three times or fewer in the past eight years (number of years that data were available). Nearly half of the sample (45 %) had been previously booked four times or more. In 12 % of cases, the individuals had been booked 11 times or more.

Table 9.9 Behavioral health services in jail for 3-month period

Total users	671
Total costs for all services	$126,572
Average amount per person per month	$188.63

Table 9.10 Current criminal charges

Current charge category	Inmates
Class C misdemeanor	27 (4 %)
Class B misdemeanor	114 (17 %)
Class A misdemeanor	127 (19 %)
State jail felony	90 (13 %)
Third-degree felony	108 (16 %)
Second-degree felony	154 (23 %)
First-degree felony	60 (9 %)
Total	680

Table 9.11 Criminal justice history ($n = 685$)

Number of previous bookings	Number in sample	Percentage in sample
0–3	377	55
4–10	226	33
11+	82	12

Table 9.12 Jail days by offense type

Charge category	No. of jail bed days
First-degree felony	118
Second-degree felony	60
Third-degree felony	41
State jail felony	36
Class A misdemeanor	10
Class B misdemeanor	5
Class C misdemeanor	2

Sentence Estimates and Criminal Justice Unit Costs

Criminal justice costs are made up of police costs per arrest, court costs, probation services, probation length, pretrial services, jail or prison costs, and jail bed days.

Jail bed days: Table 9.12 shows that the charge category with the highest number of jail bed days was first-degree felonies (118 days), followed by second-degree felonies, which occupied beds about half the amount of time (60 days).

Probation length: Although third-degree felonies ranked the highest in number of probationers (126), first-degree felonies had the longest average sentences (2,800 days) and the smallest number of probationers (9). Across all offense types, the average sentence was 1,668 days, with the shortest sentences for those charged with class B misdemeanors. Table 9.13 summarizes this information for each offense type.

Table 9.13 Probation length by offense type

Offense type	Number of probationers	Average days of sentence	Total days
First-degree felony	9	2,800	25,200
Second-degree felony	86	2,545	218,870
Third-degree felony	126	2,072	261,072
Felony—Level S	85	1,456	123,760
Class A misdemeanor	93	604	56,172
Class B misdemeanor	35	534	18,690

Table 9.14 Pretrial services

Total clients served in 2006	203
Average daily cost	$5.15
Average days in program	145

Table 9.15 Probation services: cost per day —2003 and 2004

	2003 ($)	2004 ($)
Community supervision		
State cost	1.13	1.09
Local cost	1.16	1.18
Total	2.29	2.27
Specialized caseload—mental health		
State cost	3.72	3.35
Local cost	1.16	1.18
Total	4.88	4.53

Pretrial services: Table 9.14 shows that pretrial services served a total of 203 clients in 2006; clients averaged 145 days in the program, at an average daily cost of $5.15.

Probation services: Community supervision is a deferment of adjudication or suspension of a sentence during which the defendant is subject to certain sanctions and must fulfill certain program requirements. Some offenders receive specialized caseload services instead. The main difference between community supervision and specialized caseload is the probation staff to probationer ratio. The specialized caseload has much lower ratios than the regular community supervision and results in higher costs. Table 9.15 shows that the total state and local probation costs for specialized caseloads focusing on mental health were slightly reduced to $4.53 per day in 2004 from the previous year's $4.88. This was the latest year available that had reliable cost data in Travis County. These costs likely increased, but in keeping with a conservative costs approach, these costs were left at 2004 numbers for the simulation.

Court costs: To calculate court costs, data were collected from the criminal courts, the district attorney, the county attorney, the district clerk, and the county clerk. The average court cost was $1,664 for a felony and $702 for a misdemeanor.

Table 9.16 Police cost per arrest

Average time	2.5 hr
Cost per hour (wage and benefits only)	$38
Average cost per arrest	$95

Police cost per arrest: Police costs are an important factor in criminal justice costs. Table 9.16 shows that the average police time expended per arrest was 2.5 hr. At a rate of $38 per hour, the average police cost per arrest was $95. This cost includes wages and benefits only.

Recidivism Rates

One of the anticipated outcomes of jail diversion is reduced recidivism, measured by a reduction in the number of rearrests. The task force held numerous discussions about the potential impact of the jail diversion program on recidivism rates at different points in time (e.g., one year, two years, and three years after program start). Recidivism data were available for Travis County Jails, and the estimates in the year prior to planning year had a general recidivism rate of 29 % after one year, 40 % after two years, and 43 % after three years. The TAPA Center for Jail Diversion facilitated a discussion based on local programs comparable to Travis County (programs in Connecticut and Nebraska) and estimated that a program could anticipate a 5–10 % reduction in year one, 8–12 % reduction in year two, and much smaller reductions in out years. It was determined by the task force committee to use the most conservative approach and selected the 5 % reduction in year one and the 8 % reduction in year two.

Summary of criminal justice costs by category and total costs by offense category: The criminal justice costs vary greatly between categories. State jail had a cost of $747 per offense during pretrial but a cost of $3,162 per offense during probationary or parole periods. Sentencing costs were lower, with prison time costing $1,845 per offense. Police time was valued at $98 per offense for all events. However, transport and court/trial costs fluctuated depending on the offense.

Offenses are charged at different levels, depending on the seriousness of the crime. For example, first-degree felonies had a cost of $747 for pretrial services but a cost of $9,380 for the probation/parole period. Jail time averaged $5,310 per offense. At the other end of the scale are class C misdemeanors, which have a daily cost of $747 per offense for pretrial services, no cost for probationary periods, and $90 per offense for jail time.

Calculating total criminal justice costs (for all offenses except felony 1s and misdemeanor Cs) for initial arrest only: To calculate the total criminal justice costs for an initial arrest, aspects such as police time and the average cost per arrest need to be factored in. In all offenses except first-degree felonies and class C

Table 9.17 Criminal justice costs by offense type

Charge level	Pretrial services	Probation/ parole	Jail/ prison	Police time	Court/ trial	Transportation
Unit	Day	Day	Day	Event	Event	Event
Unit cost—felony ($)	5.15	3.35	45	98	1,664	30
Unit cost—misdemeanor ($)	5.15	3.35	45	98	702.20	30
First-degree felony ($)	747	9,380	5,310	98	1,663	120
Second-degree felony ($)	747	8,526	2,700	98	1,663	120
Third-degree felony ($)	747	6,941	1,845	98	1,663	120
State-run jail ($)	747	3,162	1,620	98	1,663	90
Class A misdemeanor ($)	747	2,023	450	98	702	90
Class B misdemeanor ($)	747	1,789	225	98	702	60
Class C misdemeanor ($)	747	0	90	98	702	60

misdemeanors, the cost of the arrest and the wages of the police are all that are needed to calculate a total. Table 9.17 summarizes the individual criminal justices costs by type of offense. These costs are important variables in the model as the cost implications are different for diverted vs. not-diverted. For example, court costs could be eliminated or reduced under some scenarios, and the associated costs with recidivism would differ for each group as well.

In Table 9.18, you can see the total cost by offense category that includes all criminal justice costs factored in (police, court, pretrial, jail, probation, etc.). For example, the average cost from police pickup to sentence completion is $17,265 for the highest charge (first-degree felony), while the cost for the lowest offense category (class C misdemeanor) is $1,644.

Table 9.19 identifies the total cost per person for each category that was determined to be in the divertible to services category (as decided by the Taskforce Monitoring Committee) and provides the total cost for each offense type. The per-person and total costs decrease as the seriousness of the offense decreases.

Summary: Total Criminal Justice Costs for Sample

In order to understand the potential cost implications, the following calculations were prepared for the Taskforce Monitoring Committee based on the eligible offense categories:

- Total number divertible over a 3-month period = 593
- Projected jail days = 18,748
- Projected total criminal justice costs = $4,933,722

The question raised to the task force was how the almost five million dollars could be spent differently on a group of individuals eligible for diversion and would there be any leftover for the state. The following section on model results provides the details on different jail diversion scenarios.

Table 9.18 Total cost
by offense category
per person

Offense category	Total cost per person ($)
First-degree felony	17,265
Second-degree felony	13,801
Third-degree felony	11,361
State-run jail	7,327
Class A misdemeanor	4,057
Class B misdemeanor	3,568
Class C misdemeanor	1,644

Table 9.19 Summary of total costs by offense

Offense category	Total cost per person ($)	Potentialnumber diverted	Total cost ($)
Second-degree felony	13,801	154	2,125,277
Third-degree felony	11,361	108	1,226,983
State jail	7,327	90	659,444
Class A misdemeanor	4,057	127	515,283
Class B misdemeanor	3,568	114	406,735
Total	–	593	4,933,722

Test Model: Results

The Taskforce Monitoring Committee was interested in modeling a scenario related to the expansion of Project Recovery for homeless males (aka scenario 2). The criteria are summarized in more detail below.

As discussed in the introduction, the development of the diversions to simulate was based on the TAPA Center for Jail Diversion's Logic Model for diversion and reentry programs (see Fig. 9.9).

Diversion Scenario: Nonresidential Treatment for Defendants with Housing Stability and Co-occurring Disorders

Stage 1: The target population for this scenario was individuals with co-occurring severe mental illness, substance use disorders, and a history of multiple arrests and who did not present issues with housing stability. Housing stability was defined as a score of moderate or lower on the TRAG housing instability scale. Major depression, bipolar disorder, schizophrenia, and schizoaffective disorders were the eligible diagnoses. Three offense categories were excluded: misdemeanor C, felony 2, and felony 1 based on task force recommendations.

Stage 1	Stage 2	Stage 3

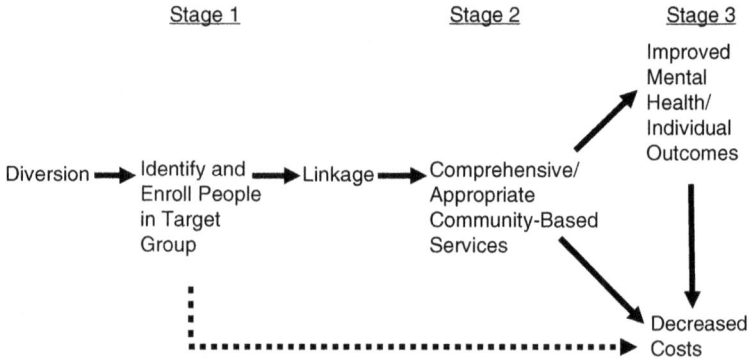

Fig. 9.9 TAPA center for jail diversion's logic model for diversion and reentry programs

Table 9.20 Divertible defendants by month and charge

Month	Felony 3	State Jail	Misdemeanor A	Misdemeanor B	Total
Total	51	79	158	134	422
Monthly average	17	26	53	45	141

Monthly averages are rounded to the nearest whole number

Two tracks were projected to serve individuals whose highest current charge was an eligible misdemeanor or felony. In the misdemeanor track, individuals would be released on bond within five days of jail admission; charges would be dismissed following six months of supervision by pretrial services. As a condition of probation, individuals processed through the felony track would be diverted within 10 days of admission; individuals would be on probation and thus supervised for two to five years. Of the total possible divertible individuals in the sample previously identified ($n=592$), 422 or approximately 141 per month were projected to meet this scenario's specific admission criteria.

Table 9.2 summarizes the eligibility criteria for the two tracks for the nonresidential treatment for defendants with housing stability and co-occurring disorders.

Table 9.20 summarizes the number of divertible defendants by month and charge.

Stage 2: Divertees would be placed on a Forensic Intensive Case Management caseload and would receive supported housing, rehabilitation, and counseling services.

Stage 3 (projected program costs and resource distribution): Assuming that 141 eligible individuals are enrolled in the diversion program in each of the 12 months following program implementation, 1,692 individuals would be diverted under this scenario. Figures 9.10, 9.11, and 9.13 display the simulated costs for this scenario. As the figures show, costs for the not-diverted group would exceed the costs of the diverted group after a two-year period. The graphs and tables in Figs. 9.10 and 9.11 show that for the first year of a jail diversion program, the diverted group is more expensive.

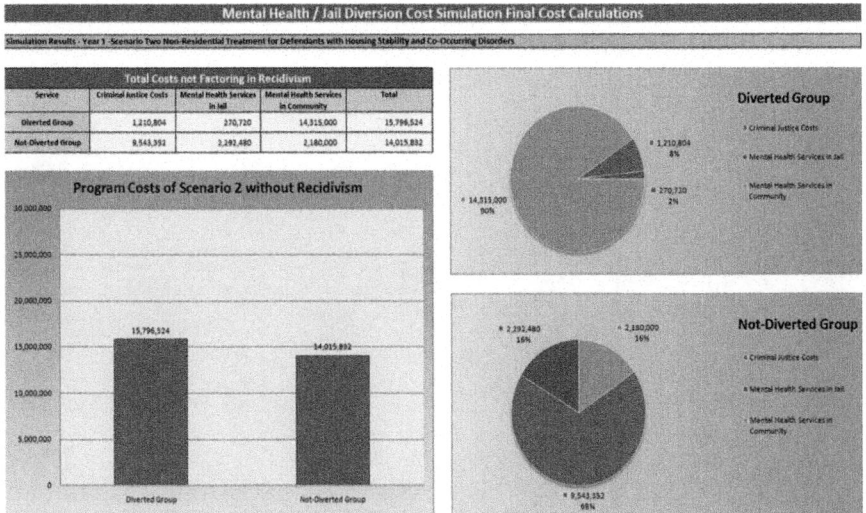

Fig. 9.10 Program costs for scenario 2, year 1 (not factoring in recidivism)

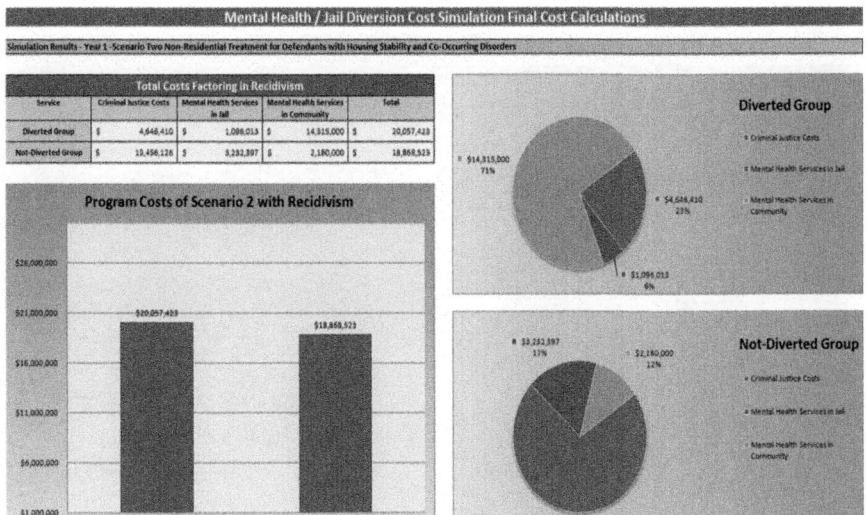

Fig. 9.11 Program costs for scenario 2, year 1 (factoring recidivism)

In year one, costs for the diverted group would total $15.8 million, compared to approximately $14 million in costs for the not-diverted group. 90.62 % of these costs would be for mental health services in the community. Of the remainder, 7.67 % would be spent on criminal justice costs, and only 1.71 % on mental health services in jail. In comparison, for the not-diverted group the majority of costs

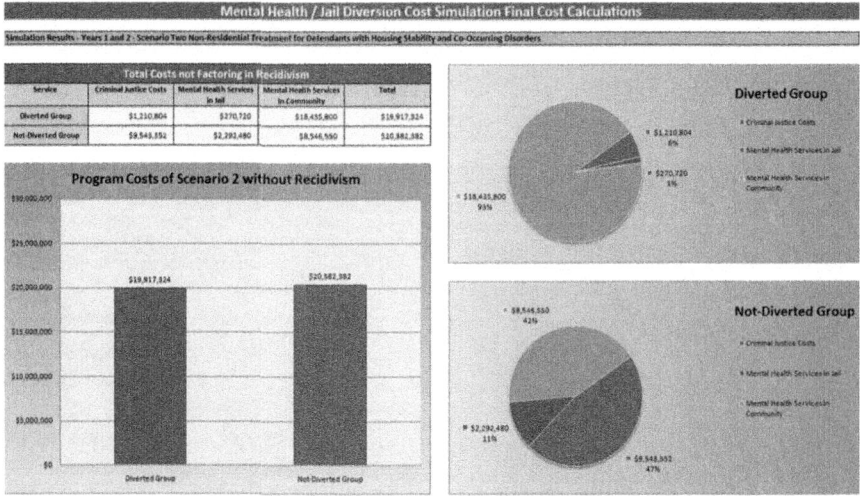

Fig. 9.12 Program costs for scenario 2, years 1 and 2 (not factoring in recidivism)

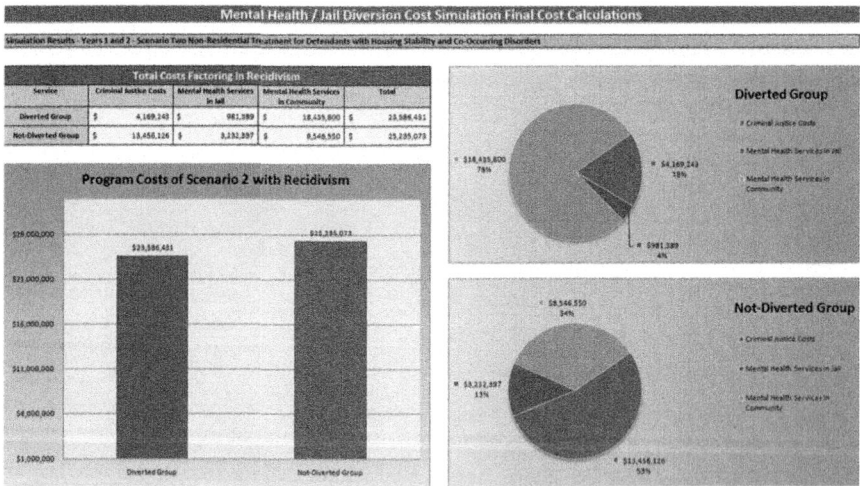

Fig. 9.13 Program costs for scenario 2, years 1 and 2 (factoring in recidivism)

(68.09 %) would be criminal justice expenses, with about an equal portion of remaining costs split between mental health services provided in the community and in jail.

Factoring in recidivism, diverted group costs would still be $1,188,900 higher in year one ($20,057,422.98) compared to the not-diverted group ($18,868,522.56). In this scenario mental health services in the community remain the majority of costs (71.37 %), and criminal justice costs for the diverted group increase to 23.17 %. For the not-diverted group, criminal justice costs remain the majority of costs, increasing to 71.32 %.

The graphs and tables in Figs. 9.12 and 9.13 look at the same group of people over two years. Note that the diverted group has now become less expensive. The diverted group is $465,058 less expensive when recidivism is not factored in and $1,648,642 when it is. It is interesting that the diversion group in this scenario took only two years to become less expensive than the not-diverted group.

Sensitivity analysis: In order to understand how the inputs would have to change in order to change the results, we conducted a sensitivity analysis on recidivism rate, transition probabilities, and charge level. The results are described below.

Recidivism rate: By changing the cost rate to 14 % for the diverted group and leaving the not-diverted group at 31 %, it is possible for the diverted group to cost less after one year of diversion implementation. This is not considered to be an impossible outcome, but would require a very successful program. This outcome might require even more intensive community services (e.g., assertive community treatment to more individuals, or more units to those who receive it), which would increase the mental health services side. Thus, the cost savings may be attenuated or washed out altogether.

Transition probabilities: Using the same slide function of the outcomes explorer, transitions (or functional level improvement) would have to increase at an 18 % rate for year one for the diverted group to cost less than the not-diverted group. The resulting transitions would also be considered to be very optimistic, based on the same comparisons to state transition data for persons receiving EBPs (the three comparison transitions included consumers in Oregon, Oklahoma, and Arizona receiving one or more EBPs).

Charge level: The first attempt for charge level to have an impact on costs after year one was not successful. This attempt involved swapping the highest misdemeanor ($n=53$) costs for the lowest felony charge ($n=26$), but that was not enough to change the outcome. However, swapping the lowest misdemeanor ($n=41$) along with the same lowest felony charge ($n=26$) resulted in year one costs being lower for the diverted group.

The results of the sensitivity analysis of key variables show that the most realistic way to change the cost outcome continues to be changes to charge level. A combination of higher charges being included and lower recidivism rates offer the best hope of reducing costs for the diverted group. Accordingly, policy makers should give careful consideration to charges included in eligibility criteria for diversion programs.

Jail days, outcomes, and costs summary: Table 9.21 provides a summary of the scenario and uses the output factoring in the conservative 5 % improvement in recidivism after year one and 8 % after year two. Net jail days saved refers to the net number of jail days avoided. Net forward steps represent the total forward steps minus the total backward steps over the number of years being simulated. The

Table 9.21 Jail days, outcomes, and costs for scenario 2

Comparison variables	Scenario	
	Years 1, 2	Years 1, 2, 3
Net jail days saved (diverted–not-diverted)	3,045	3,247
Net forward functional steps (diverted–not-diverted)	74	89
Net cost (diverted–not-diverted) ($)	991,496	643.162

Shading equals year cost savings was achieved

Table 9.22 Crimes avoided by scenario

Offense category	Nonresidential treatment/co-occurring
Third-degree felony	53
State jail	81
Class A misdemeanor	166
Class B misdemeanor	128
Total	428

higher the positive number, the greater the improvement in functional level generated. As you can see, the benefits of diversion increase with time.

Crime avoidance: An important policy discussion related to the implementation of a jail diversion program is the impact on public safety. In addition to positive consumer outcomes and cost outcomes, crime avoidance is an important community outcome. Jail diversion programs must demonstrate that the programs can be successful for participants, that there are cost savings, and that public safety increases through crime avoidance. Table 9.22 provides the number of crimes avoided by the scenario at the two-year point.

For a community attempting to implement the scenario, they could point to the potential for avoiding (based on lower recidivism rates) some serious crimes in their community. For the lowest felony (defined as state jail) and the second-lowest felony, 134 crimes would be avoided. This could be an important tool for communities to use to communicate with stakeholders about the value of this type of intervention. Jail diversion programs often avoid including defendants charged with felonies because it is politically difficult to get buy-in from all stakeholders, but with information on crime avoided, a case can be made that it is a potential threat to public safety not to provide these interventions.

Summary Test Simulation Results

This scenario suggests that cost shifting occurs over the short term (and long term) from the criminal justice system to the community mental health system, but the long-term trend is cost avoidance as individual service needs lessen over time and future criminal justice system involvement is reduced (for the recidivism scenarios). In general, if there are no low-level charges, then the model demonstrates that the first year is more expensive, but these costs are recouped after a two-year period.

However, it is important to distinguish between actual savings to a system and savings on paper. In this model, the fiscal impact is reflected at the individual level and possibly at the public level in the future. Further, achieving results predicted by the scenarios depends on accurate screening and assessment for individuals who are clinically appropriate for diversion, linking those diverted individuals to the right services at the right level, and keeping those people in services. These simulation findings are discussed in more detail in the next section, along with policy implications.

Discussion and Policy Implications

Limitations

The model could have been strengthened by testing the model in multiple counties, by drawing the Travis County sample over a longer period, and by evaluating the costs of the implemented jail diversion program against the simulation's projections. However, an earlier study of a jail diversion program documented cost trajectories and cost shifting similar to the projections reported in this paper (Ridgely et al., 2007), as did the first test of the model (Pennsylvania General Assembly, 2007). This model remains a useful tool for projecting the fiscal impact and resources required to implement a post-booking jail diversion program, even though costs may vary from place to place. That said, because the model has not been tested in different criminal justice settings—such as prisoner reentry or law enforcement-based approaches—it is not possible to draw conclusions for those settings at this time.

Study Objectives

As stated in the introduction, this chapter had two main objectives: (1) describing the development of the model and (2) demonstrating employment of that model to obtain projections for the fiscal and consumer outcomes of implementing a jail diversion program for the criminal justice system, mental health system, and total public system expenditures in a community. The section on the conceptual framework laid the foundation and justification for the model and the model inputs and provided many of the technical specifications and calculations that are used in the model. The validity of the model was demonstrated through several tests on face validity and on persons transitioning at a higher rate with the presence of an EBP.

The beta test presented as the example provided many key outputs that could be beneficial to any planner working on implementing a jail diversion program, including detailed outputs on the implications for service utilization, outcomes, and costs. One of the interesting experiences working with communities on the test of the model is how the model in itself can be a great facilitator of the planning process; going through the planning process led to the surprising result that the simulation model exercise can be important on its own even without detailed output.

While the actual output is the goal, in order to understand and plan for a system intervention, the journey to get to the results can be the most important part. This stage focuses the attention of planners, administrators, and other stakeholders on the important elements of how the program can and should function. Planning sessions can be sidetracked by competing agendas; having a process and model to follow can help keep a planning team on track for capturing the most important information in order to plan a successful intervention. This process can help avoid some of the usual "back of the envelope" or agenda-driven approaches to system planning in mental health and other human services, which do not quantitatively estimate system needs. Such approaches risk the credibility of planners, the capacity of systems to offer quality care, and the health and safety of service recipients and the communities in which they reside.

Policy Discussion

The Mental Health/Jail Diversion Simulation Model provides a tool for communities to inform the process of planning a jail diversion program with a fiscal impact assessment. The model addresses an important public policy consideration related to diversion: whether, and to what extent, jail diversion achieves current and future public-level cost savings. Through the application of the model in Travis County, Texas, it confirmed the pattern observed in prior studies of cost shifting from the criminal justice to the mental health system. Moreover, the results of these simulations provide stakeholders with insights into how eligibility criteria affect the pool of individuals who can be diverted and into the overall fiscal impact of the different groups of diverted individuals.

There are three main findings from the simulation. (1) The higher the charge category included in a diversion program, the more quickly the diverted group becomes less costly. (2) Most jail diversion programs take two or three years to show cost savings; however, by changing certain assumptions, savings may be achieved in the first year. Additionally, (3) implementing diversion programs shifts costs from the criminal justice system to the mental health system. Below, we discuss each of these findings and the resulting policy implications.

Charges to be included in jail diversion programs: In order to achieve cost savings, individuals charged with the most serious misdemeanors and low-level felonies must be included for diversion; this ensures that sufficient jail days and other criminal justice costs are avoided to create cost savings.

Of course, such decisions regarding the severity of charges to include in the program will likely be contentious and politically loaded, considering public safety and other concerns of stakeholders. Communities may balk at including individuals with felony and violent charges in diversion programs. However, individuals with co-occurring mental illness and substance use disorders with violent charges have been determined to be an appropriate target population for jail diversion.

A diversion study by Naples and Steadman (2003) found no significant differences in incidents of arrests, arrests for violent offenses, or violent acts between these types of divertees, when compared to individuals with nonviolent target offenses. Nonetheless, for a diversion program to remain politically acceptable, it must successfully prevent violent crime in particular, and demonstrating violent crimes avoided might be one strategy for acceptance. Communities will have to ensure that service packages are provided and adhered to, in order to effect the greatest chance of success for each diverted individual.

Time frame for cost savings: In all scenarios modeled, cost savings generally did not materialize for two to three years when using conservative estimates for recidivism, transition probabilities, and charge levels. Of these three variables, cost savings results were the most responsive to changes in charge level. Charge level is also the variable that is most easily changed, as intense interventions would be required to affect normal results for recidivism and transition probabilities. Programs with participants requiring expensive residential services took longer to achieve cost savings. States will need to take a long view of savings when implementing a diversion program—a challenging prospect at any time and especially in the current fiscal climate.

Cost shifting: In all scenarios, the majority of the cost burden was shifted from the criminal justice system to the community-based service system, which is already strained for resources. This presents a challenge to the community-based mental health system. However, Medicaid funding provides an opportunity to lessen this additional burden on the community mental health system and the state. To the degree that current divertees are enrolled in Medicaid, the cost of community treatment can be shared with the federal government at the current Federal Medical Assistance Percentage (FMAP) rate (varies by state).

Additionally, under the Patient Protection and Affordable Care Act (PPACA, Public Law 111–148), mental health and substance abuse services will be covered at 100 % FMAP for newly eligible individuals. However, the proportion of eligible divertees enrolled in Medicaid may be low. During the Chester County test, only 31.5 % of eligible divertees were enrolled in Medicaid. States have an interest in supporting enrollment initiatives for Medicaid-eligible populations. As such, employing a benefit-management function in the program could add to the proportion of persons under a diversion scenario with Medicaid coverage.

Additional lessons learned from both jail diversion simulations and programs include the following:

- Jail diversion interventions alone may reduce jail days, but other desired outcomes depend on access to appropriate mental health services in the community.
- A basic level of cornerstone services (e.g., housing, assertive community treatment, substance abuse services) must be provided before jail diversion can be expected to improve client outcomes.
- It takes time for a jail diversion program to become cost-effective in a system providing appropriate service packages.

Regardless of cost shifting strategies, little in the way of improved public health or safety outcomes can be achieved in the absence of community-based services. The projected consumer and cost outcomes obtained through the simulation model can be achieved only if the service packages developed by the consensus panel are actually delivered to diverted clients. A barrier to the successful implementation of jail diversion programs at the national level has been the lack of appropriate service packages delivered to diverted individuals.

One advantage of simulation is the opportunity to apply various configurations of eligibility criteria over the sample. Stakeholders can explore their options using a sample drawn from the jail—as we did in this study—or from another point in the criminal justice system (e.g., probation violators), without having to first establish the infrastructure to pilot a program. However, such a simulation bypasses the compromises and negotiations that stakeholders face while arriving at a set of eligibility criteria for a program—these criteria often represent a compromise among stakeholders with different agency missions, political realities, and operating standards. For example, a behavioral health service provider may be able to serve only individuals with illnesses that fall within state priority populations (i.e., schizophrenia, bipolar, and major depressive disorders).

Programs supported with grant funds may be limited to individuals charged with nonviolent or misdemeanor offenses, such as the Bureau of Justice Assistance's Justice and Mental Health Collaboration Program. Further, even individuals who meet the eligibility criteria may not be accepted into the program. A study of the eligibility determination and program acceptance process in post-booking jail diversion programs found that courts rejected 35 % of cases recommended for diversion by program staff (Naples, Morris, and Steadman, 2007). Similar negotiations and compromises also occurred during the Travis County simulation planning process. Some of the benefits of using the simulation model in planning a jail diversion program, as described by participants in the Chester County and Travis County projects, include the following:

- Use of the best knowledge available and expert judgment when there are gaps in data, utilization projections, and cost estimates available to local planners
- Provision of important policy and program planning insights in a timely manner—more quickly than the usual route of studying the issue for several years

- Access to projections related to the immediate consequences of changes to the service configuration and assumptions, prior to actually implementing changes
- The ability to understand and adapt to the gaps that exist in the existing mental health and substance abuse system to help with:
  - Reconciling equity and effectiveness
  - Identifying what new services to prioritize
  - Determining what existing services to phase down or out
  - Understanding the fiscal feasibility of implementing the program

Other Considerations

Jail diversion programs potentially result in broader benefits to individuals and to society as a whole than the current model can quantify. The benefits to society of individuals remaining in the community are great. The more time individuals spend incarcerated, the greater the challenge in achieving employment after leaving jail; conversely, the more time these individuals remain in the community, the more likely it is that they will become and remain employed and otherwise participate in a healthy manner in their communities. Incarcerated individuals by definition do not have the opportunity to contribute to society by advancing their own education, participating in the labor force, providing volunteer services in the community, or supporting their families and reducing intergenerational substance abuse and crime.

References

Abram, K. M., & Teplin, L. A. (1991). Co-occurring disorders among mentally ill jail detainees. Implications for public policy. *American Psychologist, 46,* 1036–1045.

Abram, K. M., Teplin, L. A., & McClelland, G. M. (2003). Comorbidity of severe psychiatric disorders and substance use disorders among women in jail. *The American Journal of Psychiatry, 160,* 1007–1010.

Bala, M. V., & Mauskopf, J. A. (2006). Optimal assignment of treatments to health states using a markov decision model: An introduction to basic concepts. *PharmacoEconomics* 24 (4):345–354.

Berk, R. A., Bond, J., Lu, R., Turco, R., & Weiss, R. E. (2000). Computer simulations as experiments: Using program evaluation tools to assess the validity of interventions in virtual worlds. In L. Bickman (Ed.), *Research design: Donald Campbell's legacy* (pp. 195–214). Thousand Oaks, CA: Sage.

Case, B., Steadman, H. J., Dupuis, S. A., & Morris, L. S. (2009). Who succeeds in jail diversion programs for persons with mental illness? A multi-site study. *Behavioral Sciences & the Law, 27,* 661–674.

Clark, R. E., Ricketts, S. K., & McHugo, G. J. (1999). Legal system involvement and costs for persons in treatment for severe mental illness and substance use disorders. *Psychiatric Services, 50,* 641–647.

CMHS National GAINS Center. (2007). *Practical advice on jail diversion: Ten years of learnings on jail diversion from the CMHS National GAINS Center.* New York: Delmar.

Cosden, M., Ellens, J., Schnell, J., & Yamini-Diouf, Y. (2005). Efficacy of a mental health treatment court with Assertive Community Treatment. *Behavioral Science and the Law, 23*, 199–214.

Cowell, A. J., Broner, N., & Dupont, R. (2004). The cost-effectiveness of criminal justice diversion programs for people with serious mental illness co-occurring with substance abuse. *Journal of Contemporary Criminal Justice, 20*, 292–314.

Frank, R. G., & Glied, S. (2006). *Better but not well: mental health policy in the United States since 1950*. Baltimore: Johns Hopkins University Press.

Hargreaves, W. A. (1986). Theory of psychiatric treatment systems. An approach. *Archives of General Psychiatry, 43*(7), 701–705.

Heeg, B. M. S., Damen, J., Buskens, E., Caleo, S., De Charro, F., & Van Hout, B. A. (2008). Modelling approaches: The case of schizophrenia. *PharmacoEconomics, 26*(8), 633–648.

Herinckx, H. A., Swart, S. C., Ama, S. M., Dolezal, C. D., & King, S. (2005). Rearrest and linkage to mental health services among clients of the Clark County Mental Health Court Program. *Psychiatric Services, 56*, 853–857.

Hiday, V. A., & Ray, B. (2010). Arrests two years after exiting a well-established mental health court. *Psychiatric Services, 61*, 463–468.

James, G. M., Sugar, C. A., Desai, R., & Rosenheck, R. A. (2006). A comparison of outcomes among patients with schizophrenia in two mental health systems: A health state approach. *Schizophrenia Research, 86*(1), 309–320.

Korte, A. O. (1990). A first order Markov model for use in the human services. *Computers in Human Services, 6*(4), 299–312.

Leff, H. S., Dada, M., & Graves, S. C. (1986). An LP planning model for a mental health community support system. *Management Science, 32*, 139–155.

Leff, H. S., Graves, S., Natkins, J., & Bryan, J. (1985). A system for allocating mental health resources. *Administration in Mental Health, 12*, 43–68.

Leff, H. S., Hughes, D. R., Chow, C. M., Noyes, S., & Ostrow, L. (2010). A mental health allocation and planning simulation model: A mental health planner's perspective. In Y. Yih (Ed.), *Handbook of healthcare delivery systems*. Boca Raton, FL: Taylor & Francis.

Miller, L., Brown, T., Pilon, D., Scheffler, R., & Davis, M. (2009). Measuring recovery from severe mental illness: a pilot study estimating the outcomes possible from California's 2004 Mental Health Services Act.

Naples, M., & Steadman, H. J. (2003). Can persons with co-occurring disorders and violent charges be successfully diverted? *International Journal of Forensic Mental Health, 2*(2), 137–143.

Naples, M., Morris, L. S., & Steadman, H. J. (2007). Factors in disproportionate representation among persons recommended by programs and accepted by courts for jail diversion. *Psychiatric Services, 58*, 1095–1101.

New Freedom Commission on Mental Health, Subcommittee on Criminal Justice. (2004). Background paper. Rockville, MD: Author. Retrieved from http://www.bipolarworld.net/pdf/CJ_ADACompliant.pdf

New Freedom Commission on Mental Health. (2003). *Achieving the promise: Transforming mental health care in America—Final report*. Rockville, MD: US Dept of Health and Human Services. DHHS Pub. No. SMA-03-3832.

Norton, E. C., Yoon, J., Domino, M. E., & Morrissey, J. P. (2006). Transitions between the public mental health system and jail for persons with severe mental illness: A Markov analysis. *Health Economics, 15*(7), 719–733.

Patient Protection and Affordable Care Act, Pub. L. No. 111–148.

Patten, S. B. (2005). Modelling major depression epidemiology and assessing the impact of antidepressants on population health. *International Review of Psychiatry, 17*(3), 205–211.

Pennsylvania General Assembly, Legislative Budget and Finance Committee. (2007). *Lessons learned from three mental health diversion and post-release programs*. Harrisburg, PA: Author.

Perry, J. C., Lavori, P. W., & Hoke, L. (1987). A Markov model for predicting levels of psychiatric service use in borderline and antisocial personality disorders and bipolar type II affective disorder. *Journal of Psychiatric Research, 21*(3), 215–232.

Pierskalla, W. P., & Brailer, D. J. (1994). Applications of operations research in health care delivery. In S. M. Pollock, M. H. Rothkopf, & A. Barnett (Eds.), *Operations research and the public sector* (pp. 469–498). New York: North-Holland.

Ridgely, M. S., Engberg, J., Greenberg, M. D., Turner, S., DeMartini, C., & Dembosky, J. W. (2007). *Justice, treatment, and cost: An evaluation of the fiscal impact of Allegheny County mental health court.* Santa Monica, CA: RAND Infrastructure, Safety, and Environment.

Shumway, M., Chouljian, T. L., et al. (1994). Patterns of substance use in schizophrenia: A Markov modeling approach. *Journal of Psychiatric Research, 28*(3), 277–287.

Solnit, A. (2004). *The costs and effectiveness of jail diversion: A report to the joint standing committee of the General Assembly.* Hartford, CT: Department of Mental Health and Addiction Services.

Steadman, H. J., & Naples, M. (2005). Assessing the effectiveness of jail diversion programs for persons with serious mental illness and co-occurring substance use disorders. *Behavioral Sciences & the Law, 23,* 163–170.

Steadman, H. J., Osher, F., Robbins, P. C., Case, B., & Samuels, S. (2009). Prevalence of serious mental illness among jail inmates. *Psychiatric Services, 60,* 761–765.

Sweillam, A., & Tardiff, K. (1978). Prediction of psychiatric inpatient utilization: A Markov chain model. *Administration in Mental Health, 6*(2), 161–173.

U.S. Department of Health and Human Services. (1999). *Mental Health: A Report of the Surgeon General.* Rockville, MD: U.S. Department of Health and Human Services, Substance Abuse and Mental Health Services Administration, Center for Mental Health Services, National Institute of Health, National Institute of Mental Health.

Chapter 10
Using Discrete-Event Simulation Modeling to Estimate the Impact of RNR Program Implementation on Recidivism Levels

April Pattavina and Faye S. Taxman

The US prison population declined by 0.3 % from 2009 to 2010. Although the decrease seems small, it was the first yearly decrease since 1972 (Guerino, Harrison, and Sabol, 2011). While this may be a hopeful sign of a trend away from the system growth of the past decades, many troubling after effects of mass incarceration will continue to remain and require sustained governmental efforts to offer viable, evidence-based interventions for offenders to help prepare them for productive lives outside of correctional control. Perhaps the most troubling prediction for those incarcerated is that many will recidivate within 3 years of release. Sixty-eight percent of those released were arrested within 3 years and 25 % had a new prison sentence (Langan and Levin, 2002). Recidivists may return to prison numerous times. This is a process Lynch and Sabol (2001) refer to as churning where offenders cycle in and out of prison. Churning translates into a public safety problem in terms of new crimes and an unfortunate future for those offenders who keep returning to prison.

High levels of incarceration and recidivism have defined the correctional landscape for years and have drawn critical attention to the way services are provided to inmates in prisons and offenders living in communities. One major challenge confronting the US corrections system is figuring out how to improve services in ways that maximize benefits for the offender and the public. As prior chapters in this book have demonstrated, the risk–need–responsivity (RNR) model has gained considerable support as a promising framework for deciding which offenders should be

A. Pattavina (✉)
School of Criminology and Justice Studies, University of Massachusetts Lowell,
One University Ave, Lowell, MA 01854, USA
e-mail: april_pattavina@uml.edu

F.S. Taxman
Department of Criminology, Law and Society, George Mason University, 10900 University
Boulevard, Fairfax, VA 20110, USA
e-mail: ftaxman@gmu.edu

F.S. Taxman and A. Pattavina (eds.), *Simulation Strategies to Reduce Recidivism:*   267
Risk Need Responsivity (RNR) Modeling for the Criminal Justice System,
DOI 10.1007/978-1-4614-6188-3_10, © Springer Science+Business Media New York 2013

served with what treatment services and how treatment services should be delivered.

While evidence-based practices are helping to guide the movement toward RNR-based programming, much less is understood about implementation. As our knowledge continues to evolve about which programs are successful at reducing recidivism and the type of offenders most likely to benefit, we must make equal efforts to understand how readily available programs are, to examine the contexts in which they may be delivered, and how best to utilize them for the purposes of maximizing a reduction in recidivism. Program matching that is guided by principles of RNR is considered the best practice for corrections (Taxman and Marlowe, 2006) and has been shown to significantly reduce recidivism in certain settings (Andrews, Bonta, and Hoge, 1990). It has also been reported that nonadherence to RNR principles in service delivery not only is ineffective but can actually be detrimental to offender treatment outcomes (Lowenkamp and Latessa, 2005). Program caliber has also been found to be an important consideration when considering treatment delivery (Latessa, Smith, Schweitzer, and Brusman Lovins, 2009).

Although there are encouraging signs that RNR treatment principles have taken hold among scholars and practitioners, a full-scale implementation of RNR would require a significant change in the way the business of treatment delivery currently works. In this chapter, we map out an implementation process and use simulation modeling to explore how the adoption of the model will impact recidivism levels. This is an important question because the outcomes of such a study could have important implications for the future of correctional programming. The results may have appeal to those interested in establishing cost–benefit estimates of programming and for those wishing to approach recidivism from a public safety perspective. Indeed, much of the empirical work needed to map the RNR process has been documented in earlier chapters in this volume. Using the synthetic offender profiles and treatment options generated by the authors' contributing work to this volume, along with historical data on correctional populations and program capacities, we will present the results from a discrete-event simulation model to determine the impact of RNR on recidivism when implemented on a national scale.

Why Simulate?

Simulation and modeling can be used in a variety of research activities with the collective goal of developing evidence-based tools and methods to assist in establishing a treatment system governed by RNR principles. For this particular chapter, simulation refers to the use of a computer-generated model to investigate the results from participation in the correctional system with a specific focus on RNR-based treatment delivery (also see Hughes, Chap. 9, this volume). Computer simulation allows the performance of the justice system to be observed over an extended period of time and under a number of different scenarios that would not be possible to test using field experiment techniques. Simulation allows us to examine key parameters and assess the impact on different outcomes.

There are a number of benefits to computer simulation for the corrections field. First, simulation can help to improve the system by informing the development of policies that govern the allocation of scarce resources. It is important to know where resources are needed and how providing additional resources in one area may affect other parts of the system. Second, simulation offers the capacity to test out the system effects of new concepts or policy changes, such as the impact of new sentencing legislation or changes in parole practices. The third benefit of simulation is that the technique provides the opportunity to acquire information without having to disrupt the actual system (Alimadad et al., 2008). In this regard, simulation is used to show the effects of interventions in a virtual setting. Simulation modeling offers an ideal environment in which to examine how the adoption of the RNR model of offender treatment will affect recidivism levels on a national scale.

Computer simulation has a long but sporadic history of use as a criminal justice planning tool. Early models date back to the late 1960s with the JUSSIM model developed by Alfred Blumstein. He brought an operations research focus to the field that integrated police, court, and corrections operations into a comprehensive criminal justice system simulation tool. The mathematical model included baseline parameters that represented the current operation of the criminal justice system. Model parameters could be altered to produce results based on planned scenarios. The JUSSIM model would estimate the changes in resource needs and costs associated with each test scenario (Blumstein, 2002). Although the JUSSIM model is no longer operational, it "had considerable value as a pedagogic device for the newly established industry of criminal justice planning" (Blumstein, 2002, p. 17).

The JUSSIM model was the first attempt to simulate the impact of planned change on the workings of the criminal justice system. Back then, simulation projects demanded considerable time and technical resources. Mainframe computers and software were the main programming resources at the time, and few had the technical skills needed to develop complex models. Moreover, the justice system data required to validate the model parameters were lacking. Thus, computer simulation projects were not common to the field. Today, the availability of powerful personal computers and desktop simulation software, along with growing archives of criminal justice system data available for model building and validation, has made computer simulation modeling more accessible to the criminal justice community.

The JUSSIM model and later more dynamic versions JUSSIM II and III have informed a more recent generation of simulation models built to reproduce how the system works for the purpose of forecasting prison populations or exploring the effects of specific policy changes. In 2007, The Pew Charitable Trusts (2007) published a report that uses simulation modeling to examine the growth in prison populations on public spending and safety. Austin and Fabelo (2004) use simulation as a tool to forecast prison populations for the purposes of examining correctional reforms and promoting accountability. Some state correctional agencies are looking to build sustainable simulation models for monitoring prison and parole operations

(Chaiken and Maltz, 2009), and state sentencing commissions have hired simulation consultants to examine changes in sentencing policies on correctional populations (Speir, Flynt, and Wright, 2008).

Computer simulation techniques vary from basic spreadsheet models to more technical models such as system dynamics models and discrete-event simulation models that may require specific simulation software and programming (Greasley, 2008). Auerhahn (2003) used system dynamics modeling to simulate the criminal sanctioning system and the consequences of criminal justice policy reform in California. She employed a model where the system is modeled through states and rates. Transition rates move offenders through various system states with possible feedback effects in the form of causal loops. A similar approach was used by Tragler, Caulkins, and Feichtinger (2001) to explore the allocation of treatment and enforcement resources for the control of illicit drug markets. System dynamic models are ambitious in scope in that they often attempt to model the workings of the entire system and are generally concerned with understanding the compositional dynamics of system populations (Auerhahn, 2008).

Discrete-event simulation (DES) is another type of computer modeling technique that has been applied in a variety of criminal justice settings. The major distinction between DES and system dynamics is that where system dynamics models capture the movement of aggregate populations, DES models are capable of assigning individual characteristics to entities or persons and using those characteristics to direct and track movement through discrete events defined in the model. Greasley (2000) has used this technique to examine the effects of changes made to the operation of an offender custody process in a UK police force. Speir et al. (2008) built a DES model to test alternative sentencing structures on correctional populations for the Alabama Sentencing Commission. Zarkin, Dunlap, Hicks, and Mamo (2005) used Monte Carlo simulation, a relative of DES, to model heroin use over the lifetime as a chronic recurring condition and estimated the benefits and costs of methadone as a treatment. Hughes (Chap. 9, this volume) uses simulation to examine various system outcomes associated with jail diversion programs.

This project uses DES modeling to estimate the impact of RNR implementation on recidivism levels. The technique is desirable for this project given the capacity to assign individual traits to inmates and then use those traits to determine treatment needs, program availability, and inmate outcomes. In the sections that follow, we will first outline and discuss the goals and objectives of the RNR simulation model. Next we present a description of the model inputs along with the underlying assumptions inherent in the model logic. This will be followed by model validation and results.

DES Model Goals and Objectives

Greasley (Chap. 3, this volume) presents a framework for developing simulation models. The first stage of model development involves explicating the goals and objectives of the project. The specific goal of RNR model is to determine how recidivism levels might change if RNR-based programming were adopted on a national level compared to a system that adheres to the status quo. The simulation objective is to translate the RNR model of treatment delivery as presented in the earlier chapters of this book into an operational simulation model intended to explore how adherence to the specifics of the RNR plan might influence the functioning of the current system and what the impact may be on recidivism outcomes at the national level. This objective has been informed by earlier chapters in this volume that describe the expectation that adopting an RNR approach will result in meaningful reduction in recidivism. In order to address the issue, the simulation model must serve two functions. First, the simulation model must incorporate two treatment options—a baseline model to represent the current system and an RNR informed treatment model. Second, the model must provide options to test the impact of different treatment programs and capacities on recidivism outcomes.

The model scope is the definition of the boundary between what is to be included in the model and what is considered external to the specification. One of the most important simulation tasks is determining the model scope or level of complexity needed to address an issue or question. It is important to note that the more questions a model is expected to answer, the more assumptions the developer must embed in the model about the operation of the criminal justice system. Additionally, as model complexity increases, so does the number of parameters that need to be estimated to conduct the simulation. The challenge is to find the right balance. A useful model will typically include only those inputs, processes, and outputs that are considered essential to the model purpose (Berk, 2008).

The RNR model developed here is designed to operate within the confines of the state prison system. It is a historical simulation model that uses existing data sources to generate model inputs and makes no attempt to predict outcomes outside the time frame for which existing data are available. The model is narrow in scope in that it is specifically designed to compare levels of recidivism predicted over time within the current system of treatment delivery with that of a model using the same admission inputs but employing an RNR-based treatment delivery model. We were not interested in using the model as a forecasting tool to predict the entire volume of prison admissions into the future nor are we introducing new programs per se. The model is designed with the intention of implementing an evidence-based RNR framework that organizes how treatment would be delivered given current levels of treatment resources and levels of incarceration. The ultimate goal is to explore how a full implementation of the RNR model would work to change one specific outcome over time: level of recidivism.

Model Inputs

Table 10.1 includes a description of the RNR simulation model inputs. We use prison admission and release data for males from the National Corrections Reporting Program (NCRP) for model inputs from 1994 through 2006. The NCRP program

Table 10.1 Nationally based estimates of model parameters for RNR discrete-event simulation

Model input	Categories	Source	Parameter
Age, crime type, and length of stay			
Age	Years	NCRP admissions	Fitted distribution
Age grouped	16–27, 28–34, 35–45, 46+	NCRP admissions	Empirical distribution
Crime type given age	Violent, property, drug, sex offense, other	NCRP admissions	Empirical distribution
LOS given age group at admission and crime type	Length of stay for current offense in months	NCRP releases	Fitted distributions
Risk assessment			
Risk given age	High, medium, low	Survey of inmates[a,b]	Empirical distribution
Needs assessment			
Primary need given risk	Criminal thinking, substance abuse, none	Survey of inmates	Empirical distribution
Substance use	Yes, no	Survey of inmates	Empirical distribution
Mental health	Yes, no	Survey of inmates	Empirical distribution
Destabilizers	Yes, no	Survey of inmates	Empirical distribution
Stabilizers	Yes, no	Survey of inmates	Empirical distribution
Recidivism and programming			
Baseline recidivism	Reincarceration within 3 years	BJS recidivism study[b]	Recidivism probability assigned to each risk–need profile
Program category assignment	A–F	Decision rule protocol[c]	Categorical assignment based on risk–needs profile
Available programs in program category and capacity	Program type	CJ-DATS: survey of programming in correctional settings[d]	Capacity based on estimates of average daily population
Recidivism reduction for programs	Percent reduction	Systematic reviews[e]	Relative to baseline

[a]See Ainsworth and Taxman (Chap. 5, this volume)
[b]Generated from synthetic data (see Bhati and Taxman, Chap. 8, this volume)
[c]See Crites and Taxman (Chap. 6, this volume)
[d]See Taxman et al. (2007)
[e]See Caudy et al. (Chap. 7, this volume)

provided admission data for 38 states covering about 93 % of the US incarcerated population.[1] Monthly counts for new court commitments were entered into the model. Figure 10.1 offers an overview of the model process. As offenders enter the system each month, the model creates profiles by assigning characteristics to each based on historical data and appropriate statistical distributions. For example, the

---

[1]Some states did not report every year. For those states, the admissions from the previous year for which data were available were applied.

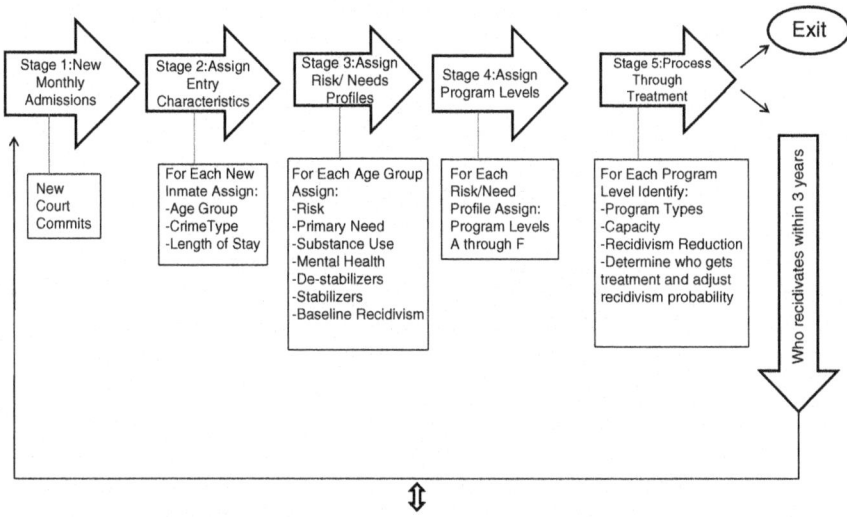

Fig. 10.1 Overview of RNR discrete-event model parameters and flow

first characteristic is age followed by crime type, both generated from the NCRP admissions data. Length of stay (LOS) was estimated using the NCRP release data. We made the assumption that a person currently being admitted by crime type would serve a similar LOS as those being released for the same crime type in the same year. LOS distributions were fitted for each crime type by age at admission and were randomly assigned to inmates admitted for the same crime type and age at admission.

We used the 1997 files to construct crime type, age, and length of stay. The year 1997 was selected since we used the first 4 years of the data as a "warm-up" period to populate a model with a critical mass of inmates. We examined the stability over time for the NCRP distributions including age, crime type, and length of stay. The yearly admission distribution for crime type was stable over the time period. The average percent for court commitments for violent crime was 21 %, property crime averaged 27 %, drug crime averaged 29 %, sex offenses averaged 6 %, and others 15 %. The variation for each of the crime types did not exceed 5 % over the time period. The inmate age at admission and length of stay categories were fairly stable over time as well with most measures having ranges that were less than 10.

The next step in the model was to assign the risk–need profiles. The development of these profiles is described in Chaps. 5 and 8. The creation of these profiles as model inputs was hierarchical in approach, with assignment of each variable category based on the values of the previous variables (Speir et al., 2008). For example, the assignment of risk for each inmate was based on the grouped age of the inmate, next is the assignment of primary need which is based on the risk assignment, etc. This hierarchical approach is useful because it will allow for the user to examine the output grouped by different characteristics.

Baseline Model

The baseline model is structured to represent treatment delivery consistent with current practices. Unfortunately there is no widely established framework or set of organizing principles that describes how treatment delivery is managed within the system (Ward and Maruna, 2007). While there is some evidence to suggest that risk assessment is increasingly being used to decide among treatment alternatives, it is not applied with any degree of consistency and validation of instruments remains problematic (see Chap. 7, this volume). What researchers in the field have found with respect to treatment delivery is that (1) there is not enough program capacity to meet the demand (Taxman, Perdoni, and Harrison, 2007; Chap. 2, this volume), (2) programs do not embrace evidence-based practices (Friedman, Taxman, and Henderson, 2007) and, (3) there is a lack of a clear connection between what offenders need and the type of treatment programs they are likely to receive (Petersilia, 2005), and (4) ending up in an ill-suited program can be detrimental to offenders (Lowenkamp and Latessa, 2005; Crites and Taxman, Chap. 6, this volume). Given that there is little theoretical or organizational guidance that currently dictates treatment program assignment, we do not attempt to invoke any logic governing the treatment delivery in the baseline model.

The issue of an undefined treatment-matching process in the current system also manifests in the data used to develop baseline recidivism probabilities. Baseline recidivism probabilities applied to each prisoner profile were generated from a nationally representative sample of released prisoners. While some of these prisoners may have had treatment in prison and benefitted from it with a reduced chance of recidivism, there is no way to link specific treatment program experiences and recidivism levels. It is therefore possible that the baseline recidivism estimates based on the BJS study may be underestimated. In light of these issues, we assume that a current systemic framework for delivering treatment is not identifiable and that any benefit of treatment that may present in a lowered chance of recidivism is likely offset by wrongly matched program assignments (or no program assignments) and therefore unlikely to have much influence on baseline recidivism levels. The baseline model was validated on inmate characteristics including age, crime type, criminal justice risk, and baseline recidivism levels by risk. One year of output from the model was compared with the values that were expected based on model inputs. For each of these characteristics, output generated from the model was consistent with what was expected given the model parameter inputs.

Treatment Model

The RNR treatment model offers specific direction about how treatment should be administered. Based on the risk–need profiles developed in Chap. 5, offenders will be assigned to specific program categories. Program categories identify the type of programming that would be most suited to the inmate profile. This process

Table 10.2 Program types and estimated capacity inputs for DES model

Program category	Program type	Program levels eligible	Estimated ADP capacity[a]	Recidivism reduction (%) Moderate quality general quality	General quality
Therapeutic community	Segregated	A, B	31,994	9	18
	Nonsegregated	A, B	9,853	8	16
Narcotic maintenance	Drug maintenance[b]	A	2,226		17
Boot camp with treatment	Boot camp with SA treatment	A	6,209	5	3
Counseling SA	5 or more hours per week	B, C, D	59,316	10	20
Mental health	Mental illness treatment[c]	C	26,142	9	17
ISP with treatment	ISP with SA treatment	B	20,133	9	18
Sex offender therapy	Sex offender therapy	C, D	9,987	16	31
Vocational	Vocational training	D, E	70,557	11	22
Employment	Work release	D, E	3,707	3	2
Educational	Education	D, E	81,672	9	18
Intermediate sanctions	Day reporting	F	741	1	2

[a]Unless otherwise noted, average daily population(ADP) estimates are based on Taxman et al. (2007), adjusted for overlapping SA treatment and serving 93 % of population
[b]Mumola (1999)
[c]Beck and Maruschak (2001)

was developed by Crites and Taxman (Chap. 6, this volume), and six program categories were identified. Program types are the specific programs that serve each program category. Caudy, Tang, Ainsworth, Lerch, and Taxman provided a list of available treatment programs, the program categories they serve, and the expected reductions in recidivism for each program type based on meta-analytic analyses (see Chap. 7, this volume).[2]

The capacity for each treatment program is a necessary input for the model. As part of the Criminal Justice Drug Abuse Treatment Studies (CJ-DATS) research cooperative, the National Institute on Drug Abuse (NIDA) sponsored the first comprehensive survey of treatment services within correctional agencies. The survey is designed to estimate national rates of treatment availability and access for offenders involved in different correctional and drug treatment programs and services (Taxman et al., 2007). This survey provided estimates for current programming options available in prisons and their respective capacities.

Table 10.2 provides information on program capacities and recidivism reductions used for the model inputs. The Taxman et al. survey (2007) identified specific

---

[2]Correctional programming can occur in different settings—prison, jail, probation/parole offices, community treatment provider, halfway house, and so on. It was not always possible to differentiate program setting for the meta-analysis (see Caudy et al., this volume).

programs that were being offered in correctional institutions along with national estimates on the number of inmates being served by that program on a given day. These program types were assigned to program categories which were matched to the program levels identified by Crites and Taxman (Chap. 6, this volume). Some of the program types were identified by Caudy et al. (see Chap. 7, this volume) as serving more than one program category. In the absence of clear indicators of program length, the assumptions underlying this process are that the programs are offered once a year. Capacity was estimated by taking the average daily population national estimates for programs included in the survey and creating a number of monthly slots available for each program. In instances where a particular program type could serve more than one program level, the number of slots available for that program was divided based on the proportion of inmates in need for each eligible program level. The treatment slot distribution across levels was calculated by first running the validated model to estimate the number of inmates that fell into each program level for a year and creating percentages based on total number of persons in all levels that could be assigned to the specific program type.

After the profiles for inmates are generated in stages 1–4, they are evaluated for treatment (see Fig. 10.1). Each month, all inmates currently in the prison system with a length of stay longer than 6 months and who had not yet been chosen for treatment would be eligible for a treatment slot in their designated program level. Inmates were randomly selected for treatment until the slots were filled for that month. In cases where inmates were eligible for more than one program, they would only be considered if they were not selected during previous months of their stay, or selected for another program in the same month. This is consistent with the assumption that inmates could receive only one program type per stay. Each inmate had a 0.75 probability of completing the program.

We ran the model with treatment assignments for a year and found that for program categories A and B, 10 % and 11 % of those in need got treatment, respectively. Program category C had about 10 % of those in need treated as well. Programs D and E had 62 % and 66 % treated, respectively. Program F (just punishment, no needs) had about 4 % of the need population treated. Clearly, programs D and E (educational and employment programs) have the most slots available and serve the highest capacity levels. Although the model only allows for one treatment program per inmate, it is possible that given the wide availability of vocational and educational programming, inmates from all program levels are receiving some type of educational or vocational treatment in the current system. It is possible that inmates may end up getting treatment from more than one program or that these programs are used to compensate for the lack of more intensive programming. The extent of each of these possibilities is unknown.

In the model, the baseline recidivism probability, measured as a reincarceration within 3 years, was applied to inmates who did not receive or complete treatment. For those who did get treatment, the recidivism reduction for the assigned program would be calculated relative to the baseline recidivism estimate initially assigned to the inmate profile created upon admission. Those predicted to recidivate would then enter back into the system within 3 years time. A weight was applied so that those who would recidivate would be more likely to come back into the system in the first year after release. Each recidivist would be aged accordingly and increased by one risk level until reaching a third level. The need profiles remain the same.

According to the BJS study, many of those who come back into the system return on a technical violation. About a quarter of the recidivists are predicted to come back for a technical violation (Langan and Levin, 2002). A separate length of stay distribution for technical violations was estimated for this group. This group would be added to the admissions for that month and would draw a length of stay from a distribution created using the NCRP data for technical violators. Other recidivists would be subtracted from the admissions (because they would be considered new court commitments and would be double counted otherwise). Those recidivists would be incarcerated for a new crime, and the associated length of stay from relevant distributions would be applied. Those who do not recidivate left the system permanently. It is also possible that an inmate may exit the system upon death in prison. This probability was calculated from the NCRP data and programmed into the model. We used SIMUL8 software package to build the DES model.

Model Output

We used the model to examine recidivism levels for the baseline and three RNR treatment scenarios. The first treatment scenario is the baseline model. It simply assigns inmate risk and need profiles, but does not consider the inmate for treatment. This is consistent with the assumption we make about an undefined treatment structure where any success is likely to be offset by poorly assigned treatment. The second scenario is structured to examine what happens to the level of recidivism if we use the current treatment options and capacity levels with moderate program quality but use RNR principles to assign the appropriate treatment. Only the process of treatment assignment is altered in this model and the recidivism reduction is applied. The third scenario adjusts the capacity for treatment. Given that RNR literature suggests that those with the higher-risk levels should be targeted for treatment, we increase the current capacity for program levels A–C by 50 %. The fourth scenario does not increase capacity from current levels, but instead focuses on program quality. The model uses the recidivism reduction for programs with general quality which are typically twice as effective as those with moderate quality. In order to populate the prison system, the models were run for 4 years to accumulate a population of about 650,000 inmates with synthetic profiles before assigning any offender to treatment programs or services.

Table 10.3 presents these results of each scenario from a single model run for each. Model 1 is the baseline model and provides the overall number of reincarcerations over the 9-year period and then is broken down by risk level and program category. The probability of reincarceration is the baseline value on that originally

Table 10.3 Total reincarcerations: 1994–2006 monthly prison admissions

Offender Risk	Model 1	Model 2		Model 3		Model 4	
	Baseline: no RNR implementation	RNR treatment, current capacity, moderate program quality	% Change from Model 1	RNR treatment, 50 % increase in capacity for program categories A–C, moderate program quality	% Change from Model 1	RNR treatment, current capacity, general program quality	% Change from Model 1
Overall	925,903	894,568	−3.4	875,095	−5.5	864,025	−6.7
Low risk	56,390	55,478	−1.6	55,556	−1.5	54,830	−2.8
Medium risk	410,472	396,492	−3.4	388,358	−5.4	382,617	−6.8
High risk	459,041	442,598	−3.6	431,181	−6.1	426,578	−7.1

assigned to the profile. These numbers are then compared with the estimates from other model scenarios. Model 2 is the second scenario where the RNR treatment assignment is activated and current capacity estimates are used to determine the number of treatment slots. This model addresses the planned scenario that asks what happens to the level of recidivism if we use the current treatment options and capacity levels with moderate program quality but use RNR principles to assign the appropriate treatment. Column 2 of Model 2 reports the percent change in the number of recidivists reentering prison. Overall the percent change is −3.4 % or a reduction of 31,335 recidivism reentry nationwide over the 9-year period. The majority of the reduction is in the medium- and higher-risk levels.

The third scenario, reported in Model 3, increases the capacity by 50 % for program levels A–C. As expected, the percentage change is larger than Model 1 with a reduction of −5.5 %. Again, the medium- and high-risk levels have the highest reduction in recidivism. Model 3 adds 19,473 fewer recidivists coming back into the system compared to Model 2. Model 4 does not change the capacity levels from the current levels, but instead applies the reduction in recidivism consistent with the general quality of programs which assumes that programs are operating at full integrity for all program levels (see Table 10.2). This model shows a reduction in the overall recidivism numbers by 6.7 %. There are sizeable differences across all risk levels which is consistent with what one would expect if programming quality at all levels is increased.

These results demonstrate that implementing an RNR model of offender treatment can be expected to have a considerable impact on levels of recidivism levels under all of the scenarios that we examined. It should be noted that the simulation model is stochastic and each run of the model may generate different results. As such, the results reported here are preliminary estimates. While the model can be run numerous times to generate results that can be statistically analyzed using means testing, the high volume of inmates being processed are unlikely to have a major impact on the outcomes. The model has flexibility in that recidivism outcomes can also be explored by program levels and profile characteristics. Inputs can also be adjusted to reflect more locally based populations.

Conclusion

The purpose of the DES simulation was to translate the RNR model of treatment delivery as presented in the chapters of this volume into an operational model intended to explore how adherence to the specifics of the RNR plan might influence the churning of offenders in the prison system and its impact on recidivism outcomes at the national level. The plan was grounded in the work presented in earlier chapters in this volume that used evidence-based research to argue that adopting an RNR approach to treatment would result in stronger reductions in recidivism than those observed in the current system. We used the synthetic offender profiles provided by the empirical work reported in earlier chapters of this book as inputs into the simulation model. The model offered two options: a baseline model to represent the current system and an RNR informed treatment model.

We were able to use simulation modeling to create a treatment environment that adheres to the RNR model. As such, we expected at the outset that when the recidivism reductions resulting from proper treatment matching were applied in the computer simulation, the model results would predict fewer returns to prison. What we could not estimate without the model execution was how much of a reduction could we expect. The simulation outputs confirm that a model which uses the current treatment capacity for a core group of treatment programs that can be found in the correctional system and assigns those programs based on an RNR model of treatment matching will reduce the number of inmates returning to prison by 3.4 % over the baseline model run for 9 years (for the purpose of building up the prison population and allowing people to process through treatment and release time). Additionally, if we modify the capacity of those programs that serve the higher-risk populations by 50 %, the reduction increases to 5.5 %. One additional test that increased the quality of the treatment programming resulted in even greater reductions in recidivism even if capacity was not expanded—recidivists were reduced by 6.7 % over the baseline model. In terms of reducing recidivism, the results are consistent with what we expected. This is encouraging for the future of RNR programming.

The results of the simulation model can be used to shed light on the need to elaborate upon the value of multiple program assignments in the context of the RNR treatment delivery model. For example, the process that informs this simulation model requires that primary needs such as criminal thinking or substance abuse are targeted for treatment for higher-risk offenders and that the specific programs designed to serve that group are expected to result in the greatest recidivism reductions. While this logic is embedded in our model, the results show that the capacity for educational and vocational programs (which are intended for lower-risk offenders in this model) is much larger than for programs at other levels. The question is how or whether high capacity levels for this type of programming should be reconciled within the RNR model. It is possible that educational and vocational programs could be viewed as sole program options for low-risk offenders and as supplementary to the recidivism reduction effects of more intense forms of substance abuse and criminal lifestyle treatment. Unfortunately, the literature is lacking in estimates of offenders assigned to multiple programs and how multiple programing may improve recidivism reduction so our model was not able to estimate the later possibility. It is probably more realistic to understand that these programs have been part of the correctional treatment inventory for a long time and may have come to be viewed as offering opportunities for inmates from all program levels and filling idle time, especially in cases where more targeted programs have limited capacity or are lacking entirely. This is a question for future research.

While a full-scale implementation may not happen in the immediate future, more state correctional systems are recognizing the importance of risk and needs assessment to manage programming for inmates. As more correctional systems incorporate RNR-based offender treatment, there may be contagion effects that may accelerate success in terms of recidivism reduction (see Caudy et al., Chap. 7, this volume). How the implementation will be integrated into the correctional system presents challenges to culture that do not readily lend themselves to mathematical structures required for computer simulation. Projects that focus changing the culture of programming in prisons should be used in conjunction with the results of simulation models.

References

Ainsworth, S., & Taxman, F. (2012). Creating simulation parameter inputs with existing data sources: Estimating offender risks, needs and recidivism. In F. Taxman & A. Pattavina (Eds.), *Simulation strategies to reduce recidivism: Risk need responsivity (RNR) modeling in the criminal justice system*. New York, NY: Springer.

Alimadad, A., Borwein, P., Brantingham, P., Brantingham, P., Dabbaghian-Abdoly, V., Ferguson, R., et al. (2008). Using varieties of simulation modeling for criminal justice system analysis. In J. Eck & L. Liu (Eds.), *Artificial crime analysis systems: Using computer simulations and geographic information systems*. Hershey, PA: IGI Global.

Andrews, D., Bonta, J., & Hoge, R. D. (1990). Classification for effective rehabilitation: Rediscovering psychology. *Criminal Justice and Behavior, 17*, 19–52.

Auerhahn, K. (2003). *Selective incapacitation and public policy*. Albany, NY: State University of New York Press.

Auerhahn, K. (2008). Dynamic systems simulation analysis: A planning tool for the new century. *Journal of Criminal Justice, 36*, 293–300.

Austin, J., & Fabelo, T. (2004). *The diminishing returns of increased incarceration*. Washington, DC: JFA Institute.

Beck, A. J., & Maruschak, L. (2001). *Mental health treatment in state prisons*. Washington, DC: Bureau of Justice Statistics, US. Department of Justice.

Berk, R. (2008). How you can tell if the simulations in computational criminology are any good. *Journal of Experimental Criminology, 4*, 289–308.

Blumstein, A. (2002). Crime modeling. *Operations Research, 50,* 16–24.

Caudy, M. S., Tang, L., Ainsworth, S., Lerch, H., & Taxman, F. (2012). Reducing recidivism through correctional programming: Using meta-analysis to inform the RNR simulation tool. In F. Taxman & A. Pattavina (Eds.), *Simulation strategies to reduce recidivism: Risk need responsivity (RNR) modeling in the criminal justice system.* New York, NY: Springer.

Chaiken, J., & Maltz, M. (2009). *California's simulation model for forecasting prison and parole populations. Presentation at the American Society of Criminology annual meeting.* Philadelphia, PA: Pennsylvania.

Crites, E. L., & Taxman, F. S. (2012). The responsivity principle-determining the appropriate program and dosage to match risk and needs. In F. Taxman & A. Pattavina (Eds.), *Simulation strategies to reduce recidivism: Risk need responsivity (RNR) modeling in the criminal justice system.* New York, NY: Springer.

Friedman, P. D., Taxman, F. S., & Henderson, C. (2007). Evidence-based treatment practices for drug involved adults in the criminal justice system. *Journal of Substance Abuse Treatment, 32*(3), 267–277.

Greasley, A. (2000). A simulation analysis of arrest costs. *Journal of the Operational Research Society, 51,* 162–167.

Greasley, A. (2008). *Enabling a simulation capability in the organisation.* London: Springer.

Greasley, A. (2012). The simulation process. In F. Taxman & A. Pattavina (Eds.), *Simulation strategies to reduce recidivism: Risk need responsivity (RNR) modeling in the criminal justice system.* New York, NY: Springer.

Guerino, P., Harrison, P. M., & Sabol, W. (2011). *Prisoners in, 2010.* Washington, DC: Bureau of Justice Statistics, US Department of Justice.

Hughes, D. (2012). A simulation modeling approach for planning and costing jail diversion programs for persons with mental illness. In F. Taxman & A. Pattavina (Eds.), *Simulation strategies to reduce recidivism: Risk need responsivity (RNR) modeling in the criminal justice system.* New York, NY: Springer.

Langan, P. A., & Levin, D. (2002). *Recidivism of prisoners released in 1994.* Washington, DC: Bureau of Justice Statistics, US Department of Justice.

Latessa, E. J., Smith, P., Schweitzer, M., & Brusman Lovins, L. (2009). *Evaluation of selected institutional offender treatment programs for the Pennsylvania Department of Corrections.* Prepared for the Pennsylvania Department of Corrections in Harrisburg, PA.

Lowenkamp, C. T., & Latessa, E. J. (2005). Developing successful reentry programs: Lessons learned from the "what works" research. *Corrections Today, 67,* 72–76.

Lynch, J., & Sabol, W. (2001). *Prisoner reentry in perspective.* Washington, DC: Urban Institute.

Mumola, C. (1999). *Substance abuse and treatment, state and federal prisoners, 1997.* Washington, DC: Bureau of Justice Statistics, US Department of Justice.

Petersilia, J. (2005). From cell to society. In J. Travis & C. Visher (Eds.), *Prisoner reentry and crime in America* (pp. 15–49). New York, NY: Cambridge University Press.

Pew Charitable Trusts. (2007). *Public safety, public spending. Forecasting America's prison population 2007-2011.* Washington, DC: Author.

Speir, J., Flynt, L., & Wright, B. (2008). *Alabama Sentencing Commission: Data analysis and simulation enhancement.* Washington, DC: US Department of Justice.

Taxman, F., & Marlowe, D. (2006). Risk, needs, responsivity: In action or inaction? *Crime & Delinquency, 52,* 3–6.

Taxman, F. S., Perdoni, M., & Harrison, L. (2007). Drug treatment services for adult offenders: The state of the state. *Journal of Substance Abuse Treatment, 32,* 239–254.

Tragler, G., Caulkins, J. P., & Feichtinger, G. (2001). Optimal dynamic allocation treatment and enforcement in illicit drug control. *Operations Research, 49,* 352–362.

Ward, T., & Maruna, S. (2007). *Rehabilitation.* London: Routledge.

Zarkin, G., Dunlap, L. J., Hicks, K. A., & Mamo, D. (2005). Benefits and costs of methadone treatment: Results from a lifetime simulation model. *Health Economics, 14,* 1133–1150.

Part IV
Conclusion

Chapter 11
Risk-Need-Responsivity (RNR): Leading Towards Another Generation of the Model

Faye S. Taxman, Michael S. Caudy, and April Pattavina

The risk-need-responsivity (RNR) framework offers great promise to the fields of community and institutional corrections. It establishes a framework for providing the appropriate type and level of responses to offenders that is grounded in empirical studies (research) and clinical science. RNR advances correctional practices in a multitude of ways, but most importantly it provides an underlying rationale for what types of treatment programs are needed and who should receive these services. By tying these decisions to improved outcomes, it provides a clearer rationale for how we address factors that contribute to criminal behavior. In the chapters of this book, we presented a conceptual framework that has allowed us to investigate prospects for the migration of the current correctional system, which presently lacks a cohesive model of treatment delivery, to one that is grounded in both empirical and clinical sciences. Moreover, we show that although there are significant gaps in the capacity of the correctional system to deliver RNR programming, we actually have many correctional resources (e.g., evidence-based reviews in support of RNR theory, offender risk and needs assessment tools, and meta-analyses that identify successful programs) needed to inform practice and implementation efforts. We have been able to show how these resources can be adapted and expanded for use in simulation models for the purpose of testing the effects of RNR programming on offender outcomes.

F.S. Taxman (✉) • M.S. Caudy
Department of Criminology, Law and Society, George Mason University,
10900 University Boulevard, Fairfax, VA 20110, USA
e-mail: ftaxman@gmu.edu

A. Pattavina
School of Criminology and Justice Studies, University of Massachusetts Lowell,
One University Ave, Lowell, MA 01854, USA
e-mail: april_pattavina@uml.edu

F.S. Taxman and A. Pattavina (eds.), *Simulation Strategies to Reduce Recidivism:*
Risk Need Responsivity (RNR) Modeling for the Criminal Justice System,
DOI 10.1007/978-1-4614-6188-3_11, © Springer Science+Business Media New York 2013

In this final chapter, we highlight six key conclusions that have emerged from the combined efforts of the authors in this volume to map out and build components of an integrated RNR model to improve offender outcomes. First, there is an expansive body of literature supporting an RNR framework of treatment and program delivery. Second, offender risk and needs assessment instruments can, with some adjustments, be used to identify primary offender risk factors and treatment needs. Third, there is currently a significant treatment gap in services necessary to address offender's primary needs, and this gap contributes to the current high rates of negative outcomes (i.e., recidivism). Fourth, meta-analyses of correctional treatment programs can be used to identify programs that significantly reduce recidivism. Fifth, simulation models that test RNR implementation scenarios on a large scale illustrate substantial reductions in recidivism. Finally, it is possible that RNR programming can be integrated into a system of treatment delivery designed for particular jurisdictions. After reviewing each of these areas, we conclude this book with recommendations for the next generation of RNR research.

Support for the RNR Framework for Offender Treatment Delivery

A strength of any empirically based framework is that, as knowledge expands, the model can be altered and modified to accommodate new findings. The RNR framework is theoretically, clinically, and empirically grounded, with an emphasis on static risk factors, dynamic need factors, and the need for programming that embraces cognitive and behavioral approaches. The RNR framework offers tremendous promise based on several key principles: (1) the risk and need of the offender should drive the type and intensity of programming needed; (2) programming should be built on a cognitive and behavioral framework which has shown to be more effective than other orientations; (3) correctional staff should be part of the treatment regime and therefore must integrate the principles of risk, need, and responsivity into their own interactions with offenders; (4) correctional organizations should model behavioral interventions in their own operations; and (5) correctional culture should embrace concepts of justice, fairness, therapeutic jurisprudence, and behavioral change to facilitate correctional programming and ultimately improve offender outcomes. All together, the RNR framework addresses the whole of the correctional enterprise as well as smaller parts such as treatment or service programs. In doing so, the framework provides a model for building a justice system that is responsive to the human service needs of offenders.

Andrews and Bonta (2010) identify the "central eight" dynamic risk factors that are related to recidivism and posit that if these factors are attended to as part of correctional programming, individual outcomes will improve. The "central eight" are as follows: a history of antisocial behavior, antisocial attitudes and cognitions, antisocial peers, antisocial values, lack of prosocial leisure or recreational times, employment or educational deficits, substance abuse, and dysfunctional families. The first four are considered the most important in terms of recidivism outcomes,

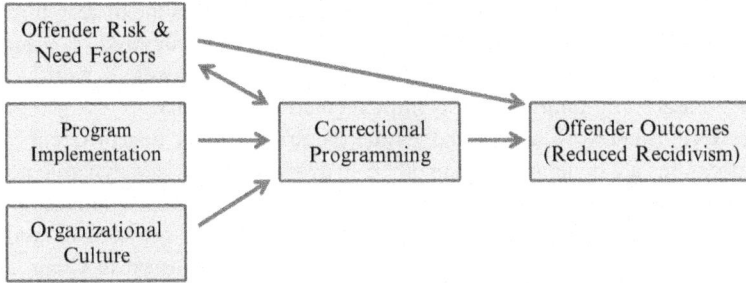

Fig. 11.1 Modified RNR conceptual model

while the later four are of "lesser importance" due to slightly smaller correlation coefficients with recidivism (pp. 498–499). The empirical evidence for each of these factors suggests a correlation with recidivism, but few studies have examined how several of these factors together affect outcomes (see Chaps. 4 and 5). Andrews and Bonta (2010) treat each dynamic need factor as a stand-alone correlate of recidivism instead of examining a spectrum of behaviors or linked conditions that include individual symptoms and characteristics. The consideration of a spectrum of dynamic needs alters the emphasis of the model and allows the RNR model to be more directly tied to responsivity (appropriate correctional programming).

The RNR framework essentially indicates that correctional programming (which is appropriate for risk and need factors) mediates recidivism-related outcomes. While the framework is premised on the direct relationship between individual-level factors (risk and needs) and recidivism, the underlying notion is that participation in appropriate levels of correctional programming will affect offender outcomes. That is, the impact of correctional programming may be moderated by offender-level risk and need factors, as shown above in Fig. 11.1. This alters the original RNR framework to focus on a slightly different empirical question: what type of correctional programming is known to impact recidivism for what type of offenders? And it assumes that changes in the offender's risk and dynamic need factors occur as a result of participation in correctional programming, which also impacts recidivism. In this conceptual model, correctional programming both directly and indirectly contributes to the recidivism outcomes.

Empirical support for this conceptual framework is derived from the large body of research and evaluation studies that test hypotheses regarding the impact of the correctional programming on offender outcomes. In one such study, Landenberger and Lipsey (2005) conducted a meta-analysis of 58 studies on the effect of cognitive-behavioral therapy (CBT) programming on recidivism for both adults and juveniles. The study reported that the recidivism outcomes varied across program features, with better results occurring when (1) the CBT program targeted higher-risk offenders, regardless of any specific need factors; (2) the CBT program included anger control and interpersonal problem solving, regardless of any particular brand of CBT program; and (3) the program was well implemented. Since the study included few individual-level characteristics of offenders other than age, ethnicity, and static risk level, it is unknown whether other demographic or dynamic need factors may affect offender outcomes as well. Overall, this study supports the underlying premise of the RNR framework; high-risk offenders in well-structured programming will make improvements when the program targets their needs as well as offers evidence-based programming that is implemented with fidelity.

The RNR framework acknowledges the empirical reality that different correctional programs are better suited than others to address the risk and dynamic needs of different profiles of offenders. The emphasis on "what type of programming works for what profile of offender?" is a question that is still being explored in studies of correctional programming and offender outcomes. As discussed in Chap. 6, correctional programming should be categorized based on the specific dynamic needs it *intends* to address, and these categories should be used to improve the "match" between offender risk and need factors and the program type that the offender is most likely to benefit from. In other disciplines, the emphasis on matching diagnostic characteristics to programming is commonplace. For example, in medicine, certain types of physical therapy are better suited for certain types of problem behaviors. Certain medications are known to treat certain conditions for women instead of men. Yet, in the correctional environment, programming tends to be more "generic" as if all offending behaviors are similar and all offenders are the same. The RNR framework offers an improvement over this "one-size-fits-all" approach by advancing the notion that programming should be tailored to meet the specific risk and need factors of the individual. However, the framework does not provide clear guidance for how the field should achieve this goal. It is for this reason that, in Chap. 6, Crites and colleagues outlined program classification criteria that focus on what risk and need factors should be targeted and how these needs should be addressed.

Group A: drug dependence on opiates, cocaine, or amphetamines. Offenders with drug dependence disorders, particularly on substances that are linked to criminal behavior (see Bennett, Holloway, & Farrington, 2008), should receive cognitive restructuring treatments focused on their drug dependence. The programs may offer cognitive-based treatments to improve decision-making, interpersonal skills, and social skills of the clientele. Once this primary criminogenic need (substance dependency) has been stabilized, other dynamic needs can be addressed.

Group B: criminal thinking/lifestyle. Offenders with a spectrum of criminal lifestyle needs (several dynamic needs relating to antisocial attitudes, values, behaviors, and social networks) should receive an emphasis on criminal thinking using cognitive restructuring techniques. These programs may also focus on building interpersonal and social skills.

Group C: substance abuse and mental health needs. Offenders who abuse drugs and alcohol but have other stabilizer-related need factors (e.g., employment issues, mental health) should be linked to programming that addresses these specific clinical needs. Once these clinically destabilizing needs have been addressed, programming should focus on interpersonal and social skill development.

Group D: social and interpersonal skill development. Offenders with few dynamic needs but other social needs (e.g., mental health, housing instability) should be linked to programming that focuses on social and interpersonal skill development. This focus is intended to address the multiple destabilizing issues.

Group E: *life skills*. Lower-risk offenders with only stabilizer-related needs (e.g., employment issues, low educational attainment) should be linked to programming that will enhance their ability to improve their overall functioning.

Group F: *punishment only*. Low-risk or low-need individuals who do not require any direct services should be designated for punishment/supervision only. These individuals do not have specific needs that can be addressed through programming. They should not be placed into programming that is overly intensive or unnecessary. Also, when programming is not available within a specific jurisdiction, it may be necessary to place offenders with certain dynamic needs in this category instead of using poorly matched programming that may exacerbate their underlying treatment needs. Punishment in this sense may include a number of options with the use of incarceration reserved for higher-risk offenders.

Building on the RNR principles, this schema provides a guide for targeting programs to different configurations of offender risk and need profiles. It is essential to consider static risk factors, the need for programming, and the intensity of programming (number of clinical hours) that may be required to realize significant impacts on recidivism outcomes. This translation of the RNR framework is based around a typology of offender profiles that focuses attention on the primary drivers of criminal behavior. It positions the offender's level of risk and type of dynamic need factors as the central determinants of the level and type of programming.

In this translational framework, there are no "lesser priority" dynamic risk factors as suggested by Andrews and Bonta (2010). Instead the emphasis is on identifying the major drivers of criminal behavior for each individual offender and tying these to evidence-based correctional programming. Given that correctional programming outcomes are highly dependent on addressing dynamic needs, the resulting system creates placement criteria for matching different offenders to different types of programming. This approach is consistent with both the clinical science literature and with focusing attention on certain factors known to affect involvement in criminal behavior.

Offender Risk and Needs Assessment Instruments Should Be Used to Identify Offender Risk Level and Primary Needs

While the RNR framework and the RNR Simulation Tools discussed in this book all stress the importance of distinguishing static risk and dynamic needs, most of the risk assessment tools available in the field fail to do so, at least not as they are currently used. This has created a controversy in the field given that a combined risk and need score is often used to identify risk level for offender classification and even sentencing decisions. This practice of combining risk and needs to calculate a global risk score does little to improve prediction and may contribute to the mismatch between offender needs and programming by overclassifying offenders as high risk (see Austin, 2006; Austin, Coleman, Peyton, & Johnson, 2003). This practice may also contribute to more severe punishments being levied against justice-involved persons who have behavioral health disorders and other treatment needs. The challenge for risk and needs assessments is how to advance attention to dynamic needs and improve offender outcomes through responsivity without overclassifying individuals with behavioral health treatment needs as high risk.

As noted above, Andrews and Bonta identified the "central eight" dynamic risk factors for recidivism. In their schema, they placed a history of antisocial behavior in the "big four" dynamic risk factors. This history of antisocial behavior is similar to, and often measured as, a history of criminal justice involvement. That is, this is a *static* risk factor that indicates not the type of offense or severity of criminal conduct but rather the number of times (and age of onset) that the individual has been involved in the justice system. Criminal justice risk has long been identified as a predictor of future criminal behavior because "the past predicts the future" (see, e.g., Gendreau, Little, & Goggin, 1996; Gottfredson & Gottfredson, 1987). The inclusion of a history of antisocial behavior as a dynamic need in the RNR framework is problematic and potentially contributes to the practice of combining risk and needs that is currently common within the field of corrections. The process of combining risk and needs has been the subject of considerable critique among scholars in recent years. For instance, in a reanalysis of the LSI-R, Austin (2006) reported that it was the criminal justice risk component, not the dynamic needs, that was predictive of recidivism.

> only a small number of the 54 LSI-R scoring items are useful and most of them are not contributing to the risk assessment process. We also found that compared to the risk groups created by the full LSI-R, the condensed instrument creates risk categories with greater distinctiveness in terms of recidivism. Not only do these items have better predictive ability, but also they reduce the "high risk" category." (Austin, 2006: doi 11/25/2012: http://www. uscourts.gov/viewer.aspx?doc=/uscourts/FederalCourts/PPS/Fedprob/2006-09/index.html)

Analysis of risk and needs assessment tools (referred to as third-generation assessment tools) tends to find that (1) dynamic need factors have lower correlations with recidivism than static risk factors; (2) other variables that are not generally included in risk assessment tools are related to recidivism such as age, gender, and educational attainment; and (3) the scoring of assessment tools that combines risk and need factors is not as efficient as scores that separate risk and needs factors (Andrews & Bonta, 2010; Austin, 2006; Austin et al., 2003; Baird, 2009; Gottfredson & Moriarty, 2006). The concept underlying third-generation risk and needs assessment tools is that the attention to both factors will improve the assessment process. But the designers of third-generation tools were considering the notion of responsivity—using the risk and needs assessment to identify the appropriate programming for a particular person—rather than prediction of recidivism risk alone. Accordingly, these instruments are often misused in the field when the inclusion of needs increases an offender's risk score.

The controversy over the inclusion of risk and need factors within risk assessment instruments has to do with both the predictive validity of the instrument and the relative role and value that dynamic risk factors contribute. Baird (2009), in his assessment of the evidence for risk assessment tools, comments:

Criminogenic Needs

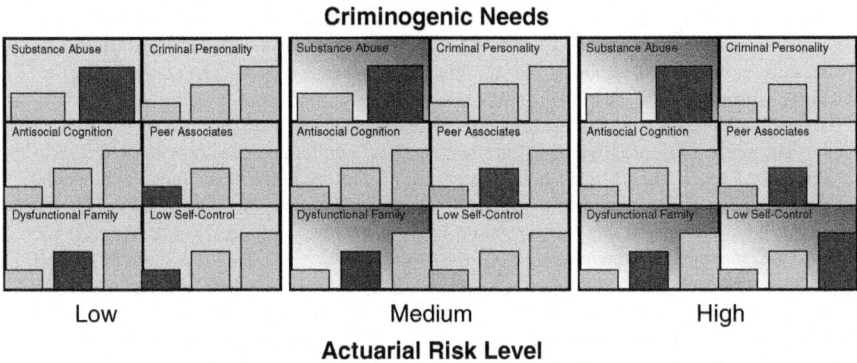

Low   Medium   High

Actuarial Risk Level

Fig. 11.2 Using actuarial risk and criminogenic needs to guide responsivity (Taxman, 2006)

Despite the inclusion of factors without significant relationships to recidivism, these risk models contain enough valid risk factors to attain, in many instances, a modest relationship with various measures of recidivism (see, for example, Flores et al., 2004). Most researchers never ask the next logical question: Would classification results improve if these non-related factors were left out of the instrument? A study of the LSI-R in Pennsylvania (Austin et al., 2003) explored this issue, and produced a dramatic improvement in accuracy using only eight of the 54 LSI-R factors.... Note that the more concise scale not only produced better separation among risk categories, it also dramatically altered the proportion of cases at each risk level, placing more cases in the moderate and low risk categories. This has substantial implications for both release decision making and allocation of resources, including staff supervision and reentry programs and services. In this instance, because the instrument is used by the parole board, the potential impact on individual offenders is especially profound. (Baird, 2009, p. 4 http://cjjr.georgetown.edu/pdfs/ebp/baird2009_Question OfEvidence.pdf)

Baird directly considers the issue that Austin (2006) and others have identified about scoring of risk and needs factors. Baird finds that a total score merely mixes apples and oranges and together it does not provide a good (statistically sound) measure of recidivism risk. In other words, a combined score of risk and need factors makes a difference in terms of how many offenders are placed in different levels of risk as well as the predictive validity of the tool. Both are critical variables that affect the practical utility of risk and needs assessment tools.

Taxman (2006) offers that risk and needs assessment should be considered separately. (Note the original design for the Wisconsin Risk and Needs Assessment Tool had two scores, one for risk and one for needs.) In *Assessment with a Flair*, Taxman argues that risk scores should be used to separate individuals into categories where more structured programming should occur, but the treatment placement should be determined by the dynamic need factor(s). This is consistent with the theoretical logic of the RNR framework.

Figure 11.2 (altered for this model) illustrates the implementation of these principles into a model. Essentially, actuarial risk level should be determined to identify what is the offender's likelihood of further criminal behavior. High-risk offenders should be targeted for treatment-based on the area (s) in which they score moderate or high on criminogenic needs. That is, the offender needs to be assessed also on the criminogenic needs to identify the drivers to their criminal behavior. The notion is that, similar to treatment placement models, actuarial risk should drive the priority for intensive control and appropriate services, with a focus on selecting programs that address multiple problem areas. "Appropriate" refers to attention to the criminogenic factors that have been identified.

The model presented in the exhibit illustrates how criminogenic factors can exist regardless of risk level. That is, a substance abuser may be low risk due to the fact that he or she does not have a history in the criminal justice system. Other criminogenic factors may exist in that low-risk person, but they are more likely to be low to moderate in severity. As the offender moves along the continuum of risk (moderate to high), then it is more likely that more severe problem behaviors may occur. This is a byproduct of the offender's inability to be a productive, contributing member of society. For example, a high-risk offender may have criminogenic needs relating to self-control, peer associates, ASPD, and substance abuse. The combined treatment and control strategies should be designed to address these issues. The model also suggests that the high-risk offender is more involved in situations, settings, and individuals that are likely to further their criminal conduct. Hence, control and treatment services should be concentrated on this individual to achieve the desired goal of reducing the risk of recidivism. (Taxman, 2006: http://www.uscourts.gov/uscourts/FederalCourts/PPS/Fedprob/2006-09/accountability.html)

In both the synthetic and the discrete simulation models described in this book, the static risk level is separated from the criminogenic needs or dynamic risk factors for treatment placement. The empirical evidence, as discussed by various scholars (see above), illustrates that static risk factors improve the predictive validity of an instrument, whereas dynamic risk factors add little, but are relevant for case planning/treatment matching. For this reason, applications of the RNR framework need to consider these separately in terms of responsivity. While Andrews and Bonta infer the importance of both risk and needs, common interpretations of the risk principle combine the two.

The finding that static risk alone serves as a more reliable predictor of recidivism than a global risk and needs score affirms the need to distinguish between static risk and dynamic needs in risk prediction and offender classification models. Following the logic of the RNR framework, static risk should be used to identify individuals in need of more intensive services and controls, while dynamic needs should be used to identify potential targets for rehabilitative interventions. Realizing this goal requires that risk assessment tools and practices distinguish static risk from dynamic needs. The RNR Simulation Tool system discussed in Chaps. 5, 6, 7, and 8 applies this logic to provide decision support tools for the field and help practitioners properly utilize the information that is collected through risk and needs assessments.

A Significant Gap in Services Necessary to Address Offender Needs Reduces Effectiveness

In Chap. 2 we established the gap between offender needs and the availability of programming for one dynamic risk factor: substance abuse. This gap is wide, with most offenders not getting services. The implication of this gap is that offenders with specific needs receive no programming, inappropriate programming, or too little "dosage" of programming to have a considerable impact on recidivism or quality of life. Combined together, this aggravates the problem of recidivism because offenders are often placed in the wrong type or intensity of programming, which results in diminished outcomes and may even be criminogenic (Andrews, 2006; see Chap. 2). The service provision gap problem is also observed for other areas of dynamic needs. For example, for criminal thinking/antisocial attitudes, there is very little direct programming that correctional agencies offer despite increased attention to this correlate of recidivism in recent years (Lipsey, Landenberger, & Wilson, 2007; Taxman, Perdoni, & Harrison, 2007). And there is frequently no direct source of funding for these programs. While correctional agencies are beginning to expand their correctional programming to include criminal thinking interventions (Lipsey et al., 2007; Polaschek, 2011), few correctional agencies routinely offer such programming (Taxman et al., 2007).

As noted throughout this volume, a large part of the reason why responsivity to offender needs has not become a more routine part of correctional practice is the lack of treatment-correctional placement criteria for offenders. Each correctional system has to develop such a process, and it needs to be engrained in sentencing patterns, probation or parole decisions, and other decision-making criteria (e.g., presentence investigations and reports, supervision case plans, correctional case plans) that dominate the criminal justice system in order to be effective. Absent such criteria, individual decision-makers can assess offenders and make placement recommendations based on their own criteria. The advantage of an evidence-based approach is that with the consensus about effective programming comes the general agreement that programs are targeting certain types of dynamic needs or drivers of criminal behavior and subsequently are more likely to improve offender outcomes. In Chap. 6 we outlined the rationale for the RNR Simulation Tool Program-Group Placement Criteria (also discussed above). This and other evidence-driven treatment matching strategies provide a rationale for the placement of offenders into different programs and services. Additional research is needed to test treatment matching strategies designed specifically for justice-involved individuals and to establish clear operational definitions of the primary drivers of recidivism that can be targeted through correctional interventions.

The various simulation projects (the "what if" expert system analyses and discrete models) that have been conducted as part of the development of the RNR Simulation Tool decision support system(s) have assisted in examining questions about the utility of using the RNR approach in assigning offenders to appropriate programming and services. We have used the simulation model approach to demonstrate the impact of the revised decision criteria in terms of offender outcomes (see Chap. 6), and we have used the flexibility and dynamic nature of the simulation models to illustrate the impact on the system over time (see Chaps. 7 and 10). Each model and approach helps to address the three types of impacts discussed in section "RNR Programming Can Lead to Fewer Recidivists: Simulation Findings and Applications" below: impact on recidivism, impact on churning through the system, and impact on the nature and types of services provided to achieve better outcomes.

Evidence-Based Reviews of Correctional Treatment Programs Can Be Used to Identify Programs that Result in Significant Reductions in Recidivism

The simulation methods described throughout this book rely upon and integrate findings from systematic and meta-analytic reviews. In Chap. 7, Caudy and colleagues document the areas where reviews of the effectiveness of correctional interventions have been conducted and report the related effect sizes. The reliance on meta-analysis and systematic reviews ensures that the best available data is used in the simulation models, and it ensures that single site studies or studies of varying rigor are not used to overstate (or understate) the potential effects of using such a program or suite of services. That is, using the best science available adds to the integrity of the simulation model.

The small to moderate effect sizes (ranging from 0 to 30 % relative reductions in recidivism risk) raise a significant question whether providing treatment programming can improve system-level offender outcomes, even when treatment programming quality is high. Austin (2009) argues that the effects of treatment and other programming are limited (looking at the absolute risk reduction numbers) and that increasing the number of offenders in programming will not have a large impact on system-level recidivism outcomes. Instead, Austin (2009) argues there is more to gain from changing policy rather than expanding treatment services. Essentially, the sentiment is that a focus on expanding treatment services, which has an overall small impact on individual-level outcomes, commands attention that would be more effectively given to altering the policies and practices that affect incarceration rates. As discussed in Chaps. 2 and 7, this argument is fostered by current correctional practices, which do not often target offenders for programming under a risk reduction rationale; offenders are frequently misplaced in programming due to limited services and the tendency to use easily accessible services. Unlike the argument put forth by Austin and others, the findings from meta-analyses and systematic reviews and the empirical research on the RNR framework lead us to have confidence that scaling up the use of appropriate treatment will have a considerable impact on recidivism. If program quality is high and a larger portion of the justice-involved population is able to access appropriate services, this will add to the potential impact on recidivism. Changing policies to decrease the size of the incarcerated population is important; however, unmet behavioral health and antisocial cognition treatment needs still represent a key problem within the criminal justice system and are a primary cause of high recidivism rates in the United States.

The controversy over the size of the effect from evidence-based programming is complicated by the poor quality of programming that prevails (Lipsey & Cullen, 2007; Lowenkamp, Latessa, & Smith, 2006). Lowenkamp, Latessa, and Smith (2006) demonstrated how program quality affects recidivism reduction outcomes where better quality programs have better outcomes than lesser quality programs. In their influential study, better quality halfway houses had more positive findings (less recidivism) than halfway houses that were poor quality and that did not embrace the risk principle of the RNR framework. The quality of program implementation is an essential feature of program effectiveness (Andrews & Dowden, 2005; Gendreau, Goggin, & Smith, 1999).

One central tenet of a therapeutic jurisprudence model is that criminal justice systems should only use programs that are known to have a positive impact on individual offenders since the justice system should ensure that the programming does not contribute to harm. That is, offenders should only be assigned to programs that improve outcomes, and assigning offenders to programs that are unlikely to provide benefits or to harm the individual is a misuse of the legal authority. Therapeutic jurisprudence experts argue that deliberately providing harmful programs is akin to providing cruel and unusual punishment because the intervention is likely to cause more harm than good (Wexler, 1993, 2000). Accordingly, under the umbrella of this tenet of therapeutic jurisprudence, it is essential that we continue to explore the relationship between program quality, program implementation, and program effectiveness in an effort to ensure that all programs offered to justice-involved individuals are capable of producing improved outcomes.

RNR Programming Can Lead to Fewer Recidivists: Simulation Findings and Applications

The results from the decision support components and the discrete event models of the RNR Simulation Tool illustrate the impact of responsivity to offender treatment needs.

Impact on Recidivism: The theoretical question of "what works for whom?" is in need of an answer. This research question has yet to be answered by the existing literature given that many studies do not target specific offender profiles or explore the impact of offender characteristics as moderators of program effectiveness. That being said, simulation modeling allows us to examine how the participation of a certain profile of offenders in a given program or service can affect outcomes.

Typically one looks at the absolute risk reduction that relates to the simple difference between the treatment and control group to determine the effectiveness of a treatment intervention. Another way of measuring treatment effectiveness is to examine the relative risk reduction that indicates the percentage change in the treatment group from the expected base rate (control group). The absolute or relative risk reduction basically creates an indicator of the size of the effect of the treatment. While these are often referred to in the field, two other issues affect the impact on recidivism: (1) the population impact and (2) program quality/implementation. Population impact is an important concept since it draws upon the notion that an intervention will have a greater impact when more of the target population is exposed to the intervention and that there is a benefit to the culture and system when the intervention is incorporated into routine practices. With an estimated 10 % of offenders currently provided access to treatment services (see Chap. 2; Taxman et al., 2007), improving access to services will have a greater impact on both individual offender and system-level outcomes. It is the latter—the correctional culture—where the greatest impact is likely to occur; when correctional agencies are more comfortable with providing quality treatment programming. Tucker and Roth (2006) note that expanding access and coverage will improve overall outcomes since a larger percentage of the offender population will be exposed to rehabilitative treatment programming. Finally, consideration of program quality/implementation issues are used in the RNR Simulation Tool models to assess the impact under different implementation scenarios.

The "number needed to treat" or NNT is another way to assess the impact of treatment. The NNT is the inverse of the absolute risk reduction and allows one to estimate the number of individuals that must be treated to prevent one negative event (i.e., one recidivist). In Chap. 7, the NNT was calculated using an estimated 0.20 effect of treatment (relative risk reduction) based on meta-analytic findings reviewed by Lipsey and Cullen (2007). The NNT for sanctions (including incarceration) was 33 people punished to prevent one recidivism event compared to 9 from rehabilitative programming (based on an estimated 0.05 effect of sanctions). According to the estimates provided in Chap. 7, by applying the risk, need, and responsivity principles developed in this book (as discussed in Chaps. 6 and 7), we could obtain an NNT of 5. That is, for every five people placed in appropriate correctional programming, this would prevent one recidivism event. For a population of 10,000 offenders, moving from 10 % (based on Taxman et al., 2007) of offenders in treatment to 50 % would result in 475 less victims of crime. As discussed by Caudy et al. (Chap. 7), making the RNR framework a staple of routine correctional practice can have a considerable population-level impact on recidivism.

Impact on Recycling Through the System: Churning through the justice system is commonplace with reported recidivism rates of around 65 % (Langan & Levin, 2002). The most costly impact of recidivism is reincarceration to prison or jail, which is generally more expensive than community-based programming (Pew Center on the States, 2011). Churning through the justice system is clearly problematic because it indicates that the punishment and/or treatment program did not achieve its stated purpose which is to reduce the likelihood of future criminal offending (except for retributive policies which are designed to provide punishment to allow the state to address offending behavior). One component of the RNR Simulation Tool estimates the impact of adhering to the principles of the RNR framework on recycling through the criminal justice system using a discrete event simulation model (see Chap. 10). This model examines the impact of providing treatment services in prison to appropriate offenders and explores the implications of providing RNR-informed treatment for prison populations.

Using reincarceration as the recidivism measure, the discrete event RNR simulation model illustrates positive impacts. Over time the findings from the discrete event model suggest that adhering to the RNR principles would result in a 3.4 % reduction in the number of inmates returning to prison nationwide. By serving higher-risk offenders, the reduced reincarceration rate would be increased to 5.5 %. By improving the quality of the programming in prison, even without expansion of capacity but solely through increased attention to matching offenders to quality programming, reincarceration rates would be reduced by 6.7 % over the baseline model. This is a conservative approach in that it assumes that one can only participate in one program in prison, and it does not consider that treatment will continue after release from prison. Meta-analytic research (see, e.g., Mitchell, Wilson, & MacKenzie, 2007) suggests an added value of involvement in continuing treatment after release. These findings further illustrate the potential impact of the application of the RNR framework on prisoner reincarceration rates across the United States.

Impact on Services Available in the System: The RNR Simulation Tool is designed to help inform justice agencies about their capacity to provide responsive treatment based on the characteristics of their offender population. As discussed in Chap. 6, the model created a taxonomy of correctional programming based on the primary treatment targets of interventions and the essential features of programs that make them more or less likely to have an impact on recidivism outcomes. In many ways, this taxonomy outlines the range and types of services that are likely to be needed in any correctional setting. The taxonomy outlines the range of programming, but the key issue is that there is likely to be a different distribution of programming in a jurisdiction based on the characteristics of their offender population and the availability of services. The goal of this portion of the tool is to help jurisdictions evaluate their program capacity and plan for future resource allocation to improve the fit between the services they offer and the needs of their justice-involved population.

The RNR Simulation Tool Expert System Can Be Adjusted to Meet the Specific Needs of a Particular Jurisdiction

The RNR Simulation Tool can assist jurisdictions with answering the question of what programming is needed and how much? The tool was designed with the highest degree of flexibility given that many jurisdictions do not have sufficient information on the dynamic needs of their offender population. There are several different approaches that allow jurisdictions to alter the inputs of the simulation model to make the tool outputs more jurisdiction specific: (1) use the national complied database (discussed in Chap. 4) as it exists to give an estimate of the distribution of profiles; (2) use the existing national database and re-weight the file (so it resembles the local jurisdiction) on key demographics such as age, gender, and perhaps ethnicity; or (3) use local data to recreate the profiles using available risk and need information. The potential impact of each of these strategies is depicted in Fig. 11.3. Each of these techniques is provided to allow for the maximum flexibility to meet the needs of the specific jurisdiction.

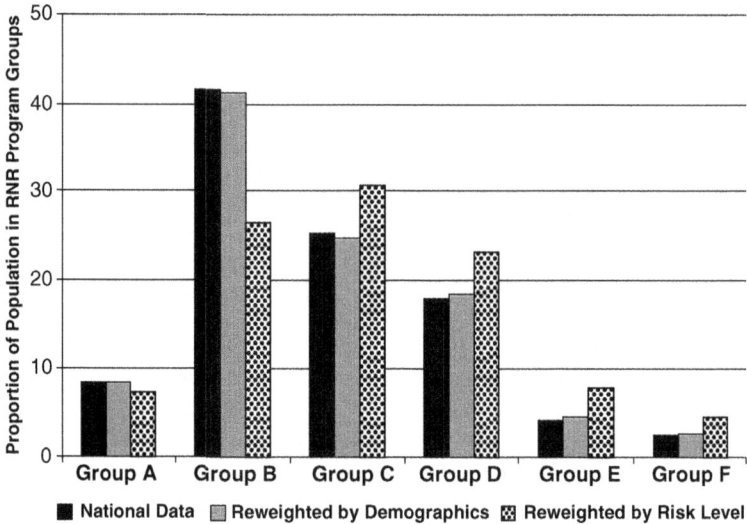

Fig. 11.3 RNR recommended program-group distributions before and after re-weighting

The distribution of programming recommended by the RNR Simulation Tool is depicted in Tables 11.1, 11.2, and 11.3. These recommended distributions are informed by nationally representative data that reflect the prevalence of both static risk and dynamic needs within the offender population. To more accurately inform practice, these capacity recommendations are disaggregated by population type with separate results reported for prisons (Table 11.1), jails (Table 11.2), and community supervision populations (Table 11.3). As displayed in the tables, it is recommended based on the RNR Simulation Tool that between 6 and 10 % of available treatment resources be allocated to address clinical dependence on criminogenic drugs (RNR Program Group A). The largest category of programming (between 40 and 45 % of programming resources) recommended by the RNR Simulation Tool model is RNR Program Group B. Group B programs target criminal thinking and/ or criminal lifestyles using cognitive-behavioral interventions. The second largest target for programming is RNR Program Category C where it is recommended that about 25–30 % of treatment resources be allocated. Group C programs target clinical destabilizers such as substance abuse (not dependence) and mental health disorders. Between 15 and 20 % of treatment resources is recommended for RNR Group D programs which target social and interpersonal skill development (e.g., education, employment, anger management programs), while less than 5 % of program resources are recommended for life skills programs (RNR Program Group E) or punishments only (RNR Program Group F).

The program groups have been designed to facilitate treatment matching and to help jurisdictions better allocate resources to reduce recidivism through responsivity to the primary treatment needs of their offender populations (see Chap. 6). The examples in Tables 11.1, 11.2, and 11.3 are based on national data from several

Table 11.1 Recommended distribution of programming for prison population based on RNR Simulation Tool data

RNR Simulation Tool Programming Group	Criminal justice risk level			
	Low	Moderate	High	Total[a]
A: Dependence on opiates, cocaine, or amphetamines	10.0	9.9	10.4	10.0
B: Criminal thinking/lifestyles	0.0	35.1	75.3	43.1
C: Substance abuse and mental health	34.4	40.4	2.6	28.4
D: Social and interpersonal instability	38.3	13.8	10.9	15.8
E: Life skills	12.8	0.0	0.8	1.7
F: Punishment only	4.5	0.8	0.0	1.0

[a]Table values represent the proportion of the population recommended for each RNR Program Group

Table 11.2 Recommended distribution of programming for jail population based on RNR Simulation Tool data

RNR Simulation Tool Programming Group	Criminal justice risk level			
	Low	Moderate	High	Total[a]
A: Dependence on opiates, cocaine, or amphetamines	6.7	6.3	6.9	6.4
B: Criminal thinking/lifestyles	0.0	42.0	74.5	41.8
C: Substance abuse and mental health	41.5	31.3	4.5	28.4
D: Social and interpersonal instability	32.8	18.8	12.3	19.6
E: Life skills	11.8	0.0	1.7	1.7
F: Punishment only	7.2	1.7	0.0	2.1

[a]Table values represent the proportion of the population recommended for each RNR Program Group

Table 11.3 Recommended distribution of programming for community supervision population based on RNR Simulation Tool data

RNR Simulation Tool Programming Group	Criminal justice risk level			
	Low	Moderate	High	Total[a]
A: Dependence on opiates, cocaine, or amphetamines	6.2	8.4	9.9	8.4
B: Criminal thinking/lifestyles	0.0	38.2	75.0	41.7
C: Substance abuse and mental health	29.7	35.8	2.9	24.7
D: Social and interpersonal instability	36.4	16.3	11.1	18.7
E: Life skills	19.0	0.0	1.1	4.1
F: Punishment only	8.7	1.2	0.0	2.3

[a]Table values represent the proportion of the population recommended for each RNR Program Group

different sources (see Chap. 5); however, these data can be adjusted to reflect the distribution of risk and need profiles within any specific jurisdiction (as discussed in Chap. 8). The translation of the RNR framework into a system-level decision support tool offers meaningful information to guide system planning efforts and to help local and state agencies build up a capacity of treatment providers to address the treatment needs of their justice-involved populations. The potential uses and implications of this feature of the RNR Simulation Tool are numerous and can have a considerable impact on recidivism rates by increasing capacity to allow more justice-involved persons to receive rehabilitative treatments.

RNR Future Research Directions

Even though the RNR framework has received a considerable amount of empirical attention and support over the last three decades (Andrews, 2006; Andrews et al., 1990; Andrews & Bonta, 2010; Andrews & Dowden, 2006; Dowden & Andrews, 1999a, 1999b, 2000), the authors of this book in their analysis of the field have identified a number of limitations of the existing RNR literature base (see Chap. 4). Our analysis, along with that of others, has explored the nuances of the framework to exalt the empirical foundations and to enhance the transportability of the framework for practical use by correctional and service organizations. In this section we highlight several key areas where further empirical research is needed to augment the RNR conceptual framework. The goal of this discussion is to provide a prospective research agenda for RNR and to facilitate further model refinements and knowledge translation of key findings.

Substance Use Disorders (SUDs) in the RNR Framework: Despite the high prevalence of substance use, mental health, and co-occurring disorders among individuals involved in the justice system (Lurigio, Cho, Swartz, Graf, & Pickup, 2003; Mumola & Bonczar, 1998; Peters & Bekman, 2007; Staton-Tindall, Havens, Oser, & Burnett, 2011; Steadman, Osher, Robbins, Case, & Samuels, 2009; Taxman et al., 2007), few justice-involved individuals are exposed to evidence-based programs or services. When justice-involved individuals are exposed to programs, the programs are often not well matched to their individual treatment needs (see Chap. 2). As discussed in Chap. 4, the RNR framework (Andrews & Bonta, 2010) does not prioritize substance use as a "big four" criminogenic need. Substance abuse is one of the central eight dynamic risk factors, but is considered to be of lesser importance than those factors related to antisocial history, peers, values, and attitudes. The omission of substance use from the list of criminogenic needs to be prioritized for treatment may be a function of the poor operational definition of this construct. That is, SUDs can vary considerably in terms of the compulsive nature and severity of the disorder, ranging from periodic use to compulsive use. The failure to consider the complexities of the drug-crime nexus and the differential impact of SUDs on recidivism is one limitation of the extant RNR literature base.

The existing literature on the drug-crime nexus needs to be extended to address key issues about the varying nature of drug use patterns in society: clinically defined drug dependence, drug abuse, recreational use, and social uses. A few unanswered questions exist given the dated literature establishing the link between opioid use and criminal behavior (see Ball, Shaffer, & Nurco, 1983; Nurco, Hanlon, & Kinlock, 1991; Nurco, Hanlon, Kinlock, & Slaght, 1984): do offenders diagnosed as drug dependent have higher recidivism rates than offenders who are classified as abusers or users only? Are some drugs more directly related to recidivism than others? Are criminal justice and/or substance abuse treatment outcomes improved when offenders are matched to levels of treatment intensity based on disorder severity?

In the RNR Simulation Tool models, we have identified clinically diagnosed substance dependence on drugs that have a stronger direct relationship with crime as a primary criminogenic need; "criminogenic drugs" include opioids, cocaine, and amphetamines. Offenders with dependence on these criminogenic substances should be prioritized for treatment and control responses because there is more direct information about relevant and effective treatment for individuals with these addiction disorders. Additional research is needed to provide further empirical support for this more specific operationalization of SUDs within the RNR framework. This reconceptualization also calls attention to the need for more evidence-based screening and assessment practices in the justice system. Prioritizing certain SUDs for treatment requires that these disorders are reliably and consistently identified within the population of offenders and that an infrastructure is in place to provide treatment services to the large portion of the justice population that needs it.

Future research should also explore the adaptability of the RNR framework for guiding substance abuse treatment case planning. The RNR framework has primarily been implemented for criminal justice populations; however, the model may have added utility for non-justice-involved individuals. Additional research is needed to better understand the transportability of the RNR principles to the substance abuse treatment field. Specifically, this research should test whether or not adherence to the RNR principles can lead to improved treatment outcomes for individuals with SUDs. Do substance users fare better when the intensity of treatment services is matched to the severity of their SUDs? Does addressing multiple dynamic needs improve treatment outcomes? Finally, does the use of cognitive-based approaches and tailoring interventions to the strengths of the individual participants improve motivation and success in substance abuse treatment?

Measurement of Dynamic Offender Needs: The RNR framework is grounded in the relationship between dynamic offender needs and recidivism. The need principle stresses that (rehabilitative) interventions should target specific offender needs that are both dynamic (amenable to change) and criminogenic (directly related to recidivism outcomes). A considerable body of empirical research in the field of criminology has been devoted to establishing risk factors for future involvement in antisocial behavior and subsequent contact with the criminal justice system. The extant research generally supports criminal history (static risk) and demographic characteristics such as age and gender as the most robust predictors of continued involvement in offending (Gendreau et al., 1996; Huebner & Berg, 2011; Makarios, Steiner, & Travis, 2010). Extensive research has also explored the relationship between dynamic offender needs (e.g., antisocial cognitions, mental health, family problems, and employment problems) and recidivism outcomes. The results of these studies vary considerably, often depending on how dynamic needs are measured as well as the study design. While some studies find support for dynamic needs as significant correlates of recidivism, the mechanisms through which these needs impact recidivism remain unclear. Once criminal history and demographics are taken into account, the relationships between dynamic needs and recidivism are often found to be weak or spurious (see Chap. 4, this volume).

One potential explanation for the inconsistent findings regarding the relationship between dynamic offender needs and recidivism is the poor measurement of these constructs. Across third- and fourth-generation risk assessments commonly used in the justice system (e.g., LSI-R, ORAS, and COMPAS), there is a lack of construct validity for many dynamic needs. The use of varying operational definitions of these constructs across tools and settings is problematic for testing the principles of the RNR framework and making generalizations across populations.

In the field, dynamic needs are often measured very differently across justice agencies depending on what assessment instruments are used. For instance, antisocial attitudes, a "big four" criminogenic need, are measured differently by the LSI-R, ORAS, and COMPAS instruments. While the LSI-R and COMPAS assessments only use attitudinal measures, the ORAS also includes behavioral measures in its operationalization of the antisocial attitudes construct. In fact, this construct is even operationalized differently across two assessment batteries within the ORAS. And while the LSI-R operationalizes this construct with only four items, the ORAS includes eight items and the COMPAS includes eleven items. Within these three risk assessments, there are four different ways to operationalize the same antisocial attitudes construct. This lack of construct validity, as well as a lack of measurement harmonization, is a barrier to the implementation of the RNR conceptual framework and limits the generalizability and transportability of research that explores the relationship between these dynamic needs and recidivism outcomes.

Future RNR research should explore the robustness (or lack thereof) of the relationship between dynamic offender needs across assessment instruments and diverse data sources. The goal of this research should be to establish standardized conceptual and operational definitions of need constructs and to establish a strong empirical literature base concerning the relationship between these needs and offender outcomes. Additional empirical attention is also need to better understand the mechanisms through which these dynamic needs impact recidivism. This is critical in light of the various instruments, the various ways in which key constructs are measured, and potential utility of each variable. This line of research is relevant to both practice and policy. Adherence to the need principle of the RNR model is only possible if needs are adequately defined and measured, and this information about individual needs is available to guide treatment matching and case planning strategies. Establishing clear definitions of these constructs and their empirical link to offender outcomes is a necessary step in the process of moving the RNR model from research into practice.

Developing and Testing Treatment Matching Strategies: The use of treatment matching strategies is scarce in the criminal justice system. More often than not, justice-involved individuals with treatment needs are assigned to correctional interventions based on programming availability, professional judgment, and/or characteristics of their instant offense. These program assignment practices are not evidence-based and often lead to a mismatch between offender treatment needs and the type or intensity of programming that is received. The RNR framework predicts that this mismatch between offender treatment need, which can be defined as a combination between static and dynamic risk, and programming is a primary cause of treatment failures and recidivism. Future research is needed to develop and test different treatment matching strategies that embrace the RNR principles and can be successfully implemented in justice settings.

This line of inquiry should explore differential offender outcomes for those who are correctly matched to levels of care and those who are not. Correctly matched treatment at the individual level should adhere to all three of the RNR principles: treatment should target high- and moderate-risk offenders, be targeted to specific criminogenic needs while also taking into account other clinically relevant offender needs, and should employ evidence-based treatment techniques such as CBT. Treatment matching strategies must also take into account other key program features to ensure that the available programming has the potential to lead to improved offender outcomes that are sustainable over time. The development of effective treatment matching strategies requires attention to key program features including dosage (frequency and duration), setting, intensity, and implementation fidelity.

Under the larger umbrella of treatment matching, the issue of program dosage is of particular salience and an area where future research is needed. Limited empirical research has explored this topic, but the research that has been done has found that dosage is an important mediator of program effectiveness (see, e.g., Bourgon & Armstrong, 2005). Based on their work assessing the effectiveness of one program within one facility, Bourgon and Armstrong (2005) suggested that the dosage of programming needed to affect recidivism varies depending on the severity of risk and needs. More specifically, they recommended that 100 hours of programming was sufficient to reduce recidivism for moderate-risk offenders with few needs, while over 200 hours of programming was needed for higher-risk or multiple need offenders. They also reported that 300 hours of programming was needed for offenders with both high static risk and multiple dynamic needs. A number of unanswered questions remain, such as whether this dosage of time can be delivered through one program or conversely via portions of several programming experiences. Future research should focus on developing a sound conceptual definition of dosage and testing the relationship between dosage and programming outcomes across a more generalizable set of programs and samples.

The development and empirical testing of treatment matching strategies is a necessary next step for the RNR framework. Most extant empirical tests of the framework have used very general definitions of "appropriately" or "inappropriately" matched treatments (see, e.g., Andrews et al., 1990). Exploring the nuances of the relationship between treatment matching, treatment dosage, treatment completion, and recidivism is essential for informing effective correctional practice. If the framework is to be successfully integrated into the field of corrections, specific, tangible guidelines need to be developed to inform practice.

Understanding the Role of Demographics in the RNR Framework: Actuarial risk assessments have been developed to be demographically neutral, as discussed in

Chap. 4. Specifically, these assessment instruments have been designed to limit the potential for extralegal bias in the prediction of risk for future offending. While excluding race and ethnicity from the risk prediction equation is important for limiting the potential for racial bias in the prediction of risk, excluding key demographics such as age and gender from the RNR framework is potentially problematic. Gender, age, and ethnicity are particularly relevant from a responsivity standpoint within the RNR framework, but they may also play an important role as key moderators of the relationships specified within the framework. The conditioning effects of age and gender in particular on the relationships between risk, needs, program outcomes, and recidivism are important avenues for future empirical investigation.

Understanding whether or not some needs are more salient as recidivism predictors for males relative to females or for younger offenders relative to older offenders is important for informing responsivity and for moving the RNR framework forward. It is also necessary for the field to continue to explore "what works for whom?" Are some correctional interventions more effective for some subgroups relative to others? How can programs be adapted to be culturally relevant and responsive to the diverse characteristics of the offender population? These are questions that warrant further investigation within the field.

Future research should focus on testing the moderating influence of demographics on the relationship between risk and recidivism, the relationship between dynamic needs and recidivism, the relationship between program participation and program success, and the relationship between program participation and recidivism. Answering these questions with empirical data will enhance the transportability of the RNR framework into everyday practice. Gaining a better understanding of what works best for whom is a critical next step for the RNR framework.

Conclusion

The RNR framework has served as a primary model for moving research into practice in the field of corrections over the last two decades. RNR offers a parsimonious conceptual framework that combines several evidence-based practices and calls attention to the need for a correctional system that is responsive to the human service needs of the offender population. While the framework has received considerable empirical attention and support, several aspects of the framework are in need of further research to advance the utility of the RNR framework to practice and policy. In this book, several refinements to the RNR framework are being used, but further work is needed. Answering the questions outlined within this chapter will advance the transportability of the framework for informing practice.

The RNR framework offers great promise for improving outcomes across the justice system, but the current evidence base tempers this promise to some degree. Continued expansion of the literature base and research underlying the RNR framework is needed. Some important directions for future inquiry include an expansion of the literature concerning the effectiveness of correctional interventions for reducing recidivism and improving offender outcomes, improved operationalization and measurement harmonization of key RNR constructs, a better understanding of the conditioning effects of age and gender on the theoretical relationships proposed within the RNR framework, exploration of the nuances of the relationship between SUDs and recidivism within the framework, and the development of evidence-based treatment matching strategies that translate the RNR principles into everyday correctional practice. Each avenues of future research has important implications for theory, practice, and policy.

References

Andrews, D. A. (2006). Enhancing adherence to Risk-Need-Responsivity: Making quality a matter of policy. *Criminology and Public Policy, 5*(3), 595–602.

Andrews, D. A., & Bonta, J. (2010). *The psychology of criminal conduct* (5th ed.). Cincinnati, OH: Anderson.

Andrews, D. A., & Dowden, C. (2005). Managing correctional treatment for reduced recidivism: A meta-analytic review of programme integrity. *Legal and Criminological Psychology, 10*, 173–187.

Andrews, D. A., & Dowden, C. (2006). Risk principle of case classification in correctional treatment. *International Journal of Offender Therapy and Comparative Criminology, 50*(1), 88–100.

Andrews, D. A., Zinger, I., Hoge, R. D., Bonta, J., Gendreau, P., & Cullen, F. T. (1990). Does correctional treatment work? A clinically relevant and psychologically informed meta-analysis. *Criminology, 28*(3), 369–404.

Austin, J. (2006). How much risk can we take—The misuse of risk assessment in corrections. *Federal Probation, 70*, 58–63.

Austin, J. (2009). The limits of prison based treatment. *Victims and Offenders, 4*, 311–320.

Austin, J., Coleman, D., Peyton, J., & Johnson, K. D. (2003). *Reliability and validity study of the LSI-R risk assessment instrument*. Washington, DC: Institute on Crime, Justice, and Corrections at The George Washington University.

Baird, C. (2009). *A question of evidence: A critique of risk assessment models used in the justice system*. Madison, WI: National Council on Crime and Delinquency.

Ball, J. C., Shaffer, J. W., & Nurco, D. N. (1983). The day-to-day criminality of heroin addicts in Baltimore: A study in the continuity of offense rates. *Drug and Alcohol Dependence, 12*, 119–142.

Bennett, T., Holloway, K., & Farrington, D. (2008). The statistical association between drug misuse and crime: A meta-analysis. *Aggression and Violent Behavior, 13*(2), 107–118.

Bourgon, G., & Armstrong, B. (2005). Transferring the principles of effective treatment into a "real world" prison setting. *Criminal Justice and Behavior, 32*(1), 3–25.

Dowden, C., & Andrews, D. A. (1999a). What works for female offenders: A meta-analytic review. *Crime and Delinquency, 45*, 438–451.

Dowden, C., & Andrews, D. A. (1999b). What works in young offender treatment: A meta-analysis. *Forum on Corrections Research, 45*, 438–452.

Dowden, C., & Andrews, D. A. (2000). Effective correctional treatment and violent reoffending: A meta-analysis. *Canadian Journal of Criminology, 42*, 449–467.

Flores, A. W., Travis, L. F., & Latessa, E. J. (2004). *Case classification for juvenile corrections: An assessment of the Youth Level of Service/Case Management Inventory (YLS/CMI), executive summary* (98-JB-VX-0108). Washington, D.C.: U.S. Department of Justice.

Gendreau, P., Goggin, C., & Smith, P. (1999). The forgotten issue in effective correctional treatment: Program implementation. *International Journal of Offender Therapy and Comparative Criminology, 43*(3), 180–187.

Gendreau, P., Little, T., & Goggin, C. (1996). A meta-analysis of the predictors of adult offender recidivism: What works! *Criminology, 34*(4), 575–607.

Gottfredson, S. D., & Moriarty, L. J. (2006). Statistical risk assessment: Old problems and new applications. *Crime & Delinquency, 52*(1), 178–200.

Gottfredson, M. R., & Gottfredson, D. M. (1987). Decision making in criminal justice (Vol. 3). New York, NY: Springer.

Huebner, B. M., & Berg, M. T. (2011). Examining the sources of variation in risk for recidivism. *Justice Quarterly, 28*(1), 146–173.

Landenberger, N. A., & Lipsey, M. W. (2005). The positive effects of cognitive-behavioral programs for offenders: A meta-analysis of factors associated with effective treatment. *Journal of Experimental Criminology, 1*, 451–476.

Langan, P. A., & Levin, D. J. (2002). *Recidivism of prisoners released in 1994*. Washington, DC: Bureau of Justice Statistics.

Lipsey, M. W., & Cullen, F. T. (2007). The effectiveness of correctional rehabilitation: A review of systematic reviews. *Annual Review of Law and Social Science, 3*, 297–320.

Lipsey, M. W., Landenberger, N. A. & Wilson, S. J. (2007). Effects of cognitive-behavioral programs for criminal offenders. *Campbell Systematic Reviews*. Retrieved from http://campbell-collaboration.org/lib/project/29/.

Lowenkamp, C. T., Latessa, E. J., & Smith, P. (2006). Does correctional program quality really matter? The impact of adhering to the principles of effective intervention. *Criminology and Public Policy, 5*(3), 575–594.

Lurigio, A., Cho, Y., Swartz, J., Graf, I., & Pickup, L. (2003). Standardized assessment of substance-related, other psychiatric, and comorbid disorders among probationers. *International Journal of Offender Therapy and Comparative Criminology, 47*, 630–652.

Makarios, M., Steiner, B., & Travis, L. F. (2010). Examining the predictors of recidivism among men and women released from prison in Ohio. *Criminal Justice and Behavior, 37*(12), 1377–1391.

Mitchell, O., Wilson, D. B., & MacKenzie, D. L. (2007). Does incarceration-based drug treatment reduce recidivism? A meta-analytic synthesis of the research. *Journal of Experimental Criminology, 3*(4), 353–375.

Mumola, C. J., & Bonczar, T. P. (1998). *Substance abuse and treatment of adults on probation, 1995*. Washington, DC: Bureau of Justice Statistics.

Nurco, D. N., Hanlon, T. E., & Kinlock, T. W. (1991). Recent research on the relationship between illicit drug use and crime. *Behavioral Sciences & the Law, 9*(3), 221–242.

Nurco, D. N., Hanlon, T. E., Kinlock, T. W., & Slaght, E. (1984) Variations in criminal patterns among narcotic addicts in Baltimore and New York City, 1983–1984. *Friends Medical Science Research Center*.

Peters, R. H., & Bekman, N. M. (2007). Treatment and reentry approaches for offenders with co-occurring disorders. In R. B. Greifinger, J. Bick, & J. Goldenson (Eds.), *Public health behind bars: From prisons to communities* (pp. 368–384). New York: Springer.

Pew Center on the States. (2011). *State of recidivism: The revolving door of America's prisons*. Washington, DC: The Pew Charitable Trusts.

Polaschek, D. L. (2011). Many sizes fit all: A preliminary framework for conceptualizing the development and provision of cognitive-behavioral rehabilitation programs for offenders. *Aggression and Violent Behavior, 16*, 20–35.

Staton-Tindall, M., Havens, J. R., Oser, C. B., & Burnett, M. C. (2011). Substance use prevalence in criminal justice settings. In C. Leukefeld, T. P. Gullota, & J. Gregrich (Eds.), *Handbook of evidence-based substance abuse treatment in criminal justice settings* (pp. 81–101). New York, NY: Springer Science + Business Media.

Steadman, H. J., Osher, F. C., Robbins, P. C., Case, B., & Samuels, S. (2009). Prevalence of serious mental illness among jail inmates. *Psychiatric Services, 60*(6), 761–765.

Taxman, F. S. (2006). Assessment with a flair: Offender accountability in supervision plans. *Federal Probation, 70*(2), 2–7.

Taxman, F. S., Perdoni, M. L., & Harrison, L. D. (2007). Drug treatment services for adult offenders: The state of the state. *Journal of Substance Abuse Treatment, 32*(3), 239–254.

Tucker, J. A., & Roth, D. L. (2006). Extending the evidence hierarchy to enhance evidence-based practice for substance use disorders. *Addiction, 101*, 918–932.

Wexler, D. (1993). Therapeutic jurisprudence and the criminal courts. *William & Mary Law Review, 35*(1), 279–299.

Wexler, D. (2000). Therapeutic jurisprudence: An overview. *Thomas M. Cooley Law Review, 17*, 125–135.

Index

F.S. Taxman and A. Pattavina (eds.), *Simulation Strategies to Reduce Recidivism:*   309
Risk Need Responsivity (RNR) Modeling for the Criminal Justice System,
DOI 10.1007/978-1-4614-6188-3, © Springer Science+Business Media New York 2013

Lightning Source UK Ltd.
Milton Keynes UK
UKOW05n0636300916

284174UK00016B/313/P